T0328633

INTELLIGENT IOT SYSTEMS IN PERSONALIZED HEALTH CARE

Cognitive Data Science in Sustainable Computing

INTELLIGENT IOT SYSTEMS IN PERSONALIZED HEALTH CARE

Edited by

ARUN KUMAR SANGAIAH

School of Computing Science and Engineering, Vellore Institute of Technology (VIT), Vellore, India

SUBHAS MUKHOPADHYAY

School of Engineering, Macquarie University, Macquarie Park, NSW, Australia

Series Editor

ARUN KUMAR SANGAIAH

School of Computing Science and Engineering, Vellore Institute of Technology (VIT), Vellore, India

Academic Press is an imprint of Elsevier
125 London Wall, London EC2Y 5AS, United Kingdom
525 B Street, Suite 1650, San Diego, CA 92101, United States
50 Hampshire Street, 5th Floor, Cambridge, MA 02139, United States
The Boulevard, Langford Lane, Kidlington, Oxford OX5 1GB, United Kingdom

Library of Congress Cataloging-in-Publication Data
A catalog record for this book is available from the Library of Congress

British Library Cataloguing-in-Publication Data
A catalogue record for this book is available from the British Library

ISBN: 978-0-12-821187-8

For information on all Academic Press publications
visit our website at https://www.elsevier.com/books-and-journals

Publisher: Mara Conner
Acquisitions Editor: Sonnini R. Yura
Editorial Project Manager: Rafael G. Trombaco
Production Project Manager: Omer Mukthar
Cover Designer: Matthew Limbert

Typeset by SPi Global, India

Contents

Contributors

Muyideen Abdulraheem
Department of Computer Science, University of Ilorin, Ilorin, Nigeria

Oluwakemi Christiana Abikoye
Department of Computer Science, University of Ilorin, Ilorin, Nigeria

Kayode S. Adewole
Department of Computer Science, University of Ilorin, Ilorin, Nigeria

Abimbola G. Akintola
Department of Computer Science, University of Ilorin, Ilorin, Nigeria

Ahmed O. Ameen
Department of Computer Science, University of Ilorin, Ilorin, Nigeria

S. Anjana
School of Computer Science & Engineering, VIT, Vellore, India

Joseph Bamidele Awotunde
Department of Computer Science, University of Ilorin, Ilorin, Nigeria

Femi Emmanuel Ayo
Department of Computer Science, McPherson University, Seriki Sotayo, Nigeria

Amos Orenyi Bajeh
Department of Computer Science, University of Ilorin, Ilorin, Nigeria

Abdullateef O. Balogun
Department of Computer Science, University of Ilorin, Ilorin, Nigeria

Muhammed K. Jimoh
Department of Educational Technology, University of Ilorin, Ilorin, Nigeria

Rasheed Gbenga Jimoh
Department of Computer Science, University of Ilorin, Ilorin, Nigeria

Raheleh Khanduzi
Department of Mathematics and Statistics, Gonbad Kavous University, Gonbad Kavous, Golestan, Iran

R. Kishore
Sri Sivasubramaniya Nadar College of Engineering, Chennai, Tamilnadu, India

K. Kumar
Vellore Institute of Technology, Vellore, Tamilnadu, India

K. Lavanya
School of Computer Science & Engineering, VIT, Vellore, India

Xu Lu
School of Automation, Guangdong Polytechnic Normal University, Guangzhou, China

Modinat A. Mabayoje
Department of Computer Science, University of Ilorin, Ilorin, Nigeria

M. Madiajagan
Vellore Institute of Technology, Vellore, India

Anand Mahendran
School of Computer Science and Engineering, Vellore Institute of Technology, Vellore, Tamil Nadu, India

Shiwen Mao
Auburn University, Auburn, AL, United States

Opeyemi Emmanuel Matiluko
Department of Computer Science, Landmark University, Omu-Aran, Nigeria

Hammed Adeleye Mojeed
Department of Computer Science, University of Ilorin, Ilorin, Nigeria

Rui S. Moreira
Intelligent Sensing and Ubiquitous Systems (ISUS) Unit at Faculty of Sciences and Technology (FCT)—University Fernando Pessoa (UFP); Laboratory of Artificial Intelligence and Science of Computers (LIACC); Institute of Systems and Computer Engineering, Technology and Science (INESC TEC), Faculty of Engineering—University of Porto, Porto, Portugal

Roseline Oluwaseun Ogundokun
Department of Computer Science, Landmark University, Omu-Aran, Nigeria

Idowu Dauda Oladipo
Department of Computer Science, University of Ilorin, Ilorin, Nigeria

Abdolmotalleb Rastegar
Department of Cartography, Golestan University, Gorgan, Golestan, Iran

Shakirat Aderonke Salihu
Department of Computer Science, University of Ilorin, Ilorin, Nigeria

Arun Kumar Sangaiah
School of Computing Science and Engineering, Vellore Institute of Technology (VIT), Vellore, India

Christophe Soares
Intelligent Sensing and Ubiquitous Systems (ISUS) Unit at Faculty of Sciences and Technology (FCT)—University Fernando Pessoa (UFP); Laboratory of Artificial Intelligence and Science of Computers (LIACC), Faculty of Engineering—University of Porto, Porto, Portugal

Pedro Sobral
Intelligent Sensing and Ubiquitous Systems (ISUS) Unit at Faculty of Sciences and Technology (FCT)—University Fernando Pessoa (UFP); Laboratory of Artificial Intelligence and Science of Computers (LIACC), Faculty of Engineering—University of Porto, Porto, Portugal

S. Sridhar Raj
Pondicherry University, Pondicherry, India

G. Sripriyanka
School of Computer Science and Engineering, Vellore Institute of Technology, Vellore, Tamil Nadu, India

Jose Manuel Torres
Intelligent Sensing and Ubiquitous Systems (ISUS) Unit at Faculty of Sciences and Technology (FCT)—University Fernando Pessoa (UFP); Laboratory of Artificial Intelligence and Science of Computers (LIACC), Faculty of Engineering—University of Porto, Porto, Portugal

Fatima E. Usman-Hamza
Department of Computer Science, University of Ilorin, Ilorin, Nigeria

Xuyu Wang
California State University, Sacramento, CA, United States

Chao Yang
Auburn University, Auburn, AL, United States

Yujing Zhang
School of Automation, Guangdong Polytechnic Normal University, Guangzhou, China

Foreword

An intelligent Internet of Things (IoT) computing system for personalized health care is an evolving discipline that addresses vital research challenges in the medical field. The Internet of Medical Things (IoMT) is an equivalent mechanism involving the connection of health-care equipment to the Internet to provide live communication between physicians and patients. Computational intelligence (in the form of machine learning techniques, optimization, and learning algorithms) is progressively being applied to solve problems in the health-care domain, due to its continuing evolution and the processing capabilities applied to deal with dynamic and voluminous data. Moreover, as health care moves toward data-driven personalized health-care approaches via IoT systems, it is imperative to develop more sophisticated information-based computing intelligence techniques and appropriate frameworks to derive knowledge from data collected, to allow for high-quality health-care delivery.

The extension of IoT sensors to include devices in the medical domain (thus creating IoMT) has improved the quality of personalized health-care services. Recently, the huge volume of big data generated by IoMT sensing devices in the health-care environment has been a major concern. This has created several challenges, including the identification of effective techniques to mine this huge amount of available data. The aim is therefore to determine the optimum applicability of various new technologies in the development of smart systems, wearable biosensors, and to cope with the existing massive sensitive health data toward the goal of providing successful and sustainable medical care solutions across the globe.

The book *Intelligent IoT Systems in Personalized Health Care* highlights the insight of the advanced methodological, technological, and scientific approaches, in light of the development of information and communication technology-based intelligent systems, implementing the use of IoMT sensors for personalized health care, through an array of new technologies comprising artificial intelligence, big data, analytics, wireless communications and networking, embedded systems, wearable computers, etc. This book represents a collection of 11 chapters of real significance in the field, covering various computationally intelligent technologies and their applications in IoT computing, designed to address the need for a seamless integration of computing systems for personalized health care. The book delivers an excellent review of various areas of intelligent decision support and health-care analytics in depth and caters to the needs of researchers and practitioners in the field.

To my knowledge, this is the first attempt to create such a comprehensive collection of IoMT applications, to create a transition in the design and development of computing paradigms for ambient intelligence applied to human health. I expect that the audience will find the book a key resource for a better understanding of *Intelligent Computing Systems*, through IoT platforms for personalized health care. This book is a key realization in measurement in this field and will serve to assist to address, unify, and advance the key research challenges seen in this field across the scientific community.

I congratulate the editors and authors on their valuable work and accomplishment in generating this important work, and I expect that the readers will find the book a valuable source of inspiration to advance their research and to expand professional activity in the field.

Ken Grattan
University of London, London, United Kingdom

Preface

In the modern era of computing, Artificial Intelligence (AI) with the Internet of Things (IoT) in biomedical engineering and informatics plays an important role in personalized health care. With the availability of low-cost sensing in potential computing and medical resources everywhere, there is great scope toward health consciousness and a medical society. Consequently, the IoT offers tremendous advantages/contributions in diversified research areas in health care and technology, and hence this book has focused on health-care computing through wearable biosensors.

In all realms of our daily lives, computers are integrated through wearable biosensors and generate huge information, which creates challenges to our ability in processing enormous bioinformatics data. It will be transforming a meaningful knowledge, especially patient information, medical diagnosis, monitoring, and tracking management on health issues, assisting physicians in clinical decision, early detection of infectious diseases, disease prevention, rapid analysis of health hazards, and so on. Emerging research trends in the construction and deployment of such wearable sensors, actuators, IoT, and AI and machine learning algorithms have revealed the growing significance, potential capability, and the unique benefits that can be brought into personalized intelligent health-care and rehabilitation systems. The new era of convergence of AI and machine learning techniques (supervised-unsupervised and reinforcement learning) with reference to the Internet of Medical Things (IoMT) quality of data and services for industrial applications has three main components: (a) intelligent devices, (b) intelligent system of systems, and (c) end-to-end analytics.

In this book, the chapters would focus on recent developments on integrating AI and machine learning methods, advanced data analytics, and optimization opportunities to bring more computer IoMT data and services, which lead to a new and promising health-care strategy transform. This book aims to bring together researchers and practitioners to address several research and achievements in the field of wearable sensor-based intelligent IoT systems for personalized intelligent health-care and rehabilitation applications.

1 Organization of the book

The Volume is organized into 11 chapters and the following is a brief description of each chapter:

Chapter 1 focuses on identifying the major goals and benefits of integrating IoT architectures for personal home health-care environment. The authors Rui S. Moreira et al. pinpoint home e-health demonstrative use case scenarios, where the benefits of using cloud, edge, and fog architecture types become evident, and the concurrent combination of such IoT architectures inevitable.

Chapter 2 by Chao Yang et al. presents the AutoTag system for unsupervised respiration rate estimation and real-time apnea detection with commodity radio-frequency identification (RFID) Tags. The AutoTag system has been demonstrated by experiments in typical health-care environments. The proposed system also utilized an unsupervised learning, and thus has the desirable advantage of not requiring labeled medical data, making it inexpensive and easy to deploy.

Chapter 3 by Raheleh Khanduzia et al. presents a novel cooperative hierarchical protection-interdiction-allocation model for a real-world health network, where clinics, hospitals, and medical centers are subject to the cooperative attacks and defensive operations. This chapter investigates a new model of the interdiction median problem with protection on a health-care network called cooperative hierarchical and interdiction median problem with protection (CHIMP).

Chapter 4 by Sridhar Raj et al. discusses the Internet of medical things (IoMT) devices and functionalities. The architecture of applying deep learning to the IoMT base is addressed in this chapter. The technical and market level challenges of the incorporating IoMT devices are included.

Chapter 5 by Kayode S. Adewole et al. addresses in detail the fundamental concepts of cloud computing and IoT applications within the context of health-care data management. The cloud-based framework for cardiovascular disease (CVD) prediction and diagnosis has been proposed to address some of the research issues highlighted in the chapter. The framework comprising both cloud architecture and IoMT for personalized e-healthcare has been implemented for the prediction and early diagnosis of CVD.

Chapter 6 by Sripriyanka et al. gives a clear picture on the review of IoMT technologies and services in remote health-care monitoring. This hybrid theoretical framework that discussed the numerous security and privacy solutions to the IoMT industry is addressed in this chapter.

In Chapter 7, the author Amos Orenyi Bajeh et al. illustrate the application of computational intelligence (CI) models in IoMT big data for heart disease diagnosis in personalized health care. This chapter examines the recent applications of CI in the form of machine learning techniques, optimization, and fuzzy algorithms. In addition, application of big data analytics in predicting heart diseases based on IoMT is reviewed.

Chapter 8 presented by Xu Lu and Yujing Zhang proposes the human anteflexion angle recognition algorithm based on edge detection and feature point extraction. The adaptive Gaussian filter is proposed in this chapter, which is based on pixel point gray value and local mean difference, and an improved Sobel template is used to calculate the gradient amplitude and direction, which improves the antinoise performance of the algorithm.

Chapter 9 by Joseph Bamidele AWOTUNDE et al. investigates the fundamental concepts of genomic data and machine learning within the context of prediction and classification in diabetes mellitus (DM) management. The use of genomic data and its applications in health-care practices is a new and fast-growing trend in health-related fields.

Chapter 10 by Anjana and Lavanya analyzes the suicide rate data with the tree type visualization taking World Health Organization (WHO) region and country as standards for the evaluation. The analysis of the data was done by creating a rule-based decision tree from a set of decision rules over the indexed data model represented in a tree template.

Chapter 11 emphasizes the possible communication and protocol architecture of IoT health-care systems. It also gives an overview of various applications of IoT in health care along with the several challenges such as privacy issues, security requirements, and security attacks and then a short discussion about the possible solutions.

2 Audience

The intended audience of this book includes scientists, professionals, researchers, and academicians, who deal with the new challenges and advances in the specific areas mentioned above. Designers and developers of applications in these fields can learn from other experts and colleagues through studying this book. Many universities have started to offer courses on IoT at the graduate/postgraduate level in information technology and management disciplines. This book starts with an introduction to IoT and intelligent computing paradigms and its application on the health-care domain, hence suitable for university-level courses as well as for research scholars. Their insightful discussions and knowledge, based on references and research work, will lead to an excellent book and a great knowledge source.

Arun Kumar Sangaiah[a] and Subhas Chandra Mukhopadhyay[b]
[a]Vellore Institute of Technology (VIT), Vellore, India
[b]Macquarie University, Macquarie Park, NSW, Australia

Acknowledgments

The editors would like to acknowledge the help of all the people involved in this book project and, more specifically, the contributing authors and reviewers that took part in the review process. Without their support, this book would not have become a reality.

First, the editors thank each one of the authors, who contributed their time and expertise to this book.

Second, the editors acknowledge the valuable contributions of the reviewers regarding the improvement of quality, coherence, and content presentation of chapters. We deeply appreciate the comments of the reviewers who helped us refine the context of this book. Most of the authors also served as referees who are highly appreciated for their double task.

Finally, our gratitude goes to our friends and colleagues, who were so generous with their encouragement, advice, and support.

Arun Kumar Sangaiah[a] and Subhas Chandra Mukhopadhyay[b]
[a]Vellore Institute of Technology (VIT), Vellore, India
[b]Macquarie University, Macquarie Park, NSW, Australia

CHAPTER ONE

Combining IoT architectures in next generation healthcare computing systems

Rui S. Moreira[a,b,c], Christophe Soares[a,b], Jose Manuel Torres[a,b], and Pedro Sobral[a,b]

[a]Intelligent Sensing and Ubiquitous Systems (ISUS) Unit at Faculty of Sciences and Technology (FCT)—University Fernando Pessoa (UFP), Porto, Portugal
[b]Laboratory of Artificial Intelligence and Science of Computers (LIACC), Faculty of Engineering—University of Porto, Porto, Portugal
[c]Institute of Systems and Computer Engineering, Technology and Science (INESC TEC), Faculty of Engineering—University of Porto, Porto, Portugal

1. Introduction

The exponential growth in the semiconductor domain favored the explosion of cost-effective sensor-based systems, empowered by standard wireless communication technologies such as Bluetooth, LoRA, ZigBee, and 3G/4G, commonly targeted to ubiquitous intelligent applications and services. Suddenly, the Internet became the backbone for merging millions of things (aka motes or smart objects), aiming particularly to cope with: massive scale, heterogeneity, and interoperability, simplified integration, easy development, and the deployment of smart things.

The seminal term of the Internet of Things (IoT) was firstly introduced in 1999 by Kevin Ashton [1]. This concept has been evolving since then, especially in the last 15 years, driven by an enormous number of both enterprise and academic projects [2]. Such projects focus on allowing the establishment of smart interactions between people and surrounding objects and services, enhancing the emergence of context-aware Ubicomp systems and applications. IoT became synonymous with a global infrastructure enabling advanced services for interconnecting physical and virtual things, anchored on standard interoperable networking technologies. This was particularly true for smart home and personal healthcare settings.

The architectural paradigms for building IoT-based systems followed both, a spatial and temporal evolution, from earlier larger-scale cloud-based approaches, mainly devoted to store, process, and visualize data, to the more

Intelligent IoT Systems in Personalized Health Care
https://doi.org/10.1016/B978-0-12-821187-8.00001-0

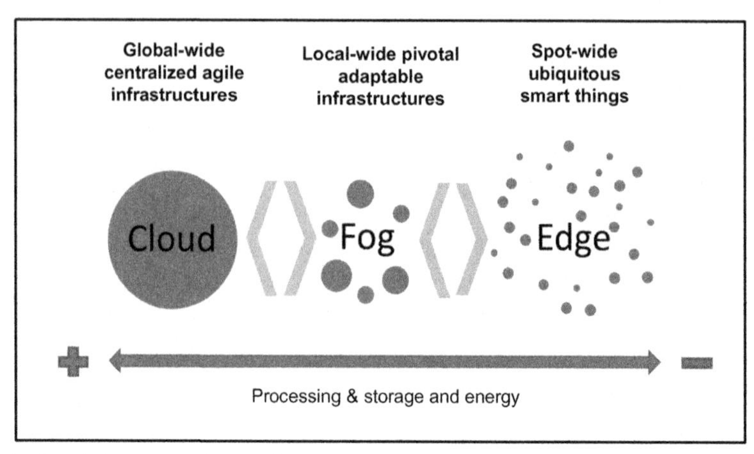

Fig. 1 Cloud, fog, and edge architectures coverage.

recent smaller-scale edge and afterward fog computing complementary architecture paradigms, more tailored for real-time local processing and privacy concerns (see Fig. 1). All of these architecture styles are typically bonded to IoT projects, and De Donno et al. [2] provide an interesting illustrative temporal evolution of these paradigms, placing edge and cloud architectures emergence before the more recent fog computing architectures. Each of these architectural approaches provides a contrasting granular scale, though being complementary to the goals and potential application scenarios.

Cloud computing has been the most predominant paradigm since 2008, though edge and more lately fog computing approaches had been gaining more predominance, especially considering the past 5 years. Some authors consider the fog computing architecture as a subset of the edge computing paradigm. In such cases, the former is seen as an instance of edge computing, mostly because of both advocates the transference of computing and processing to the edge of the network, near real-world objects or things that are able to perceive and act upon the surrounding environment. In this work, however, fog computing is considered a first-class paradigm, that is, a refinement of the edge computing architecture. A fog architecture addresses concrete requirements and plays the role of a middle layer, gluing remote cloud services to local smart edge devices.

In short, the cloud-based architectures offer global-wide centralized agile facilities, focusing on more transversal aspects of IoT systems. The fog-based architectures provide local-wide pivotal adaptable infrastructures, tackling issues more related to response time, energy autonomy, and security concerns. Finally, the spot-wide edge-based architectures embody devices with sensing and acting behaviors, thus transforming regular objects into smart things. The following bulleted paragraphs enumerate some of the major characteristics of each of these IoT-based architecture paradigms:

1. IoT cloud as an architecture designed to enable centralized management and provision of physically hyperlinked context-based services:
   - Interconnecting physical (world objects) and virtual (information) things, based on ubiquitous on-demand interoperable communication technologies.
   - Allowing shared usage of pools of scalable and configurable computing resources for data collecting, storage, processing, analysis, visualization, etc.
   - Promoting rapid and efficient use/release of resources with minimal management effort and interaction.
   - Leveraging the connectivity and the integration of heterogeneous user-centric things.
2. IoT fog as an architecture designed to enable complex distributed services, close to the end devices with the following goals:
   - Reducing communication latency by placing fog devices located closer to end devices.
   - Preserving end devices battery, since the communication time is shorter, hence end devices can turn off their radios and go into deep sleep much faster.
   - Enabling real-time services because the fog device (with better processing resources than end nodes) may perform sensor fusion and trigger faster responses to emergency situations.
   - Increasing local processing in fog devices is not usually restricted to energy spending and processing power, hence they can locally support more processing intensive tasks on behalf of end devices.
   - Improving user privacy since raw user data is never sent to the cloud, rather it is processed locally in local fog devices.
   - Improving ubiquitous system scalability, alternatively to centralized cloud services, which may not scale properly to the massive number of expected IoT devices in the near future.
3. IoT edge as an architecture designed to enable computation in the end nodes (aka motes or things):
   - IoT end devices (things) become presenting increasingly processing and storage capabilities, thus being capable of, independently, transforming raw data into more elaborated contextual information.

- Typically smart end devices are freed from the burden of constantly streaming sensor raw data, allowing them to implement a more aggressive energy management strategy and improve their energy consumption and autonomy.
- Smart objects present a more autonomous behavior, promoting easy distribution less dependency of remaining Internet infrastructure, thus easing their use, deployment, integration, and fault tolerance.

In the scope of this book, we argue that the combination of IoT-based architectures, provides killer application features, briefly enumerated above, for several medical care areas, in general. Personal home care settings, in particular for this chapter, are increasingly becoming core application areas since they provide quite a lot of suited scenarios for applying the key characteristics and benefits of each IoT architecture paradigm.

As illustrated throughout this chapter, most of the application examples focus on providing data collection and monitoring solutions for several physicals, health, and environment aspects, personal activity recognition, prediction or detection of dangerous or emergencies, support treatment adherence by controlling, for example, medication intake, promoting, and baking physical or fitness activities, etc. [3]. The continuous emergence of sensors and actuators embodied with wireless capabilities have been enabling its application on manufacturing multiple smart personal and health-related devices. Such cost-effective devices empower the arising of various monitoring and supporting IoT solutions for manifold home ambient-assisted living scenarios. This document aims to organize and emphasize the characteristics and benefits of each IoT architecture paradigm with concrete application use cases highlighting the characteristics of each approach.

The chapter is organized as follows: Section 2 summarizes the major goals of cloud computing architectures, describes relevant-related work on could-computing for e-health, and presents some demonstrative IoT cloud-based e-health use case scenarios; Sections 3 and 4 follow the same rationale, respectively, for the use of fog-computing and, afterward, edge-computing architectures; these sections advocate not only the broad benefits of each IoT computing architecture but also demonstrate, in practice, with the use of several e-health scenarios, the need for combining the three approaches, which makes inevitable sense. Final concluding remarks and future work are presented in Section 5.

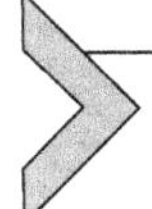

2. Cloud computing architectures

2.1 Major goals of cloud computing

Cloud computing is an architecture model tailored for enabling ubiquitous and on-demand access to a shared pool of configurable and adaptable computing resources, for example, data storage and monitoring, device and systems automation and integration, data mining, visualization, analysis, etc. Cloud-based IoT systems emerged mainly to simplify the rapid provision and release of solutions, with minimal management efforts and interactions with service providers. Moreover, IoT cloud platforms provide valuable application-specific facilities, both for vertical and transversal domains.

A cloud-based architecture is tailored for the provision of an highly scalable infrastructure (i.e., hardware and software) for supporting an agile and adaptable ecosystem of services [2]. Pivotal key features of such infrastructure are typically:

- the ability to automate the real-time provision of on-demand resources for masses of hosted IoT application;
- the facilities to provide and sync shared resources and services among concurrent agents of IoT applications;
- the faculty of supporting several communication technologies and networking protocols for IoT application (e.g., MQTT, CoAP, AMPQ, etc.); and
- the capacity to measure and account, in real time, the required resource usage for each individual IoT application, thus allowing specific and fare pay-per-use business models.

The raising needs for cloud-based architectures was endorsed by several IoT cloud services providers that tuned their platforms for leveraging market suitable applications. Currently, dozens of IoT cloud platforms compete in the market, some are interesting open-source projects (e.g., KAA, Open Remote, ThingSpeak, etc.) while most of the enterprise solutions. The focus of these IoT cloud platforms is summarized in four core categories presented in Table 1. This classification does not intend to be closed, rather presents a high-level view of existing cloud-based platforms, aggregated by their major characteristics or concerns for supporting IoT applications.

The classification of cloud frameworks/providers, also presented in Table 1, is not to be taken strictly since generically many of them possess

Table 1 Summary of major IoT cloud core categories and example providers.

Core goals of cloud-based frameworks	Cloud frameworks/providers
Hardware and apps integration and management: simplify device configuration, integration, system automation and management	For example, KAA, Carriots, Temboo, SeeControl, SensorCloud, Etherios, Xively, etc.
System and heterogeneity management: promote agile development and deployment of diverse and disparate applications	For example, thethings.io, Exosite, Open Remote, Axeda, Oracle IoT cloud, etc.
System deployment and monitoring management: empower cost-effective management of data storing and monitoring	For example, Nimbits, ThingWorx, Echelon, ThingSpeak, etc.
Data management, analytics and visualization: provision of tools for data monitoring, mining, and dashboard visualization	For example, Jasper, Plotly, GroveStreams, IBM Watson, Google IoT, AWS IoT, MS Azure, etc.

more than one of the listed major characteristics. However, most of the solutions, although covering several application aspects, give more emphasis to certain cloud functionalities typically associated with their business models. Given the wide set of existing cloud-based solutions, what criteria can or should be used to select the most suited framework or provider? There are a few steps or criteria, listed below, that may be followed to select an IoT cloud platform:

1. Check specific requirements:
   - Endure data storage, monitoring and management, and security and scalability
   - Support for particular devices or functionalities (hardware and software)
   - Uphold set of integrated services and platforms
   - Underpin for analytics and data visualization
2. Pricing model:
   - Evaluate and compare open-source versus paid PaaS platforms
   - Free open-source platforms
   - Pay-as-you-go model (charged for the resources you actually consume, e.g., AWS IoT Core)
   - Pay a subscription model (billing flat fee per month, e.g., Salesforce)

3. Availability of a free tier and effective technical support:
   - Existence of a free trial period (e.g., Oracle has no free option)
   - Ability to test a prototype, that is, freely run a simple project, for example, AWS offers free tier option although with restrictions)
4. In house development team experience:
   - Balance project requirements and team knowledge with platform supported tools
   - Evaluate support for development languages, hardware device platforms, and involved technologies (e.g., communication protocols, wireless technologies, etc.)

2.2 Related work on cloud architectures for e-health

Cloud computing has been the prevalent paradigm when covering major IoT applications in the e-health area. In particular, personal healthcare places several challenges and opportunities for IoT architectures [4, 5]. Many cloud-based projects address multiple aspects of remote monitoring [6], home-assisted living, and managing quality of life while reducing healthcare costs. Such projects focus on coping with current and future limitations of health systems, due to critical shortages of medical staff, increase population aging, and associated widespread of chronic illnesses and medical conditions requiring continuous monitoring.

In particular, personalized e-health provides a rich field for physiological monitoring (e.g., blood pressure, body temperature, heart rate, sugar level) [7], real-time remote monitoring [8] of activities of daily living (ADL) and emergency care, pattern analysis, and machine learning (ML) classification for early diseases diagnosis [9], and even support for treatment adherence, and fitness and sport promotion. Therefore, cloud-based architectures play a key role due to their ability to support: (i) real-time information gathering; (ii) long-term and continuous data collection facilities; and (iii) information completeness and tools for mining and visualizing (cf. support for medical staff diagnoses and caretaker information) [10]. In fact smart cloud-based personal e-health systems extend health care from traditional clinical facilities to home settings, by allowing 24/7 physical monitoring, reducing labor costs, and increasing efficiency [11, 12].

Several earlier IoT projects focused precisely on physiological monitoring. For example, the Code-Blue proposes a wireless sensor network for assisting the triage process and monitoring victims in emergencies and disaster scenarios. The caregivers accessed query graphical user interfaces to access data obtained from sensors [13]. The AMON project embedded many sensors (cf. blood pressure, pulse oximetry, ECG, accelerometer, and skin temperature) into wrist-worn devices directly connected to a telemedicine center via a GSM network, allowing direct contact with the patient [14]. AlarmNet focuses on continuously monitoring and ambient-assisted living of independent home residents. The system integrates information from sensors in the living areas as well as body sensors. It features a query protocol for streaming online sensor data to user interfaces [15]. The Mobile Cloud for Assistive Healthcare (MoCAsH) is a mobile could infrastructure providing portable sensing integrated with context-aware middleware for automating deployment and collaboration of agents [16].

More recent projects on personal health care for smart environments propose low-cost bodycentric systems for gathering information about the user's living environment (e.g., temperature, humidity, etc.) together with data about the human behavior. This monitoring info about the user's health is intimately bonded with cloud-based remote assistance activation [17]. Another project, named CUIDATS, provides an RFID-based and wireless sensor networks (WSNs) monitoring system able to track the location, movement, and vital signs of patients (e.g., blood pressure, heart rate, temperature, etc.) through the use of wearable devices monitoring [18]. In [19], the authors combine the use of cloud together with wireless ZigBee-based body sensor networks to enable a reliable collection stream of health data. UbeHealth is a ubiquitous healthcare framework combining data collected from edge devices, to be used on cloud-based high-performance computing infrastructures on which deep learning and big data algorithms are able to run with on demand resources [20]. This framework uses deep learning algorithms to predict network traffic and implement network quality of service (QoS), which is used by Cloudlet and network layers to optimize data rates, data caching, and routing decisions. The major focus of this project addresses network QoS parameters challenges (e.g., latency, bandwidth, energy consumption, etc.) to be faced by the next generation of global smart cities networked healthcare systems.

2.3 IoT cloud-based e-health use case scenarios

This section presents three personal homecare use case scenarios where cloud-based architecture fits the needs of the applications, namely: (i) support for data storing and dashboard visualization of personal health parameters and real-time activity registration; (ii) manage services integration for monitoring medicine intake and assisting personal treatment adherence; and (iii) sustain the registration of contextual functional information about home appliances for interference detection through pattern analysis.

In these scenarios, there is a need for external storage and computing facilities that are not available directly in the IoT devices. Moreover, such scenarios may gain also from combining information from several similar home settings. Thus, a cloud-based architecture enables to store and process information gathered from a panoply of devices and home settings. Moreover, a cloud solution is capable of actively monitoring patients and commercial off-the-shelf (COTS) devices, and also providing the means and services for integrating these devices in a home environment.

2.3.1 Use case I: Personal health data storage and visualization

Project [21] purposes the use of a ZigBee ad hoc network for collecting personal health parameters and ADLs from in house inhabitants. The proposed system allowed, for example, real-time monitoring of elderly people in a nursing house. Each elderly user carries an Arduino Wee-based mote coupled with a ZigBee shield, a temperature body sensor, and a three-axis accelerometer. The mote was able to produce real-time series of elderly ADLs, such as sitting, lying, walking, running, and falling.

The collected data for each individual were conveyed to a gateway which then transmitted it, via a 3G link, to a cloud server. The server provides services for storing, visualizing, and posterior analysis by doctors and nurses (see Fig. 2). The could service enables also triggering alarm events for notifying noteworthy or emergency situations (e.g., inactivity, falls, etc.) to day-to-day remote caretakers. Furthermore, ML techniques could be used to enable the detection of unusual or abnormal patterns of ADL.

2.3.2 Use case II: Personal treatment adherence via drug dispenser

The integration and orchestration of ubiquitous devices and mobile apps through cloud services provide endless advantages on concrete application use cases. For example, in [22], the authors advocate the usefulness of cloud services for integrating a smart drug dispenser (DD) together with an Android mobile app, for controlling and improving personal treatment adherence. The goal was to combine the use of a smart DD with a personal ubiquitous assistant mobile app to help and alleviate the problems of lack of memory and disorientation associated with aging, mostly on elderly people.

This project proposes an Arduino-based medicine dispensary prototype, embodied with a Bluetooth shield for connecting and synchronizing

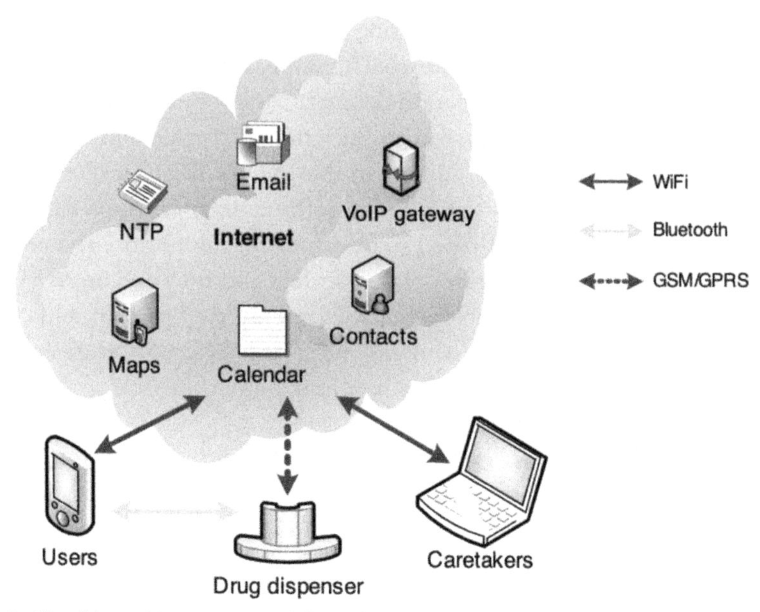

Fig. 2 Cloud-based integration of drug dispenser devices.

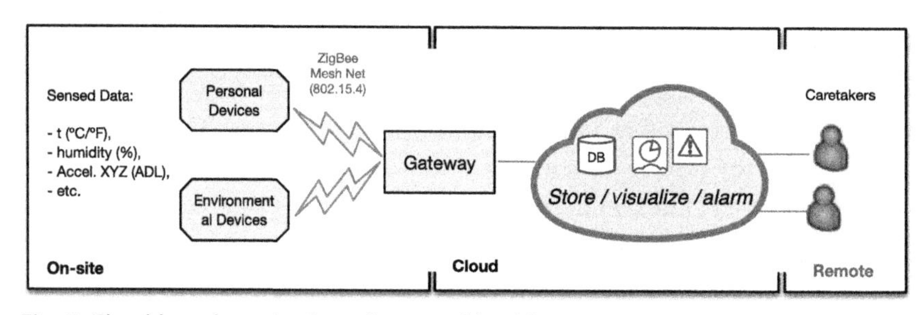

Fig. 3 Cloud-based monitoring of personal health parameters and ADLs.

medicine intake information with a personal assistant mobile app. The smartphone app and the DD were synced to manage personal information and alerts, by taking advantage of a wide range of services distributed by the cloud (cf. calendar, location, maps, NTP, SMS, e-mail, etc.) as pictured in Fig. 3.

The DD prototype was built on an Arduino Mega platform with a set of shields and specific hardware (cf. GSM/GPRS shield, Bluetooth BlueSMiRF shield, LCD, LED, push-button, and buzzer). During the initialization stage the DD synchronized the date/time with the smartphone, via Bluetooth or SMS. In the latter case, the dispensary sends an SMS to the smartphone and waits for the automatic response, using the date/time of the messaging service to update its own date/time. Afterwards it syncs its own calendar events (i.e., medicine intake moments) with the Google calendar. After the synchronization period, the DD displays information of next medication intake on the LCD. When reaching the scheduled intake moment the DD rotates the medicine dispensing cup and triggers the buzzer alarm beeps. The user must press the push button to stop the beep and acknowledge the intake. The DD then rotates to the resting position. Whenever the medicine is not taken the DD times out and sends an SMS to the caregiver, alerting for the missed dosage.

The personal ubiquitous assistant app focuses on: monitoring the location of the elderly and interacting with the DD. Among other features, it was possible to: view the current location of the elderly person and the location of caregiver contacts; show the return path back home or to the caregiver current location; show and manage the list of scheduled daily activities; send alerts by SMS when the security perimeter is exceeded or a scheduled event is missed; maintain a location history path; configure the application, namely the refresh rate of the map, the selection of caregivers, the definition of the security perimeter; and activate the sending of alerts by SMS, etc.

This project demonstrates the integration and orchestration of Ubicomp systems through convenient cloud services. Both the DD device and the mobile app interact with services available on the cloud to combine several useful features into people's daily lives. The cloud-based architecture promotes and facilitates the integration of devices, making global management and coordination easier and more serviceable.

2.3.3 Use case III: Interference detection between personal health devices

Nowadays, COTS devices and applications are pivotal in the massive deployment of pervasive computing technology around home settings. Moreover, the integration of these systems in the same household may result in unplanned functional interactions between users and their personal home devices (cf. entertainment, communication, and health-related devices). Such unplanned interactions are a serious concern when, for example, a phone or a TV set interferes with the behavior of a drug dispensary, accounting for medicine intake. Such functional interference problems are tackled in [23] with a novel graph-based approach for representing the expected behavior of COTS devices and their interactions. Both the process of extracting the Graph of Expected States (for each COTS system) and the runtime checking of unexpected behavior are provisioned by could-based services running the appropriate algorithms.

Every COTS device is planned to work properly on its own, however, when placed together with other existing systems that they may interfere. Imagine, for instance, a security system monitoring house window breaks during owners' absence, and an home automation system that may be configured to open windows for ventilation purposes. Both systems will work properly in isolation but may cause unexpected behaviors on each other when deployed together (since window opening for ventilation may be detected as a break-in). This issue is known as functional interference and can be tackled by comparing the expected behavior of each COTS device against the runtime observed behavior of those devices.

In [24], the expected behavior of COTS devices is modeled using Graphs of Expected States (GoES). These graphs can be automatically extracted through the use of unsupervised ML algorithms (e.g., association-based Closet+, cluster-based k-medoids), which was the case for the GoES of both a DD and a Phone/VoIP system [25]. The GoES are then extended with the representation of possible interferences among node states and environment medium. Since the GoES extraction algorithms may be computing intensive, their runs are usually performed on cloud premises. Another cloud service was used for monitoring the runtime behavior of COTS devices and build the Graph of Observed States (GoOS). Then a pruning algorithm compares both graphs (GoES and GoOS) to detect unconformities between the expected and current behavior of observed COTS system [24]. When some unexpected behavior is detected between two devices or between devices and users, some preplanned events may be triggered to signal and even correct such occurrences (e.g., prevent opening windows when the security system is on; disconnecting the TV set when DD is beeping for alerting medicine intake, etc.).

In this scenario, the cloud architecture allowed storing the state samples for all COTS devices, and then use them to build the GoES and also compare these with real-time GoOS. Moreover, the use of cloud for storing knowledge about multiple COTS devices deployed in different house settings, could be useful in more situations. For example, to diminish the installation bootstrap time of several COTS devices on new home settings, by reusing GoES from similar house environments, that is, the initial selection of COTS devices and respective GoES could be facilitated by organizing house profiles (i.e., number of inhabitants, activity habits, location, set of deployed COTS devices, etc.). This way any new installation could be booted from configurations used on similar houses, thus diminishing the time for learning new state patterns.

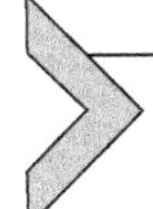

3. Fog computing architectures

3.1 Major goals of fog computing

Cloud computing is omnipresent and plays an important role in today's world of the IoT. The IoT local infrastructure is, most of the time, tailored to accomplish mainly the basic functions of sensing, acting, and communicate or relay raw data to the network. Hence, certain features, such as storing or processing data, are most of the time absent locally, being associated exclusively with the cloud where data will be processed and stored for future analysis. Frequently, IoT devices communicate directly with a cloud server using technologies such as GPRS or NB-IoT or through a LoRa or Sigfox gateway, enabling data reports to the cloud via the Internet.

Although cloud computing brings more processing power and storage, some disadvantages are obvious when is taken into consideration the increasing number of IoT devices, which have to do the heavy lifting of reporting collected data. This communicating process consumes a considerable amount of energy. If the IoT device is battery powered, it will be negatively impacted. Besides energy consumption, scalability is another concern. Since IoT devices generally do not have great processing power or storage, the cloud infrastructure must be able to serve all the devices connected to the system, and the more devices are in use the more cloud resources are required to support it. The IoT devices often are responsible for the production of streams of continuous data and this puts some pressure into the existing storage capabilities in cloud computing that have to adapt to the amount of data been generated.

Fog computing [26] is the distribution of cloud computing resources to small agents closer to the IoT devices, in the edge of the network where they are deployed, taking on the role of cloud computing in a smaller scale. As such, these devices can take the responsibility for storing, processing, and reporting information generated by the IoT devices. This creates a dynamic where the IoT device does not necessarily know the existence of a cloud server and it does not need to. The IoT device reports data more efficiently since it has to send data to another much closer device, resulting in battery life improvements for the IoT end node. Since the fog device stores and processes data from all nearest IoT devices, only processed data or a summary of

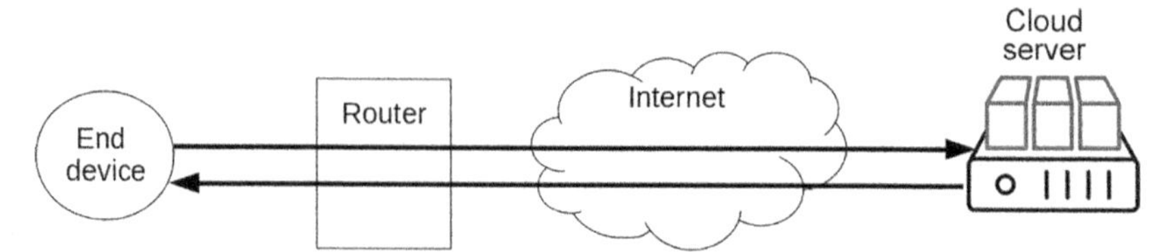

Fig. 4 Communication in cloud-based IoT systems.

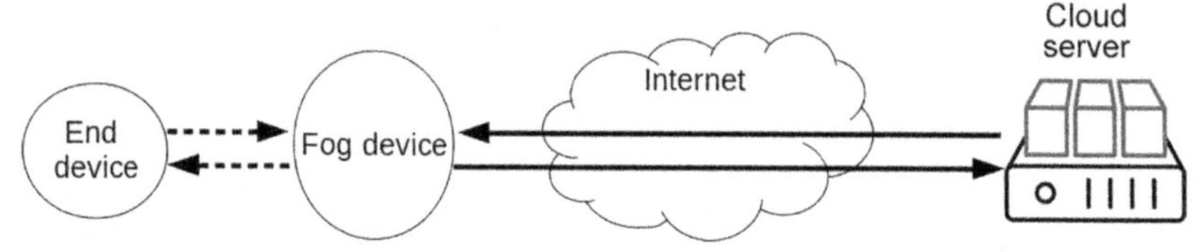

Fig. 5 Communication in fog-based IoT systems.

it is, most of the time, sent to the cloud, saving storage space and simultaneously improving the privacy levels for IoT environments.

Traditionally, in cloud-based IoT systems, the application logic is implemented between end devices and the cloud server, using a router to provide Internet connectivity (Fig. 4). The router is only responsible for packet forwarding at the network layer without any intervention in the application protocol. On the other hand, in fog-based IoT systems, the application protocol operates between the end device, the fog device, and the cloud server (Fig. 5).

From the end device perspective, the existence of the fog device enables a significantly lower communication latency. As such, the end device can save battery by turning off the radio interface and going into a deep sleep much faster and more often [27]. From the cloud server perspective, the raw data processing performed on the fog device can save storage space and improve scalability. This is of paramount importance to cope with the expected exponential growth on the number of connected IoT devices, each one endlessly generating vast amounts of sensor data. Fog computing handles this problem by saving most of the generated data locally and only sending partial or summarized processed data to the cloud. However, the original raw data can still be kept in the network-accessible fog device. Data integrity and fault tolerance are important requirements, so the data replication between the fog and cloud layers is important [28]. For the system as a whole, the existence of the fog device enables use case scenarios where local sensor fusion is desired and fast feedback to urgent events is of major importance. The fog device is usually not restricted in processing power and energy spending like the end devices. So, it is a suitable place to perform computationally intensive tasks on sensor data like pattern recognition, deep learning, sensor fusion, etc. Since it is located close to the end devices it can also react in near real-time to emergencies, reducing the system response time. Cloud-based systems struggle to meet the real-time requirements of several application scenarios due to unpredictable communication latency and server load.

3.2 Related work on fog architectures for e-health

One typical application scenario pointed for fog computing is in e-health, where real-time decisions are required. For instance, a monitored user can have his/her information sent to a nearer (fog) device that can store and analyze data and determine if the user is in a critical health condition and trigger necessary reactions. In a cloud-based architecture, the latency involved in waiting for a response could be fatal. As Monteiro et al. [29] suggest, using a fog architecture can improve response time and enable the use of deep learning algorithms to determine the cause of the emergency and act upon it.

Another system devoted to e-health in a smart-home context is SPHERE [30]. It was deployed in the real world in 30 homes in Bristol, UK, and focus on three sensing areas: environmental, video, and wearable. They use a home gateway to aggregate and process all the sensor data. No information is sent to the Internet without the user consent. The system generates rich datasets where ML algorithms can be used to improve the understanding of user behaviors in ambient-assisted living scenarios. Their experience demonstrates the need for user-centered design, capital to gain user acceptance. They also state that real-time data transmission to a remote location is not practical so the system has to be designed accordingly, that is, with a local home gateway.

Tewell et al. [31] also present a monitoring system devoted to ambient-assisted living using small low-cost devices in a smart home. They monitor "meaningful activities" that they define as "physical, social and leisure activities... that provide emotional, creative, intellectual and spiritual stimulation." Besides common electronic sensors, they also use Bluetooth beacon sensors to provide location data. Their target is helping people affected with cognitive disabilities (dementia, Parkinson's disease) in daily living activities. They use a single manufacturer sensor kit (Xiaomi) stating that it provides a LAN API for local sensor data processing without having to send it to cloud servers. All processing is done on the premises reducing the latency and ensuring data privacy.

Quintana and Favela [32] describe a project to provide augmented reality annotations to assist persons with Alzheimer's disease or memory problems and their caregivers. They tag the important objects in the smart home and provide help in the form of audio, text, or images when a mobile phone is used to identify the tags. They use computer vision algorithms to process the images gathered from the mobile phone on a server to identify the user location and the tags present on that location. Their architecture requires a powerful server located close to the client to reduce the system latency and provide near real-time operation.

3.3 IoT fog-based e-health use case scenarios

Use case I: Reference architecture for ambient-assisted living

One of the challenges that ambient-assisted living seeks to solve is how to improve the quality of life of people with special needs, such as the elderly, in their homes, so that they can carry out their daily activities with some degree of independence.

This use case presents a reference architecture for an AAL smart system [33]. The architecture is presented in Fig. 6. It is composed by three modules, whose tasks are to automate and standardize the collection of data, to relate and give meaning to that data, and to learn from it. It enables real-time response to emergencies and also a long-term analysis of the user's daily routine useful to induce healthier lifestyles.

The automation module handles the communication with different sensors, actuators, and existing applications over the network, as well as the interaction with different personas and the remaining modules. The knowledge representation module performs inference rules based on data input from the automation module (cf. sensor values, persona activities, etc.). The ML module is able to classify user activity patterns by processing input features also received from the automation module.

The implementation of the automation module was based on the environment automation framework OpenHAB.[a] OpenHAB is an open-source software that centralizes communications with many heterogeneous types of technologies and protocols, commonly used for automation and IoT. This broker allows the integration and management of different devices and their

[a] See https://www.openhab.org.

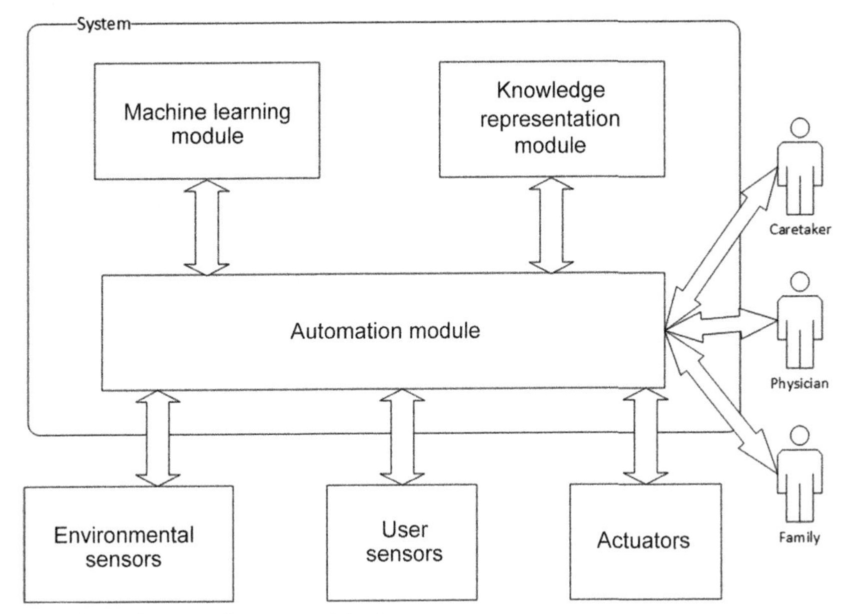

Fig. 6 Smart home reference architecture.

data. The OpenHAB facilitates considerably the effort to deal with a vast variety of protocols and home automation devices.

The primary goal of knowledge representation module is to provide a semantic representation of data received from the automation module. This representation is achieved using an Ontology Language, OWL,[b] and implemented by Protégé[c] and Jena[d] frameworks. The OWL allowed the creation of SWRL[e] rules to infer complex activities based on a temporal sequence of simple activities.

The system can identify daily living activities with different levels of complexity using a temporal logic. Examples of those activities are

- *Simple activities* that result directly from a sensor trigger, like dismissing an alarm from a DD to acknowledge a medicine intake.
- *Complex activities* that require the participation of different sensors according to a temporal logic. As an example, detecting that the user is watching TV in the living room, requires notifications from several sensors and devices in a specific order.

[b] See https://www.w3.org/2001/sw/wiki/OWL.
[c] See https://protege.stanford.edu.
[d] See https://jena.apache.org.
[e] See https://www.w3.org/Submission/SWRL/.

- *Detecting abnormal behaviors* like inconsistent sensor readings when considering the activity that the user is performing. As an example, a high heart rate value may be consistent with the user doing exercise but not with watching TV.

3.3.1 Use case II: Monitoring daily living activities through computer vision

Image sensors can be very helpful in the e-health context in general and in the ambient-assisted living scenarios in particular. Electronic sensors have some limitations because we can only retrieve sensor data from objects or places where they are installed on. Image sensors, on the other hand, can be placed in strategic locations in the household providing a broad view over the rooms that require monitoring. These sensors, when coupled with powerful deep learning models on computer vision, allow the retrieval of important information from the environment. However, running those models, usually requires computational resources not available in common IoT end devices. Transmitting the video feed to a cloud server for processing is not an option because of obvious privacy issues, network latency, and system scalability. This is the perfect ground for the fog computing paradigm. Placing a fog device with suitable resources to run the models inside the premises, ensures that the video feed never leaves the residence, and enables a near real-time follow-up of the user activities. Only these high-level activities may be reported to the outside world without compromising the user's privacy. In this scenario, we present a project using computer vision, deep neural networks, and sensor fusion to monitor daily living activities. This project extends the reference architecture presented in the previous scenario and has the following requirements:

- identify persons of interest in a household with several individuals;
- follow them throughout the household recognizing the room where they are present combining image and other electronic sensors;
- recognize their posture (lying, sitting, standing); and
- determine if the user is moving or not.

The monitoring application processing pipeline is presented in Fig. 7. Each phase is detailed in the following steps:

1. A frame is captured from the image sensor and a deep learning model (GluonCV[f]) searches the image for person shapes. If the detection is

[f] See https://gluon-cv.mxnet.io.

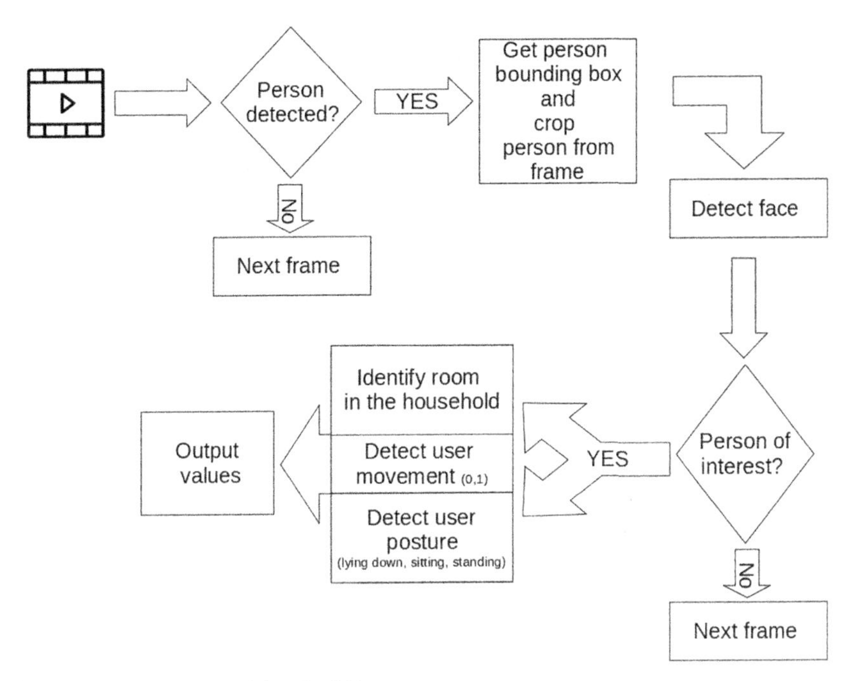

Fig. 7 Using computer vision in AAL.

successful, their bounding boxes are sent to the next phase, otherwise, the frame is discarded.

2. The person's face is isolated from their bounding boxes using a face recognition algorithm.[g]
3. Each face is compared with a person's of interest database for a match through the OpenCV[h] local binary patterns histogram algorithm. If no match is found, the frame is discarded.
4. In the case of a positive identification, three threads work on the frame to determine:
   (a) the room in the household where the person is present using the Inceptionv3 [34] deep neural network,
   (b) the person's posture (lying down, sitting, standing), also through Inceptionv3, and
   (c) if the person is moving or still using SqueezeNet [35].

As we are working with equipment that may have limited resources, the neural network model used has to consider that restriction. Squeezenet

[g] See https://github.com/ageitgey/face_recognition.
[h] See https://opencv.org.

model was an obvious choice to determine user movement since the learning process requires a small dataset and the model can classify images in less than 500 ms with an accuracy of 85%. For bigger datasets inception v3 was chosen, as the training is quite fast, and its classification takes between 1 and 2 s with an accuracy between 80% and 90%. When the three modules finish their classifications, the outputs (location, posture, and movement) are used to create an event that is stored for future analysis. We plan to merge this project in the reference architecture for ambient-assisted living presented in the previous scenario. Using the temporal logic and ML, we intend to search for (short- and long-term) patterns on the user behavior. Those patterns are useful to follow the user's health condition.

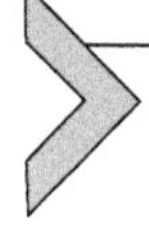

4. Edge computing architectures

4.1 Major goals of edge computing

Today's world is deeply connected by computer networks and dominated by the ubiquitous computing paradigm. More and more devices, typically used in daily life, are linked to the Internet. Devices, such as smartphones, smartwatches, smart lights, smart thermostats for automatic temperature control, video cameras, set-top boxes, televisions, or yet self-driving cars, and many other examples of smart sensors and smart actuators, have easy access to the Internet. The overwhelming growth of these IoT ecosystems has led to the emergence of edge computing.

The edge computing architecture represents, fundamentally, the relentless shift of critical aspects of processing, analysis, and storage of data to the network edge devices, well closed to the end users. It is a form of distributed computing and offers a new perspective when compared to other computing paradigms such as cloud computing and fog computing.

Very often, for instance, in smart health or real-time monitoring, the *things* produce an impressive amount of continuous data that need to be transmitted over the network backhaul, adequately processed and stored, while ensuring privacy protection, and short response delays, unveiling the weaknesses of the more conventional paradigm of cloud computing to cope with these scenarios. Typically, smart applications, the ones that make use of artificial intelligence or ML, deal with vast amounts of data, which becomes costly to send or store in a centralized cloud infrastructure.

By definition, edge computing intents to reduce the loop between end devices, where the applications, users, sensors, and actuators reside, and the place where the processing of the response occurs as presented in Fig. 8.

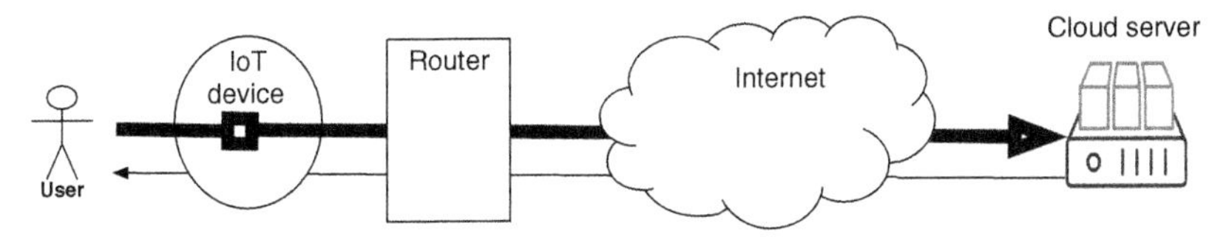

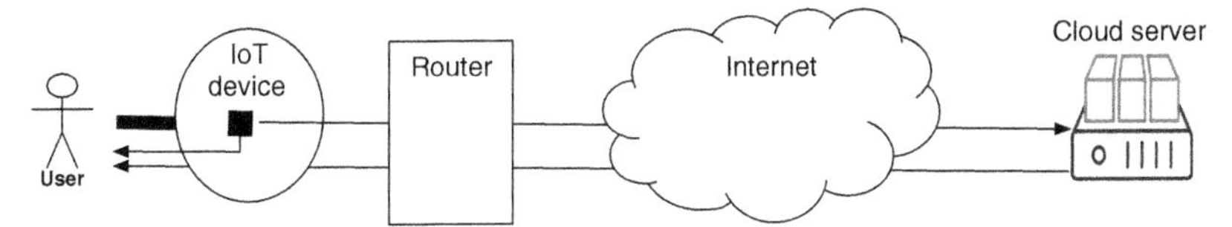

Fig. 8 Cloud versus edge data loop.

Edge computing presents some particular characteristics, suitable to deal with the novel reality of the pervasiveness of IoT devices in daily life, producing massive amounts of data:

- faster response times in real-time or near real-time applications;
- less dependency on limited bandwidth communication channels;
- greater privacy and security levels;
- greater control over data generated in end devices; and
- lower costs of transmission since less data needs to be transmitted remotely, that is, more sensor-derived data is used in the end device.

Edge computing, as an architecture, has been gaining importance. This growth in importance, in an era of cloud computing, is related to the increasingly important role that IoT currently plays and its challenges. Although, currently, in 2020, roughly 80% of data processing and analysis still takes place in the cloud, in data centers, and centralized computing facilities, and only 20% in smart connected objects and computing facilities close to the user, it is estimated that by 2025 these proportions are set to change markedly [36]. This foreseen evolution is also tightly connected with the emerging of artificial intelligence everywhere. Edge AI and edge analytics are some important trends in the ever-flourishing area of artificial intelligence [37].

4.2 Related work on edge architectures for e-health

In recent years, there has been a lot of work, developed by the scientific community, concerning the edge computing topic. According to Merenda et al. [38], 6342 papers were reported from a Scopus query using edge computing as the search term. The application of edge computing to health care is a very important topic, well covered in literature, with several published surveys, such as Ray et al. [39] or Abdellatif et al. [40].

In the vision of the authors of this chapter, there is an important distinction between fog and edge computing, with practical impacts on the way the solutions are designed and deployed. In edge computing, smart objects, most of the time with very useful processing, storage, and communication capabilities, are deployed at the very edge of the network, including, for instance, judicious management of energy consumption [41]. With respect to fog computing, the prevalent idea is the offloading of the cloud infrastructure to computing resources located somewhere in the local area network where smart objects reside, but not necessarily the objects themselves [42]. This point of view is shared by Shi et al. [43] who acknowledge that edge computing is interchangeable with fog computing, but edge computing focuses more on the side of the things, while fog computing focuses more on the infrastructure side.

In practice, the concepts of edge and fog computing are closely related as observed in works such as the fog-based IoT-healthcare solution structure presented by Mahmud et al. [44]. In their work, the authors explore also the integration of cloud-fog services in interoperable healthcare solutions extended upon the traditional cloud-based structure.

The concepts of health monitoring systems and health smart home (HSH) are gaining more and more important as is well described by Mshali et al. [45]. With the aging of population happening faster than ever before, it urges to deploy truly HSH systems to permanently monitor and evaluate any health condition that the elderly subject may have, as well as monitoring how they carry out their daily activities. Only that way will it be possible to reduce the pressure placed on the conventional existing health system as Mshali et al. [45] emphasizes.

A significant difference exists between e-health raw data and contextual information. Raw data such as a person's vital signs, or environmental parameters, such as temperature, humidity, and sound, or personal movements is useless if not adequately interpreted and understood. Context information is generated by processing raw data and it refers to extracting high-level information such as behavior patterns, or a subject's activity [45]. Edge computing in healthcare envisions that a substantial part of the contextual information is performed at edge devices.

4.3 IoT edge-based e-health use case scenarios

The appearance of open-source computing hardware projects, aka *Do It Yourself* (DIY), such as Arduino[i] in 2005 and Raspberry Pi[j] in 2012, just to mention two of the most popular communities, fostered the development of innovative ideas in IoT. The advances in low energy computing platforms and in the available sensor ecosystem are also important key factors. In a white paper, Cisco company emphasizes the explosion of IoT in the first decade of the 21st century, with the things/people ratio growing from 0.08 in 2003 to 1.84 in 2010 [46].

The revolution of IoT is also hitting the healthcare area. One important application scenario in health care is real-time monitoring of daily activity, particularly, the falling detection. This is a well-known health problem among elder people which gains even more importance with the generalized aging of the world population since most of the countries in the world are experiencing growth in the number and proportion of older persons in their population. In this type of health applications, solutions based on temporal signals produced by accelerometer sensors are commonly used, as the work presented in [47] or the review described in [48].

4.3.1 Use case I: Personal body devices for monitoring ADLs

The monitoring of daily activity is a classical example where edge computing can be applied and where many projects can be identified. Nevertheless, the following two projects present an interesting chronological evolution of edge computing application. In the first one [49], an Arduino-based hardware platform is used to implement a rule-based system to detect activity, running on the Arduino microcontroller. The second one [50] uses an Android smartphone and ML approach to infer, directly on the device, the activity being performed by the smartphone user.

A three-axis accelerometer can produce a steady and considerable stream of data. In [49], a sampling period of 100 ms was selected to acquire three accelerometer values, producing 10 measures per second and per accelerometer axis. For each temporal window of 1 s, the variance for each axis is calculated and then the average of those three variances gives origin to the value for activity indicator (VAI). Based on this VAI index, the activity of the user is determined using heuristics. The VAI values are used to identify one among six possible user activities.

[i] See https://www.arduino.cc/.

[j] See https://www.raspberrypi.org/.

Those activities are ordered by VAI according to the rule: $VAI(\{Standing, Sitting, LayingDown\}) < VAI(Walking) < VAI(Running) < VAI(Falling)$. Postprocessing based on a lookup table is applied to fine tune the activity detection. Each entry of the table has, as input, five consecutive activities and, as output, one possible activity to represent the central point of those activities. The corporal device (the edge device) sends to the gateway a message with the activity detected, avoiding the congestion of the network with the raw accelerometer signal.

Another possible approach for the problem of user activity detection is to model it as a classification problem and apply ML techniques to solve it. This was done by Torres et al. [47] and Vale [50]. The activity detection module is one of the components of the broader personal activity detection system (PADS) described in the work of Vale [50]. The activity detection module runs in an Android smartphone and is capable of logging new data for increasing the dataset to better train the model (see Fig. 9). The five activity types considered were standing, walking, running, laying down, and falling. For the deployment of the ML classifier to the Android smartphone, an unofficial and reduced version of Weka data mining software [51] was used. The Android device application is capable of producing one activity classification as a standalone service or of sending the activity detected to a central service.

In the performed tests, a sampling rate of 10 Hz was used to acquire raw signals from the accelerometer, gyroscope, and magnetometer to be used in the training set. All those three sensors were used in the classification to

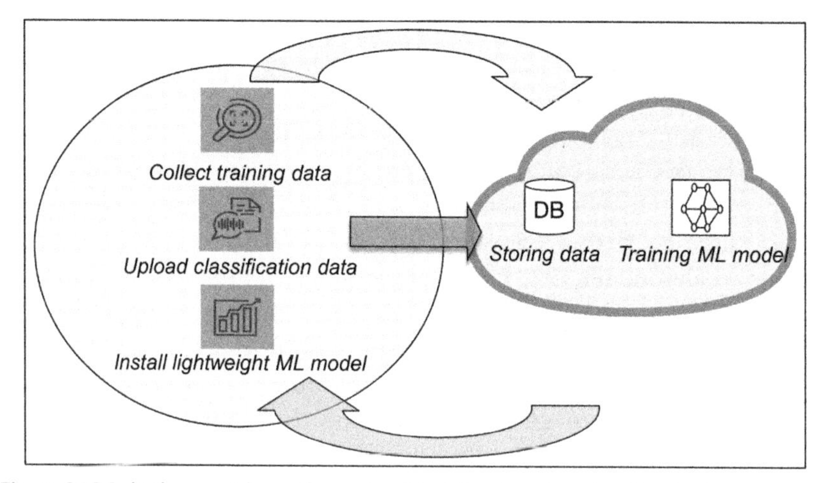

Fig. 9 PADS deployment loop (from cloud training to edge classification).

Table 2 Confusion matrix for LogitBoost classifier trained with 766 instances.

	Classified as				
Event	**A**	**B**	**C**	**D**	**E**
A	100%	0%	0%	0%	0%
B	0%	98.5%	1.5%	0%	0%
C	0%	2%	97.5%	0%	0.5%
D	0.7%	1.4%	0%	94.3%	3.6%
E	0%	3.8%	7.7%	23.1%	65.4%

Notes: A: standing, B: walking, C: running, D: laying down, E: falling

improve the accuracy. The acquired training set was composed by a total of 766 instances gathered from two users wearing the device near their chest. Each instance corresponds to a sampling window of 1 s of measures. Following some tests performed by Torres et al. [47], the classification algorithm selected was the LogitBoost [52].

The preliminary results obtained in the evaluation performed (see Table 2) were very promising. The model validation used was 10-fold cross-validation. Looking at the results obtained a slight contamination between D (laying down) and E (falling) activities can be observed. This effect can be easily corrected with a more robust training set of falls.

5. Conclusion

Currently, the world is faced with the challenge to continuously improve the quality of life of the population and, simultaneously, to integrate, in daily living objects, the tremendous technological innovations invading all the areas of society. On one side, we notice the escalating need for home and ambulatory healthcare solutions, to compensate the limits of present health systems that are unable to cope with the ever-rising needy population and with people's preferences for predominantly living and being cared for at home. On the other side, technological advances allowed the profusion of the smart object's paradigm, everywhere, anytime, and in every area, bringing technologies, such as artificial intelligence, at one's fingertips.

The healthcare stakeholders trust and welcome the evolution of IoT. The acceptance and recognition by end users, physicians, regulatory bodies, and remaining healthcare community in general, of the benefits of IoT in-home health care, follows along the fulfillment of the technical solutions and architectures proposed by IoT community. In this scope, three architectural paradigms for the deployment of IoT solutions are becoming de facto standards, as addressed throughout this chapter. Each of these three architectural categories, namely cloud, fog, and edge computing, have particularities that can be combined to obtain the best of all worlds, on providing real intelligent and sensing ubiquitous e-health systems to the world population.

The unprecedented capacity for data storage and processing, interconnection and integration of devices and users, and data mining and reasoning on top of globally collected data, is unique aspects of cloud infrastructures. Fog architectures, tightly bounded to edge solutions, guarantees real-time and privacy requirements, so important when dealing with real persons, particularly in health care. Finally, edge computing, ease the deployment of an ever-growing population of smart sensing and actuating objects, ranging from simple wearable computing devices to sets of local cooperating infrastructures. Therefore, the benefits and expectations of combining these three IoT-based paradigms is growing as demonstrated by some of the example uses cases presented in this chapter. Much more examples are being exploited in the IoT community and the future seams promising.

References

[1] K. Ashton, That "Internet of Things" Thing, RFID J. 22 (2009) 97–114. https://www.rfidjournal.com/that-internet-of-things-thing.

[2] M. De Donno, K. Tange, N. Dragoni, Foundations and evolution of modern computing paradigms: Cloud, IoT, Edge, and Fog, IEEE Access 7 (2019) 150936–150948.

[3] D.A. Gandhi, M. Ghosal, Intelligent healthcare using IoT: a extensive survey, in: 2018 Second International Conference on Inventive Communication and Computational Technologies (ICICCT), 2018, pp. 800–802.

[4] S.C. Mukhopadhyay, N.K. Suryadevara, Internet of things: challenges and opportunities. in: Internet of Things Smart Sensors, Measurement and Instrumentation, 2014, pp. 1–17, https://doi.org/10.1007/978-3-319-04223-7_1.

[5] S.B. Baker, W. Xiang, I. Atkinson, Internet of things for smart healthcare: technologies, challenges, and opportunities, IEEE Access 5 (2017) 26521–26544.

[6] N. Xu, A survey of sensor network applications, IEEE Commun. Mag. 40 (2002) 102–114.

[7] N. Kumar, IoT architecture and system design for healthcare systems, in: 2017 International Conference on Smart Technologies for Smart Nation (SmartTechCon), 2017, pp. 1118–1123.

[8] H.S. Ng, M.L. Sim, C.M. Tan, C.C. Wong, Wireless technologies for telemedicine. BT Technol. J. 24 (2006) 130–137, https://doi.org/10.1007/s10550-006-0050-9.

[9] S. Durga, R. Nag, E. Daniel, Survey on machine learning and deep learning algorithms used in Internet of Things (IoT) healthcare, in: 2019 3rd International Conference on Computing Methodologies and Communication (ICCMC), 2019, pp. 1018–1022.

[10] K. Venkatasubramanian, G. Deng, T. Mukherjee, J. Quintero, V. Annamalai, S.K.S. Gupta, Ayushman: a wireless sensor network based health monitoring infrastructure and testbed, in: V.K. Prasanna, S.S. Iyengar, P.G. Spirakis, M. Welsh (Eds.), Distributed Computing in Sensor Systems, Springer, Berlin, Heidelberg, 2005, , pp. 406–407.

[11] J.A. Stankovic, Q. Cao, T. Doan, L. Fang, Z. He, R. Kiran, S. Lin, S. Son, R. Stoleru, A. Wood, Wireless sensor networks for in-home healthcare: potential and challenges, in: Workshop on High Confidence Medical Devices Software and Systems (HCMDSS), 2005.
[12] A.D. Jurik, A.C. Weaver, Remote medical monitoring. Computer 41 (4) (2008) 96–99, https://doi.org/10.1109/MC.2008.133.
[13] D. Malan, T. Fulford-Jones, M. Welsh, S. Moulton, CodeBlue: an ad hoc sensor network infrastructure for emergency medical care, in: International Workshop on Wearable and Implantable Body Sensor Networks, 2004.
[14] U. Anliker, J.A. Ward, P. Lukowicz, G. Troster, F. Dolveck, M. Baer, F. Keita, E.B. Schenker, F. Catarsi, L. Coluccini, A. Belardinelli, D. Shklarski, M. Alon, E. Hirt, R. Schmid, M. Vuskovic, AMON: a wearable multiparameter medical monitoring and alert system, IEEE Trans. Inf. Technol. Biomed. 8 (4) (2004) 415–427.
[15] A.D. Wood, G. Virone, T.-V. Doan, Q. Cao, L. Selavo, Y. Wu, L. Fang, Z. He, S. Lin, J.A. Stankovic, ALARM-NET: wireless sensor networks for assisted-living and residential monitoring, Technical Report CS-2006-11Department of Computer Science, University of Virginia, 2006.
[16] D.B. Hoang, L. Chen, Mobile cloud for assistive healthcare (MoCAsH), in: 2010 IEEE Asia-Pacific Services Computing Conference, 2010, pp. 325–332.
[17] S. Amendola, R. Lodato, S. Manzari, C. Occhiuzzi, G. Marrocco, RFID technology for IoT-based personal healthcare in smart spaces, IEEE Internet Things J. 1 (2) (2014) 144–152.
[18] T. Adame, A. Bel, A. Carreras, J. Melià-Seguí, M. Oliver, R. Pous, CUIDATS: an RFID-WSN hybrid monitoring system for smart health care environments. Futur. Gener. Comput. Syst. 78 (2018) 602–615, https://doi.org/10.1016/j.future.2016.12.023.
[19] M.M. Hassan, K. Lin, X. Yue, J. Wan, A multimedia healthcare data sharing approach through cloud-based body area network. Futur. Gener. Comput. Syst. 66 (2017) 48–58, https://doi.org/10.1016/j.future.2015.12.016.
[20] T. Muhammed, R. Mehmood, A. Albeshri, I. Katib, UbeHealth: a personalized ubiquitous cloud and edge-enabled networked healthcare system for smart cities, IEEE Access 6 (2018) 32258–32285.
[21] P. Gonçalves, J. Torres, P. Sobral, R.S. Moreira, Remote patient monitoring in home environments, in: Proceedings of the 1st International Workshop on Mobilizing Health Information to Support Healthcare-Related Knowledge Work, MobiHealthInf 2009 in Conjunction With BIOSTEC 2009, 2009, pp. 87–96.
[22] R.S. Moreira, F. Silva, J.M. Torres, P. Sobral, C. Soares, Sistema ubíquo de acompanhamento pessoal, in: CISTI'2012 (7th Iberian Conference on Information Systems and Technologies), 2012.
[23] C. Soares, R.S. Moreira, R. Morla, J. Torres, P. Sobral, Prognostic of feature interactions between independently developed pervasive systems, in: PHM 2012–2012 IEEE International Conference on Prognostics and Health Management: Enhancing Safety, Efficiency, Availability, and Effectiveness of Systems Through PHM Technology and Application, Conference Program, 2012.
[24] C. Soares, R.S. Moreira, R. Morla, J. Torres, P. Sobral, A graph-based approach for interference free integration of commercial off-the-shelf elements in pervasive computing systems. Futur. Gener. Comput. Syst. 39 (2014) 3–15, https://doi.org/10.1016/j.future.2013.12.035.
[25] R.S. Moreira, R. Morla, L.P.C. Moreira, C. Soares, A behavioral reflective architecture for managing the integration of personal Ubicomp systems: automatic SNMP-based discovery and management of behavior context in smart-spaces. Pers. Ubiquit. Comput. 20 (2016) 229–243, https://doi.org/10.1007/s00779-016-0901-4.

[26] P. Bellavista, J. Berrocal, A. Corradi, S.K. Das, L. Foschini, A. Zanni, A survey on fog computing for the Internet of Things. Pervasive Mobile Comput. 52 (2019) 71–99, https://doi.org/10.1016/j.pmcj.2018.12.007.
[27] A. Mebrek, L. Merghem-Boulahia, M. Esseghir, Efficient green solution for a balanced energy consumption and delay in the IoT-Fog-Cloud computing. in: 2017, pp. 1–4, https://doi.org/10.1109/NCA.2017.8171359.
[28] J. Grover, R.M. Garimella, Reliable and fault-tolerant IoT-Edge architecture, in: 2018 IEEE Sensors, 2018, pp. 1–4.
[29] K. Monteiro, É. Rocha, É. Silva, G.L. Santos, W. Santos, P.T. Endo, Developing an e-health system based on IoT, fog and cloud computing, in: 2018 IEEE/ACM International Conference on Utility and Cloud Computing Companion (UCC Companion), 2018, pp. 17–18.
[30] T. Diethe, M. Holmes, M. Kull, M.P. Nieto, K. Sokol, H. Song, E. Tonkin, N. Twomey, P. Flach, Releasing eHealth analytics into the wild: lessons learnt from the SPHERE project, in: Proc. of the 24th ACM SIGKDD International Conference on Knowledge Discovery & Data Mining, Association for Computing Machinery, New York, NY, 2018, , pp. 243–252. ISBN: 9781450355520https://doi.org/10.1145/3219819.3219883.
[31] J. Tewell, D. O'Sullivan, N. Maiden, J. Lockerbie, S. Stumpf, Monitoring meaningful activities using small low-cost devices in a smart home. Pers. Ubiquit. Comput. 23 (2019)https://doi.org/10.1007/s00779-019-01223-2.
[32] E. Quintana, J. Favela, Augmented reality annotations to assist persons with Alzheimer's and their caregivers. Pers. Ubiquit. Comput. (2013) 1105–1116, https://doi.org/10.1007/s00779-012-0558-6.
[33] A. Miguez, C. Soares, J. Torres, P. Sobral, R.S. Moreira, Improving ambient assisted living through artificial intelligence. in: Proc. of WorldCIST 2019, 2019, pp. 110–123, https://doi.org/10.1007/978-3-030-16184-2_12.
[34] C. Szegedy, V. Vanhoucke, S. Ioffe, J. Shlens, Z. Wojna, Rethinking the inception architecture for computer vision, CoRR abs/1512.00567 (2015). http://arxiv.org/abs/1512.00567.
[35] F. Iandola, S. Han, M. Moskewicz, K. Ashraf, W. Dally, K. Keutzer, SqueezeNet: AlexNet-level accuracy with 50x fewer parameters and <0.5 MB model size, ArXiv, (2017)abs/1602.07360.
[36] European Commission, White paper on artificial intelligence—a European approach to excellence and trust., European Commission, Brussels, 2020.
[37] Gartner, 5 Trends Appear on the Gartner Hype Cycle for Emerging Technologies. (2019)https://www.gartner.com/smarterwithgartner/5-trends-appear-on-the-gartner-hype-cycle-for-emerging-technologies-2019/ (Accessed 1 March 2020).
[38] M. Merenda, C. Porcaro, D. Iero, Edge machine learning for AI-enabled IoT devices: a review. Sensors 20 (9) (2020) 2533, https://doi.org/10.3390/s20092533.
[39] P.P. Ray, D. Dash, D. De, Edge computing for Internet of Things: a survey, e-healthcare case study and future direction. J. Netw. Comput. Appl. 140 (May) (2019) 1–22, https://doi.org/10.1016/j.jnca.2019.05.005.
[40] A.A. Abdellatif, A. Mohamed, C.F. Chiasserini, M. Tlili, A. Erbad, Edge computing for smart health: context-aware approaches, opportunities, and challenges. IEEE Netw. 33 (3) (2019) 196–203, https://doi.org/10.1109/MNET.2019.1800083.
[41] B. Gomes, N. Melo, R. Rodrigues, P. Costa, C. Carvalho, K. Karmali, S. Karmali, C. Soares, J.M. Torres, P. Sobral, R.S. Moreira, A power efficient IoT edge computing solution for cooking oil recycling, in: Proc. of WorldCIST 2020, Budva, Montenegro, 2020.

[42] P. Costa, B. Gomes, N. Melo, R. Rodrigues, C. Carvalho, K. Karmali, S. Karmali, C. Soares, J.M. Torres, P. Sobral, R.S. Moreira, Fog computing in real time resource limited IoT environments, in: Proc. of WorldCIST 2020, Budva, Montenegro, 2020.
[43] W. Shi, J. Cao, Q. Zhang, Y. Li, L. Xu, Edge computing: vision and challenges. IEEE Internet Things J. 3 (5) (2016) 637–646, https://doi.org/10.1109/JIOT.2016.2579198.
[44] R. Mahmud, F.L. Koch, R. Buyya, Cloud-Fog interoperability in IoT-enabled healthcare solutions. in: ACM International Conference Proceeding Series, 2018. https://doi.org/10.1145/3154273.3154347.
[45] H. Mshali, T. Lemlouma, M. Moloney, D. Magoni, A survey on health monitoring systems for health smart homes. Int. J. Ind. Ergon. 66 (2018) 26–56, https://doi.org/10.1016/j.ergon.2018.02.002.
[46] D. Evans, The internet of things: how the next evolution of the internet is changing everything, Cisco White Paper, April 2011.
[47] J.M. Torres, P. Sobral, R.S. Moreira, R. Morla, An adaptive embedded system for physical activity recognition, in: Proc. of CISTI 2012–7ª Conferência Ibérica de Sistemas e Tecnologias de Informaç ao, 2012.
[48] R.S. Moreira, J.M. Torres, P. Sobral, C. Soares, Intelligent sensing and ubiquitous systems (ISUS) for smarter and safer home healthcare, in: A.K. Sangaiah, S. Shantharajah, P. Theagarajan (Eds.), Intelligent Pervasive Computing Systems for Smarter Healthcare, first ed., John Wiley & Sons, Inc., 2019, , pp. 1–36
[49] P. Gonçalves, J. Torres, P. Sobral, R.S. Moreira, Remote patient monitoring in home environments, in: Proceedings of the 1st International Workshop on Mobilizing Health Information to Support Healthcare-Related Knowledge Work, MobiHealthInf 2009 in Conjunction with BIOSTEC 2009, 2009.
[50] S. Vale, PADS Personal Activity Detection System (Ph.D. thesis), University Fernando Pessoa, 2013, http://hdl.handle.net/10284/3825.
[51] E. Frank, M.A. Hall, I.H. Witten, The WEKA Workbench. Online Appendix for "Data Mining: Practical Machine Learning Tools and Techniques" fourth ed., Morgan Kaufmann Publishers, Inc., 2016
[52] J. Friedman, T. Hastie, R. Tibshirani, Additive logistic regression: a statistical view of boosting (with discussion and a rejoinder by the authors). Ann. Stat. 28 (2) (2000) 337–407, https://doi.org/10.1214/aos/1016218223.

CHAPTER TWO

RFID-based unsupervised apnea detection in health care system

Chao Yang[a], Xuyu Wang[b], and Shiwen Mao[a]
[a]Auburn University, Auburn, AL, United States
[b]California State University, Sacramento, CA, United States

1. Introduction

Nowadays, smart health care has attracted increasing interest [1–4]. Rather than being treated after getting sick, it is more desirable to detect early and prevent diseases by monitoring people's vital signs on a daily basis. Unfortunately, traditional sensors for vital sign monitoring, such as a capnograph and a sphygmomanometer [5], are unwieldy for day-round monitoring, especially when the patient is sleeping. Moreover, the abnormality diagnosis of vital signs may consume considerable resources and efforts at medical institutions. Therefore, autonomous, low-cost, unobtrusive vital sign monitoring systems with abnormality detection capability are highly desired for IoT-based smart health care, which can benefit many people for monitoring health conditions in their daily life and in their houses.

Breathing rate is one of the key human vital signs. Irregular breathing patterns could be an indicator of unhealthy human health conditions. For example, one of the breathing disorders is obstructive sleep apnea, which can imply serious health problems including high blood pressure, heart disease, and sudden infant death syndrome (SIDS) [6]. Considering the mobility and flexibility of RF devices, wireless signals are widely used in smart health-care systems to monitor human vital signs. The low-cost and near-field features of passive RFID tags have triggered great interest in applying them for breath monitoring and health sensing. Since the movement of the human chest and heart can slightly affect the propagation of RF signals, the signal of breathing and heartbeat can be reconstructed by analyzing the variations in received RF signals.

The Radar technique is a straightforward way to identify the fluctuation in the human chest caused by respiration, because Radar can directly monitor the distance variations between the human chest and the device antenna.

Intelligent IoT Systems in Personalized Health Care
https://doi.org/10.1016/B978-0-12-821187-8.00002-2

One of the representative works in this category leverages an FMCW Radar to monitor the respiration rate and heart rate for multiple users simultaneously [7]. Furthermore, other Radar-based systems have also been proposed to measure human respiration, including Doppler Radar [8] and ultra-wideband (UWB) Radar [9]. These Radar-based systems can accurately detect vital signs, and the influence caused by surroundings can be mitigated. However, since the special hardware is essential to such systems, the cost of such systems is usually high, thus hampering their wide deployment such as in homes.

To achieve low cost and ease of deployment for vital sign monitoring, WiFi-based techniques have been utilized to measure both the human breathing rate and heart rate. The human breathing rate and heartbeat can be extracted by analyzing the variations in WiFi channels, for example, the RSS as in Ubibreathe [10]. However, the Ubibreathe system requires the patient to stand between the transmitter and the receiver, while some other CSI-based techniques have no such strict requirements. Different from RSS-based systems, CSI can provide fine-grained channel information, the variations for each channel in WiFi transmission, which means CSI can achieve higher resolution and sensitivity than RSS for monitoring human vital signs. One of the CSI-based techniques leverages the amplitude of CSI to monitor breathing and heartbeat when the patient is sleeping [11]. In addition to amplitude, the CSI phase information can also carry the human respiration signal [12]. Furthermore, the TensorBeat system can estimate respiration rates for multiple persons in a small space [13], by applying tensor decomposition to the collected CSI phase data. Although WiFi-based techniques can measure human vital signs with off-the-shelf WiFi devices, the accuracy is easily affected by the surrounding environment, because of the broadcasting nature and the relatively long range of WiFi transmissions.

To address this issue, RFID-based systems, such as TagBreathe, are developed to track human respiration by analyzing the RFID response data collected at an RFID reader [14]. Since the passive UHF RFID tags are low in cost and are easily attachable to the human body, the RFID-based system can monitor human vital signs at a low cost and is resilient to interference (e.g., other movements) from the environment.

Another existing system, called AutoTag [15], utilizes an unsupervised recurrent variational autoencoder model for respiration rate estimation and abnormal breathing detection with off-the-shelf RFID Tags. To mitigate the influence caused by channel hopping, the system proposes a novel technique to map the phase values from different channels to one reference channel. Since FCC requires RFID hop to a different channel every 200 ms and the CSI is channel dependent, the typical RFID-based sensing system can hardly be used for realtime sensing applications. Rather than offline calibration employed in other systems, the AutoTag system can enable real-time phase calibration, which is amenable to dynamic environments. Furthermore, compared with the method used in TagBreathe, the system incurs much lower delay, because grouping data from all channels, as in TagBreathe, is a time-consuming process.

Moreover, the AutoTag system incorporates an unsupervised deep learning approach for apnea detection for measuring results, which can autonomously detect abnormality in human respiration. Deep learning has gained great interest in the wireless networking community and has been successfully applied to solve many wireless networking problems [16–25]. With the advantage of unsupervised learning, labeled medical data is not necessarily required, which is usually very costly and time-consuming to obtain. Recently, a recurrent variational autoencoder model has been employed for sequence modeling and human motion synthesis. Inspired by these prior works, an enhanced recurrent variational autoencoder model is developed for apnea detection in AutoTag, where the abnormality can be detected by evaluating the similarity between the sampled breathing signal and the signal reconstructed by the deep learning model. The proposed method is superior to typical energy threshold-based methods, especially when the testing environment is not completely stationary. This is because the method can easily distinguish nonperiodic signals from normal periodic signals by learning the features of the normal breathing signals, while the energy-based method only considers apnea as a relative weak signal compared with the normal cases.

We summarize the three main contributions of this work as follows.

- To the best of our knowledge, the AutoTag system is the first apnea detection system incorporating an enhanced recurrent variational autoencoder model. The proposed scheme is an unsupervised learning approach, with the desirable advantage of not requiring costly labeled medical data.
- The system also proposes a novel technique to address the frequency hopping offset, which is a real-time calibration. The proposed scheme is simple but effective in mitigating the frequency hopping offset, thus enabling many real-time sensing applications for FCC-compliant RFID readers and tags.

- The prototype AutoTag system is designed, which is composed with (i) signal extraction, (ii) data calibration, and (iii) respiration monitoring modules, and evaluates the system in two different representative health-care environments. We present our experimental results that validate the efficacy of the proposed AutoTag system.

The remainder of this chapter is organized as follows. The preliminaries of RFID-based sensing is discussed in Section 2. The AutoTag system design is presented in detail and evaluated in Section 3. We conclude this chapter in Section 4.

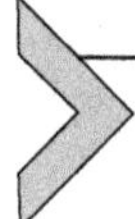

2. Preliminaries and challenges for RFID sensing systems

According to FCC regulations, UHF RFID readers should use channel hopping to avoid co-channel interference. The spectrum from 902.5 to 927.5 MHz is partitioned into 50 nonoverlapping channels, and the reader remains on each channel for 200 ms. Usually such frequency hopping introduces an additional phase offset in the RFID response signal, causing large errors in RFID-based sensing.

According to the manual of RFID reader, e.g., Ref. [26], the phase ϕ of the received RFID response signal can be expressed as

$$\phi(f_i, d) = mod\left(\frac{2\pi f_i d}{c} + \delta_T + \delta_R + \delta_{Tag}, 2\pi\right), \tag{1}$$

where d is the total distance from the reader's antenna to the tag and then back to the reader antenna, f_i is the frequency of channel i, c is a constant representing the speed of light, and δ_T, δ_R, and δ_{Tag} are the phase offsets caused by the transmitter circuit, the receiver circuit, and the tag's reflection characteristics, respectively. From the equation we find that the tag-to-reader distance (d) can be reflected by the sampled phase. Once the tag is attached on the human chest, the respiration could be monitored by the variation of d. However, for Impinj R420, a commodity RFID reader, the phase offset between two adjacent channels that it hops to is not a constant, even though the distance d remains the same, as found in our experimental studies. Since the three offsets in Eq. (1) are irrelevant to the distance d, we can lump the three offsets into a single variable δ_i for each channel i. The phase $\phi(f_i, d)$ for channel f_i under round-trip distance d can thus be expressed as

$$\phi(f_i, d) = mod\left(\frac{2\pi f_i d}{c} + \delta_i, 2\pi\right). \tag{2}$$

The main challenge for extracting the breathing signal from the RFID-phase measurements is how to mitigate the discontinuity in phase data, which is caused by channel hopping. Because of the channel hopping effect, the captured phase information cannot be directly used for detecting the respiration signal. In Fig. 1, we plot the uncalibrated phase data received from one of the reader antennas for a duration of 28 s. It can be observed that when the reader hops around various channels (200 ms on each channel), the measured phase data exhibits a wide range of variations. Furthermore, there is a large offset incurred in the phase data when the frequency hopping happens. It is thus highly challenging to extract the weak respiration signal from such uncalibrated data. In order to effectively extract the breathing signal from the data, the raw phase data should be calibrated first to facilitate the extraction of the respiration signal.

One way to eliminate the channel hopping influence, as proposed for the TagBreathe system [14], is to group the signals collected from the same channel and to use the estimated displacement in each channel to track the breathing signal. As discussed earlier, this method may not work well for RFID systems in the United States, since the reader must hop among 50 different frequencies, following the FCC requirement. Fig. 2 plots the change of channel index in a period of 30 s. We can see that it takes about 10 s for the antenna to hop through all the 50 channels. Thus, the TagBreathe method will take a very long time to collect and group multiple phase readings on the same channel, leading to an extremely long delay in respiration measurement with FCC-compliant readers.

To address the extreme phase distortion caused by channel hopping among 50 different frequencies, the Tagyro system calibrates phase values

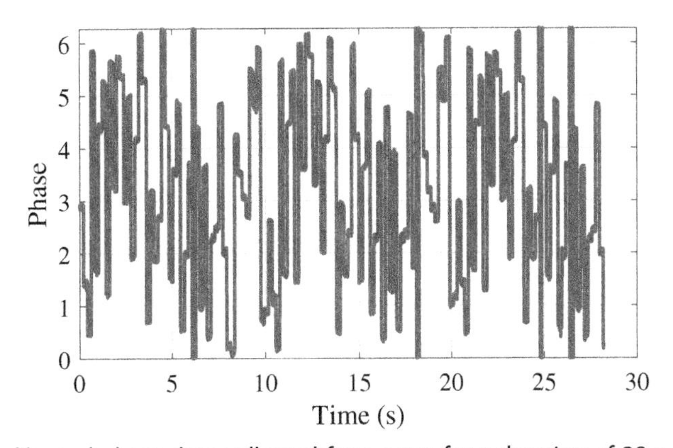

Fig. 1 Uncalibrated phase data collected from a tag for a duration of 28 s.

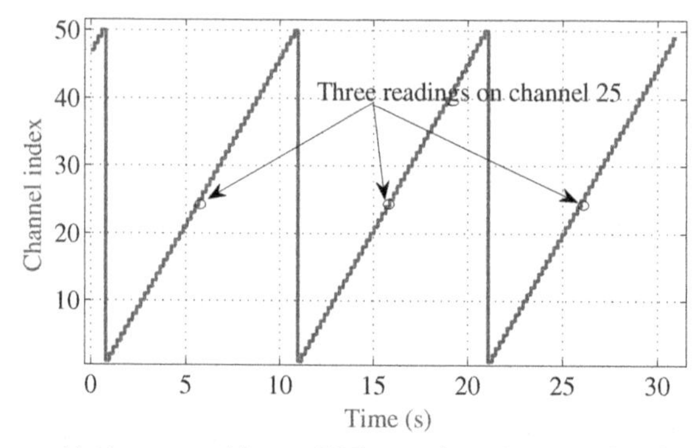

Fig. 2 The channel indexes used by an FCC-compliant RFID reader during a period of 30 s.

collected from all channels based on one reference channel [27]. Specifically, the Tagyro technique first measures the phase offset δ_i for all the 50 channels. Then the phase offset introduced by channel hopping can be removed by subtracting the phase offset δ_i in each channel. This method is suitable for a static environment, but it may not be effective for tracking the human breathing signal and apnea, where the tags are mounted on the human body and move as the patient breathes. This is because the wireless channel will change if the patient moves (even slightly). The movement causes an additional offset in δ_i, so that the estimated phase offset $\hat{\delta}_i$ does not match the real-time δ_i after the small movement.

Fig. 3 plots the calibrated phase data obtained with the Tagyro method. It can be observed that the breathing signal can be detected in the beginning

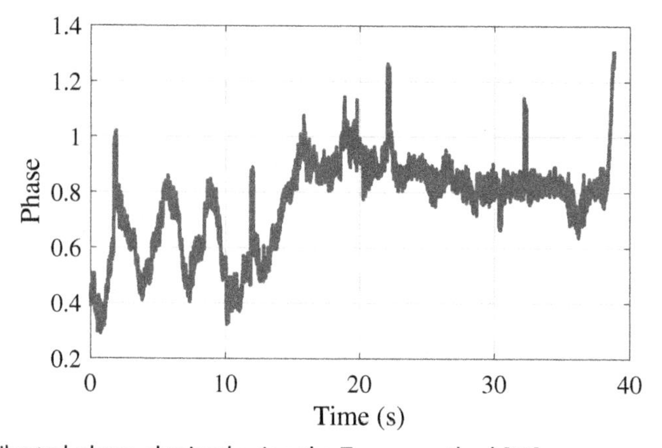

Fig. 3 Calibrated phase obtained using the Tagyro method [27].

(i.e., the first 15 s), because the initial phase offset is correct. After the first 15 s, the breathing signal cannot be detected because the channel hopping effect cannot be perfectly mitigated. To continuously eliminate the frequency hopping effect, a new method is proposed in the AutoTag system, to update and remove the phase offset δ_i in real time for breathing and apnea detection.

3. The AutoTag system

3.1 AutoTag system design

The AutoTag system is designed to measure human respiration and detect breathing abnormalities, such as apnea, with multiple RFID tags attached to the patient's body (i.e., clothes). As given in Eq. (1), the collected phase information is indicative of the round-trip distance d between the reader antenna and the corresponding tag. When the patient breathes, the distance d changes slightly with the chest movements. Thus, by analyzing the phase variations collected from the tags attached to the human body, the system can obtain the periodic breathing signals caused by chest movements. However, for accurately measuring the human respiration rate and precisely detecting apnea, several challenges should be addressed, such as mitigating the channel hopping effect, the sensitivity divergence for different tags, and dealing with the interference from surroundings. To address these issues, the system incorporates three modules in the AutoTag system, including (i) signal extraction, (ii) data calibration, and (iii) respiration monitoring, as illustrated in Fig. 4.

In the *signal extraction* module, the phase data is extracted from three tags attached to the human body, using a directional antenna at the reader. With the sampled phase data, in the following *data calibration* module, the channel hopping influence is firstly removed from the RFID reader. Then whether the monitored patient is moving or not is also detected based on a threshold-based method, i.e., movement detection. After that, the system removes the DC component from the selected signal to eliminate the impact of small movements of the patient. Then tag selection is implemented to choose the tag with the strongest signal strength. Finally, downsampling and filtering are implemented to obtain the final respiration signal. In the *respiration monitoring* module, it adopts a recurrent variational autoencoder for detection of abnormalities such as apnea. Our approach is an unsupervised learning scheme, which has the great advantage of not requiring labeled medical data, which is extremely costly to collect. The respiration rate can also be

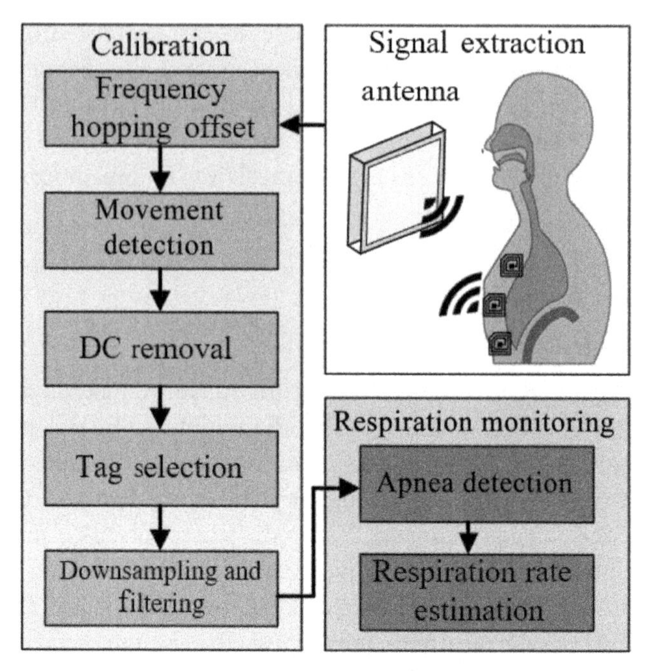

Fig. 4 The AutoTag system architecture, which includes signal extraction, data calibration, and respiration monitoring.

estimated with a peak detection method when the patient is breathing normally. The detailed design of each module will be introduced in the remainder of this section.

3.2 Signal extraction

As shown in Fig. 4, the first module is used for extracting low-level phase readings from received tag responses. The movements of the patient's chest and abdomen, induced by breathing, cause the tag-reader antenna distance to vary with human respiration. The time-varying distance translates to the time-varying phase in the tag response signal, which is indicative of the respiration signal. To increase the system's robustness, three passive UHF RFID tags are attached to the upper body of the patient. The RFID reader uses a directional antenna to transmit the RF interrogating signals to the tags and to read low-level data from the backscattered signals from the tags, which includes the time stamp, phase, received signal strength indicator (RSSI), and Doppler shift.

For most RFID systems, the collected RSSI data usually has a very low resolution, and the signal-to-noise ratio (SNR) of Doppler shift is usually low. Thus, these two types of information are not very helpful for detecting the respiration signal. In AutoTag introduction, we focus on the collected phase information from RFID responses for respiration rate estimation and apnea detection.

3.3 Data calibration

3.3.1 Mitigating the frequency hopping offset

The captured phase signal is firstly unwrapped to remove the offset introduced by the modulo operation in Eq. (2). With the modulo operation, a slight change in the real phase may lead to a large jump in the received phase signal. For instance, a small change of the real phase from 0.1π to -0.1π will cause the received phase to change from 0.1π to 1.9π. Since the sampling rate of the reader is usually higher than 100 Hz for each tag in our system, the interarrival time of phase samples is usually smaller than 0.01 s. Assuming that the phase value change of two back-to-back readings is smaller than π under such a small interarrival time, $\pm 2\pi$ can be added to recover the original phase value when the change exceeds π. Note that frequency hopping can also cause big variations between two back-to-back phase samples, and unwrapping will only be used for consecutive phase samples collected from the same channel. After the unwrapping operation, the phase samples from each channel are smoothed.

The second step is to splice the smoothed phase signals from all the 50 channels into a single-phase signal, by mitigating the frequency hopping offsets. The key is to translate the phase signal from the next channel that the reader hops to, to a transformed phase signal using the previous channel as a reference with real-time calibration. As illustrated in Fig. 5, the phase signal from the previous channel, e.g., channel i, can be written as

$$\phi(f_i, d) = mod\left(\frac{2\pi f_i d}{c} + \delta_i, 2\pi\right). \tag{3}$$

Then the reader hops from channel i to channel $(i + 1)$. Similarly, for the next channel that the reader hops to, we have

$$\phi(f_{i+1}, d) = mod\left(\frac{2\pi f_{i+1} d}{c} + \delta_{i+1}, 2\pi\right). \tag{4}$$

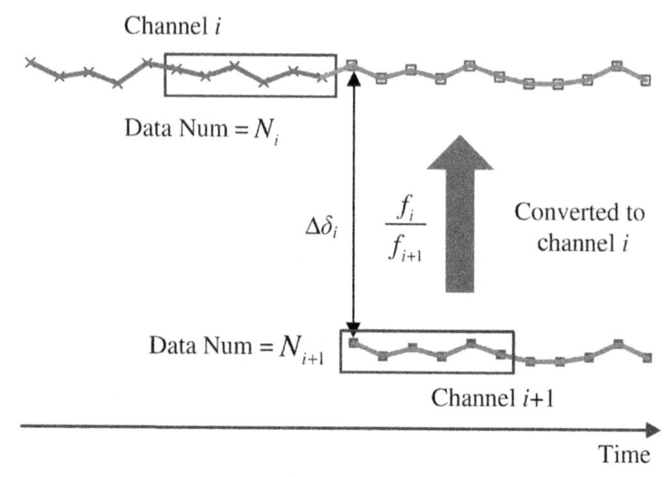

Fig. 5 Illustration of the proposed frequency hopping offset mitigation scheme.

For simplifying notation, ignore the modulo operation for now. Then the channel $(i + 1)$ phase signal is multiplied by a factor of (f_i/f_{i+1}), to have

$$\begin{aligned}\hat{\phi}(f_{i+1}, d) &= \phi(f_{i+1}, d) \times \frac{f_i}{f_{i+1}} \\ &= \frac{2\pi f_i d}{c} + \delta_{i+1} \times \frac{f_i}{f_{i+1}} \\ &\doteq \frac{2\pi f_i d}{c} + \delta_i + \Delta\delta_i,\ i = 1, 2, \ldots, 49.\end{aligned} \tag{5}$$

Therefore, we have

$$\hat{\phi}(f_{i+1}, d) = \phi(f_i, d) + \Delta\delta_i, i = 1, 2, \ldots, 49, \tag{6}$$

where

$$\Delta\delta_i \doteq \delta_{i+1} \times \frac{f_i}{f_{i+1}} - \delta_i, \tag{7}$$

represents the transformed phase offset, as illustrated in Fig. 5, which can be easily estimated as follows.

Note that the frequency for each channel is given, so we firstly multiply the phase $\phi(f_{i+1}, d)$ collected from channel $(i + 1)$ with a ratio f_i/f_{i+1}. The next value required to be calibrated is $\Delta\delta_i$ as in Eq. (7). In AutoTag, only three tags are attached to the human body. The overall sampling frequency of the reader is 600 Hz, and the sampling rate for each tag is more than

100 Hz. It takes only <1 ms for the reader to hop from one channel to another. Due to the high sampling rate and the small channel hopping time, it is reasonable to assume that the antenna-to-tag distance and the surrounding environment remain the same during channel hopping. Under this assumption, the difference between the phase data before channel hopping and the transformed phase data after channel hoping is all caused by $\Delta\delta_i$ along with thermal noise.

To mitigate the influence of thermal noise, a Hampel filter is applied to the signal read from each channel. The system applies a sliding window of 20 samples and a threshold of 0.01 for the Hampel filter. Next, the system computes the difference between (i) the average of the last six phase readings in the previous channel i (denoted as $\phi(f_i, d, k)$) and (ii) the average of the first six transformed phase readings in the present channel $(i + 1)$ (denoted as $\hat{\phi}(f_{i+1}, d, k)$), as an estimate for the transformed phase offset $\Delta\delta_i$. As shown in Fig. 5, and also from Eq. (6), we have

$$\hat{\Delta\delta}_i \approx \frac{1}{6}\left(\sum_{k=N_i-5}^{N_i} \phi(f_i, d, k) - \sum_{k=0}^{5} \hat{\phi}(f_{i+1}, d, k)\right), \tag{8}$$

where N_i is the total number of phase readings collected from channel i.

After compensating for the frequency hopping offset, the phase samples on the present channel $(i + 1)$ can be approximated as in Eq. (6). Fig. 6 plots the calibrated phase samples after removing the frequency hopping offset. It can be seen that after calibration, the collected phase data shows a visible

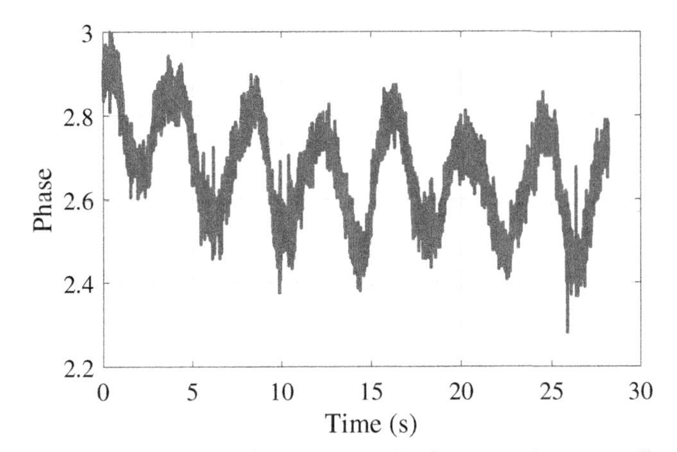

Fig. 6 The resulting phase data after removing the frequency hopping offset.

periodic respiration signal, which is completely missing in the uncalibrated phase data in Fig. 1 (for the same period of 28 s).

3.3.2 Movement detection

After mitigating the frequency hopping offset, the next step is to detect whether the patient is moving. Note that the breathing signal is very weak compared to other types of body movement. Therefore, only the signals collected while the patient is in a stationary state can be used for the following process, to avoid the interference introduced by large movements of the human body.

Movement detection is accomplished with a threshold-based method. In particular, a sliding window is applied to the collected phase data as shown in Fig. 7. For each window, the system calculates the total mean absolute deviation of the phase samples from all tags, denoted by T, as

$$T = \frac{1}{|W|} \sum_{j=1}^{N} \sum_{k \in W} |\phi^j(k) - \mathbb{E}(\phi^j(k))|, \tag{9}$$

where W represents the set of all the phase readings in the sliding window, $|\cdot|$ is the cardinality, N is the number of tags, and $\phi^j(k)$ is the kth phase sample obtained from tag j. If the patient is not stationary, the phase values will exhibit big changes (due to the large changes in d caused by body movements). Thus, by setting a threshold on T, the considerable movements of the patients are detected. In AutoTag, the threshold is set as 0.9 and window size is set as 6 s for movement detection.

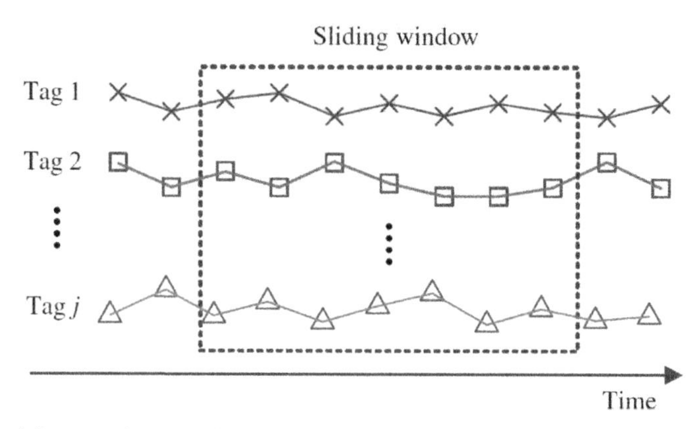

Fig. 7 A sliding window on phase readings from the tags.

3.3.3 DC removal

Although the frequency hopping offset can be effectively mitigated using a reference channel, the initial offset on the reference channel is still a random value that introduces a random DC component in the phase data. To remove the random DC component, as well as interference from other movements from the environment, the system passes the phase signal through a detrending operation.

Specifically, another Hampel filter is used, whose window size is 2000 and threshold is 0.001, to obtain an estimate of the DC component (i.e., the trend). Finally, the calibrated phase signal is obtained by subtracting the trend from the filtered signal. The calibrated signal is plotted in Fig. 8 for the same period of 28 s. It can be seen that the calibrated signal is now centered at zero, like a periodic AC signal.

3.3.4 Tag selection

The experiments result shows that each tag has a different sensitivity to the human breathing signal from other tags, which depends on the slightly different tag parameters and the different propagation environment (e.g., distance or angle) of each tag.

In particular, the angle between the tag and the reader antenna is a factor that causes the difference in the tag's sensitivity. To obtain the most sensitive signal, the AutoTag measures the signal strength using the average absolute deviation within a certain window size of 6 s (see Eq. 9). The tag with the largest signal strength will be chosen for the remaining processing steps.

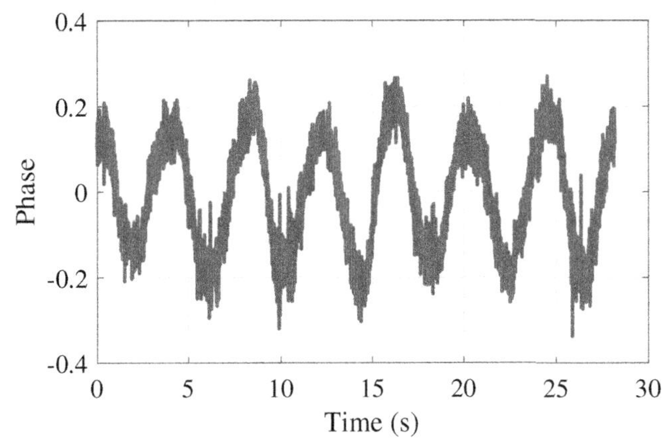

Fig. 8 The calibrated phase signal after removing the DC component with a Hampel filter.

3.3.5 Downsampling and filtering

Due to the high sampling frequency, i.e., about 600 Hz with one reader antenna, the signal needs to be downsampled to reduce the computational complexity. In AutoTag, the signal is downsampled with a factor of 10 before feeding to the respiration monitoring module. In addition, there are still some false peaks introduced by thermal noises, which affect the accuracy of peak detection for respiration rate estimation. Note that the normal human respiration rate is usually much lower than 0.5 Hz. Thus, the system applies a low-pass filter and chooses a cutoff frequency of 0.5 Hz to the downsampled signal, in order to remove the remaining high-frequency noise. The downsample and filtered phase signal is plotted in Fig. 9, which is quite smooth now. It is then used as input to the next module for the following apnea detection and respiration rate estimation tasks.

3.4 Apnea detection and respiration rate estimation

3.4.1 Recurrent variational autoencoder for apnea detection

For the purpose of accurate respiration detection, we introduce a recognition model as an approximation to the intractable true posterior, where the parameters are not computed from some closed-form expectation, but are learned from the calibrated data. To detect apnea, the main idea of AutoTag is to incorporate a *variational autoencoder* to compute the difference between the sampled signal and the reconstructed signal within a time window. Note that this is an unsupervised learning approach [28, 29], with the desirable advantage of not requiring labeled medical data that is hard or costly to obtain. If the computed difference is smaller than a given threshold, the

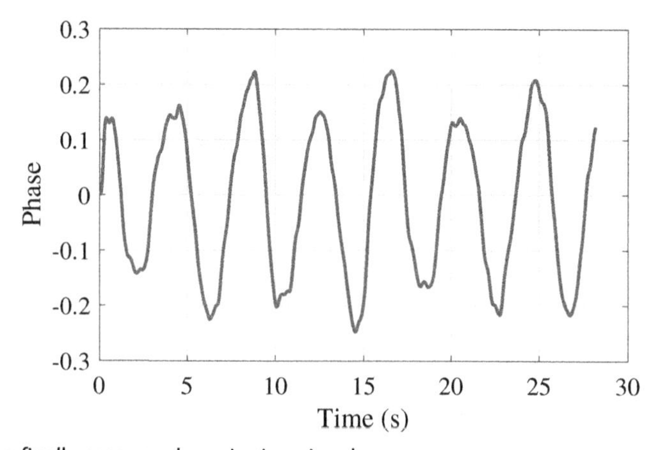

Fig. 9 The finally recovered respiration signal.

sampled signal is regarded as a regular breathing signal, from which the respiration rate can be estimated; otherwise, the signal sequence is regarded to be abnormal, and apnea is detected. Since the energy-based threshold has been applied for movement detection in the earlier stage (see Section 3.3.2), small breathing signals can be detected now at this stage.

The proposed recurrent variational autoencoder for unsupervised respiration abnormality detection is illustrated in Fig. 10.

The variational autoencoder model is applied to obtain a reconstructed version of the input signal. This model is to maximize the marginal likelihood given below.

$$p_\theta(x) = \int p_\theta(x|\, z)p(z)dz, \tag{10}$$

where x, z, and θ are the observed variables, the latent random variables, and the set of parameters, respectively; $p(z)$ is the prior over the latent random variables z; and $p_\theta(x\,|\,z)$ is the conditional probability, representing an observation model under the parameter set. Usually $p_\theta(x)$ is intractable due to the integral operation. Although the Monte Carlo sampling method can be used

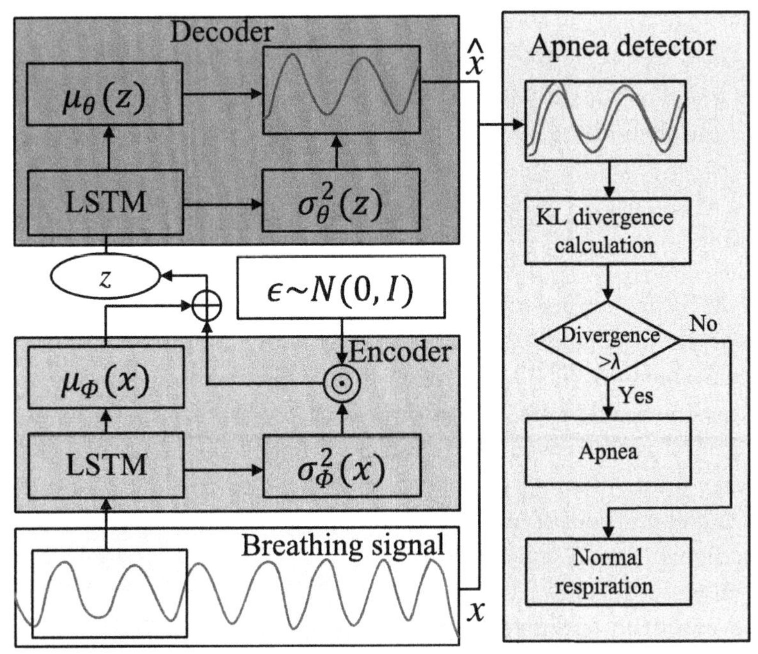

Fig. 10 Architecture of the proposed recurrent variational autoencoder for unsupervised apnea detection.

to solve the problem, it usually incurs a considerable computational cost even for small-sized datasets. The variational autoencoder model utilizes the variational approximation $q_\phi(z|x)$ instead of the true posterior $p_\theta(z|x)$. Specifically, the variational autoencoder model has $q_\phi(z|x)$ with parameter set ϕ as encoder and $p_\theta(x|z)$ with parameter set θ as decoder. According to Jensen's inequality, the variational autoencoder model can achieve the optimal values for sets ϕ and θ. This is achieved by maximizing a lower bound on the log-likelihood, given by [28]

$$\max \mathcal{L} = -D_{KL}\left(q_\phi(z|x)\|p(z)\right) + \mathbb{E}_{z\in q_\phi\left(z|x\right)}[p_\theta(x|z)], \tag{11}$$

where D_{KL} represents the KL divergence. In Eq. (11), $-D_{KL}(q_\phi(z|x)\|p(z))$ represents the regularization over the latent variables z, while $\mathbb{E}_{z\in q_\phi\left(z|x\right)}[p_\theta(x|z)]$ is the autoencoder. The latent variables z are sampled from $q_\phi(z|x)$, and the reconstructed signal $\hat{x}$ can be sampled from $p_\theta(x|z)$.

To reduce the training overhead, the variational autoencoder model utilizes the reparametrization technique. With this technique, the latent vector z is computed from the mean vector $\mu_\phi(x)$ and the variance vector $\sigma_\phi^2(x)$, as

$$z = \mu_\phi(x) + \sigma_\phi(x) \odot \epsilon, \tag{12}$$

where $\epsilon \in \mathcal{N}(0, 1)$ (i.e., a standard Gaussian distribution) and $\odot$ represent the element-wise product operation. The lower bound on the log-likelihood, i.e., $\mathcal{L}$, can then be approximated as follows.

$$\mathcal{L} \approx \frac{1}{2}\sum_{j=1}^{J}\left(1 + \log\left(\sigma_j^2(x)\right) - \mu_j^2(x) - \sigma_j^2(x)\right) + \frac{1}{M}\sum_{l=1}^{M}\log p_\theta(x|z_l), \tag{13}$$

where M is the number of samples in z, and J is the cardinality of z.

Next, the data samples within a time window are processed by a long short-term memory (LSTM) network. LSTM belongs to the class of recurrent neural networks (RNN) that can effectively handle time series data. It can also deal with the vanishing gradient problem that exists with RNN. Besides, consider that the major difference between the normal breathing signal and apnea signal can be reflected by both periodicity and the shape of the signal, which can be considered as long-range dependency and short-range dependency, respectively. These long-range dependencies can be effectually captured by LSTM, because the data in a time series can be stored or deleted by a nonlinear gate in each unit. Thus, in the proposed AutoTag system, the LSTM network is utilized to encode the respiration signal sequence within a time window. Then its outputs are used to obtain estimations for the mean vector $\mu_{\phi}(x)$ and the variance vector $\sigma_{\phi}^2(x)$ using two linear modules, respectively. The sampled z is fed to another LSTM network for decoding the estimated mean and variance vectors. Eventually, a reconstructed respiration signal is achieved with the same time window.

Once the reconstructed respiration signal is obtained, a KL divergence-based method is proposed for apnea detection. Specifically, the original signal and the reconstructed signal in the same time window should be normalized, in order to ensure that both the apnea signal and the respiration signal are within the same amplitude range. The similarity between the original signal and the reconstructed signal is then calculated in the form of KL divergence. Since the proposed neural network is well trained by a large amount of normal breathing signal, the KL divergence between input signal and reconstructed signal is very small when the signal is sampled during normal breathing. In contrast, the KL divergence is very large when the input signal includes abnormal respiration (apnea), because the network did not know the features of these abnormalities. Finally, the collected values of KL divergence calculated for normal breathing and apnea, respectively, and a threshold λ are properly chosen to determine whether the signal in this time window is for apnea or normal respiration. Specifically, if the KL divergence is greater than the threshold, apnea is detected in this time window; otherwise, the signal is for normal respiration.

The proposed deep learning-based approach has two desirable features. First, the recurrent variational autoencoder is an unsupervised learning approach. Therefore, there is no need to collect labeled data for regular and abnormal respiration signals, which could be costly and time-consuming. Second, the proposed method can learn the periodic features of respiration signal in offline training, rather than simply detecting the strength of the breathing signal. Therefore, this method is superior to the energy threshold-based method, especially when the patient moves.

3.4.2 Peak detection for breathing rate estimation

Finally, the peak detection technique [12] is used to estimate the interval between two neighboring peaks, when a regular respiration signal is detected. Although most of the noise coming from the environment has been removed at this stage, some false peaks may still exist. False peaks are usually relatively smaller than the peaks of the respiration signal, but they still could affect the accuracy of peaks detection.

To mitigate the influence of such false peaks, the system executes a special peak detection algorithm with a small sliding window of data points, instead of on the entire sampled signal sequence. In AutoTag, a window size of 11 data points is used. Only if the center data point of the sliding window is the maximum among all the data points within the window, the center data point will be identified as a peak. Then the mean value of all the peak intervals is calculated to approximate the period of the respiration signal τ. Finally, the respiration rate can be computed as $60/\tau$ breaths per minute (bpm).

3.5 Prototyping and experiment results

To evaluate the proposed AutoTag system, a prototype system utilizes Impinj R420 as the RFID reader to collect phase information from ALN-9740 tags. To be FCC-compliant, the circular polarized antenna equipped in our system hops among 50 channels ranging from 902.5 to 927.5 MHz, and remains on each channel for 200 ms. The user interface and signal processing are implemented with an MSI laptop equipped with an Nvidia GTX1080 GPU and an Intel(R) Core(TM) i7-6820HK CPU. A software named Tagsee [30] is also implemented in our system to collect data from the reader using the Low-level Reader Protocol (LLRP), which can extract useful low-level data from received tag responses, including time stamp, Doppler shift, RSSI, and phase value. All collected sampled phases are processed in the MATLAB installed in the laptop.

Extensive experiments are conducted involving four volunteers. The experimental results in two different environments are presented in this section. As illustrated in Fig. 11, the test settings include a 5.6 m × 7.5 m lab, which is cluttered with tables, chairs, and computers, and a 2.4 m × 20.0 m empty corridor with no obstacles or no moving persons in Broun Hall in the Auburn University Campus. In the lab setting, the multipath effect is obviously larger than that in the corridor setting because of more reflections caused by desks and computers. All the volunteers are tested in three cases: (i) sitting in a chair and breathing normally, (ii) sitting in the chair and holding their breath, and (iii) moving randomly while breathing normally. For respiration rate estimation, the cumulative distribution functions (CDFs) of estimation errors are shown for performance validation. For apnea detection, the following two performance metrics are used:

- True Negative (TN) rate: this is the success rate when a regular respiration signal is successfully detected;
- True Positive (TP) rate: this is the success rate when apnea is correctly detected.

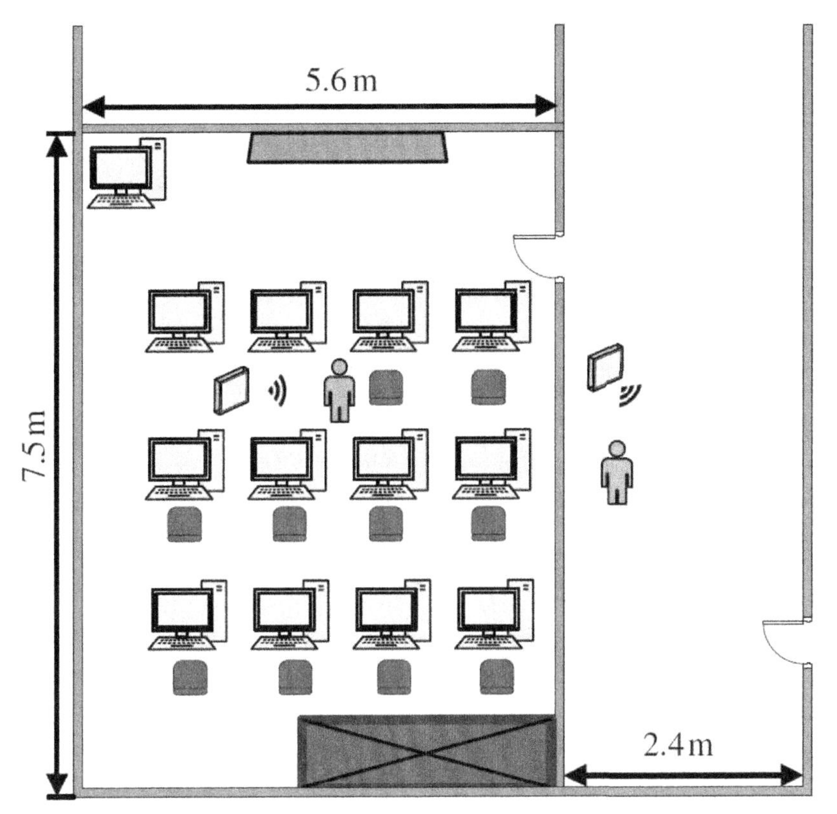

Fig. 11 Experimental environments for validating the performance of AutoTag: (i) A cluttered computer laboratory; (ii) an empty corridor.

For a normal breathing scenario, the accuracy of respiration rate estimation is evaluated under the two settings. The NEULOG Respiration Belt sensor wrapped on the volunteer's chest is applied to measure the ground truth. The CDFs of estimation errors are plotted in Fig. 12 for the lab and corridor experiments. The results show that the maximum errors are 0.462 and 0.326 bpm, while the median errors are 0.105 and 0.093 bpm for the lab and corridor settings, respectively. The maximum and median errors in the lab setting are both larger than the corresponding error in the corridor setting. In addition, more than 50% of the estimation errors obtained in the corridor setting are smaller than those obtained in the lab setting. These indicate that the multipath effect does affect the accuracy of respiration rate estimation. Furthermore, it can be seen that the median errors in the lab and corridor tests are 0.104 and 0.0925 bpm, respectively. The close median

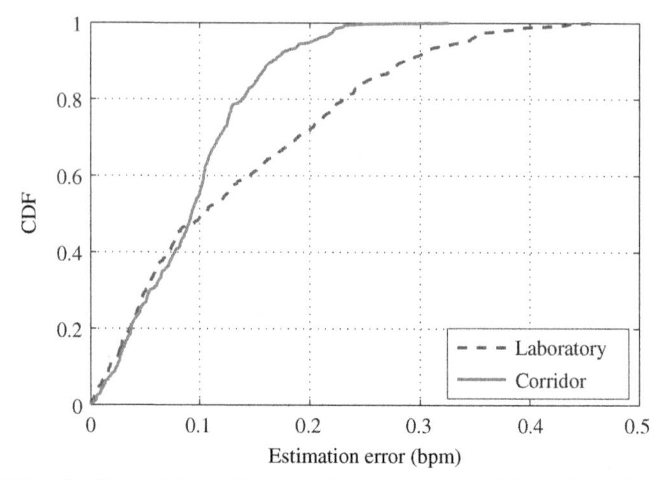

Fig. 12 CDFs of estimated breathing rates in the computer lab and corridor scenarios.

errors indicate that the effect of multipath is not substantial. Since a directional antenna is used by the reader, the backscattered tag response in the line-of-sight (LOS) path is the dominant component. Thus, the multipath effect is not serious and the respiration rate estimation is accurate both in the corridor and laboratory environments.

4. Conclusions

This chapter presented the AutoTag system for unsupervised respiration rate estimation and detection of apnea in real time with commodity RFID Tags. The AutoTag system incorporated a novel technique to effectively address the effect of frequency hopping offset for RFID systems that comply with FCC regulations, and thus can be used for many RFID applications with real-time requirements. For the signal extraction module, we extracted phase information from commodity RFID readers. Besides, the Calibration module was implemented by frequency hopping offset, movement detection, DC removal, tag selection, downsampling, and filtering. The proposed system also utilized an unsupervised learning approach, thus having the desirable advantage of not requiring labeled medical data, making it relatively inexpensive and easy to deploy. The superior performance of the AutoTag system was demonstrated by experiments in typical health-care environments. For the current system, the user is required to be in a stationary state when being monitored. Thus, the proposed system could be furthered developed by removing the interference of the user movement, so that the user could move casually while the system is monitoring.

Acknowledgment

This work was supported by the US NSF in part under Grant Number CNS-1702957, and by the Wireless Engineering Research and Education Center (WEREC) at Auburn University, Auburn, AL, United States.

References

[1] C. Yang, X. Wang, S. Mao, AutoTag: recurrent vibrational autoencoder for unsupervised apnea detection with RFID tags, in: Proc. IEEE GLOBECOM 2018, Abu Dhabi, United Arab Emirates, Dec 2018, pp. 1–7.
[2] O. Boric-Lubeke, V. Lubecke, Wireless house calls: using communications technology for health care and monitoring, IEEE Microw. Mag. 3 (3) (Apr. 2002) 43–48.
[3] X. Wang, X. Wang, S. Mao, RF sensing for Internet of Things: a general deep learning framework, IEEE Commun. Mag. 56 (9) (Sept. 2018) 62–69.
[4] X. Wang, R. Huang, S. Mao, Sonarbeat: sonar phase for breathing beat monitoring with smartphones, in: Proc. ICCCN 2017, Vancouver, Canada, July/Aug, 2017, pp. 1–8.
[5] M.L.R. Mogue, B. Rantala, Capnometers, J. Clin. Monit. 4 (2) (1988) 115–121.
[6] C. Hunt, F. Hauck, Sudden infant death syndrome, Can. Med. Assoc. J. 174 (13) (2006) 1309–1310.
[7] F. Adib, H. Mao, Z. Kabelac, D. Katabi, R. Miller, Smart homes that monitor breathing and heart rate, in: Proc. ACM CHI'15, Seoul, Korea, April 2015, pp. 837–846.
[8] P. Nguyen, X. Zhang, A. Halbower, T. Vu, Continuous and fine-grained breathing volume monitoring from afar using wireless signals, in: Proc. IEEE INFOCOM'16, San Francisco, CA, Apr 2016, pp. 1–9.
[9] J. Salmi, A.F. Molisch, Propagation parameter estimation, modeling and measurements for ultrawideband mimo radar, IEEE Trans. Microwave Theory Tech. 59 (11) (2011) 4257–4267.
[10] H. Abdelnasser, K.A. Harras, M. Youssef, Ubibreathe: a ubiquitous non-invasive wifi-based breathing estimator, in: Proc. IEEE MobiHoc'15, Hangzhou, China, June 2015, pp. 277–286.
[11] J. Liu, Y. Wang, Y. Chen, J. Yang, X. Chen, J. Cheng, Tracking vital signs during sleep leveraging off-the-shelf WiFi, in: Proc. ACM Mobihoc'15, Hangzhou, China, June 2015, pp. 267–276.
[12] X. Wang, C. Yang, S. Mao, PhaseBeat: exploiting CSI phase data for vital sign monitoring with commodity WiFi devices, in: Proc. IEEE ICDCS 2017, Atlanta, GA, June 2017, pp. 1–10.
[13] X. Wang, C. Yang, S. Mao, Tensorbeat: tensor decomposition for monitoring multi-person breathing beats with commodity WiFi, ACM Trans. Intell. Syst. Technol. 9 (1) (2017) 8:1–8:27.
[14] Y. Hou, Y. Wang, Y. Zheng, Tagbreathe: monitor breathing with commodity RFID systems, in: Proc. IEEE ICDCS 2017, Atlanta, GA, June 2017, pp. 404–413.
[15] C. Yang, X. Wang, S. Mao, Unsupervised detection of apnea using commodity RFID tags with a recurrent variational autoencoder, IEEE Access J. 7 (1) (2019) 67526–67538.
[16] Y. Sun, M. Peng, Y. Zhou, Y. Huang, S. Mao, Application of machine learning in wireless networks: key technologies and open issues, IEEE Commun. Surv. Tutorials 21 (4) (2019) 3072–3108.

[17] T. Zhang, S. Mao, Energy-efficient power control in wireless networks with spatial deep neural networks, IEEE Trans. Cogn. Commun. Netw. 6 (1) (2020) 111–124.
[18] K. Xiao, S. Mao, J. Tugnait, TCP-Drinc: smart congestion control based on deep reinforcement learning, IEEE Access J. 7 (1) (2019) 11892–11904.
[19] M. Feng, S. Mao, Dealing with limited backhaul capacity in millimeter wave systems: a deep reinforcement learning approach, IEEE Commun. Mag. 57 (3) (Mar. 2019) 50–55.
[20] Z. Whang, S. Mao, W. Yang, Deep learning approach to multimedia traffic classification based on QoS characteristics, IET Netw. 8 (3) (2019) 145–154.
[21] X. Wang, X. Wang, S. Mao, CiFi: deep convolutional neural networks for indoor localization with 5GHz Wi-Fi, in: Proc. IEEE ICC 2017, Paris, France, May 2017, pp. 1–6.
[22] W. Wang, X. Wang, S. Mao, Deep convolutional neural networks for indoor localization with CSI images, IEEE Trans. Netw. Sci. Eng. 7 (1) (2020) 316–327.
[23] X. Wang, L. Gao, S. Mao, S. Pandey, CSI-based fingerprinting for indoor localization: a deep learning approach, IEEE Trans. Veh. Technol. 66 (1) (2017) 763–776.
[24] X. Wang, L. Gao, S. Mao, CSI phase fingerprinting for indoor localization with a deep learning approach, IEEE Internet Things J. 3 (6) (2016) 1113–1123.
[25] X. Wang, L. Gao, S. Mao, BiLoc: bi-modality deep learning for indoor localization with 5GHz commodity Wi-Fi, IEEE Access J. 5 (1) (2017) 4209–4220.
[26] Impinj Support Portal, Low level user data support, in: Impinj Speedway Revolution Reader Application, 2013 Available from: https://support.impinj.com.
[27] T. Wei, X. Zhang, Gyro in the air: tracking 3D orientation of batteryless internet-of-things, ACM GetMobile 21 (1) (2017) 35–38.
[28] D.P. Kingma, M. Welling, Auto-encoding variational bayes, in: arXiv, May 2014 preprint arXiv:1312.6114.
[29] O. Fabius, J.R. van Amersfoort, Variational recurrent auto-encoders, in: arXiv, June 2015 preprint arXiv:1412.6581.
[30] L. Yang, Y. Chen, X.-Y. Li, C. Xiao, M. Li, Y. Liu, Tagoram: real-time tracking of mobile RFID tags to high precision using COTS devices, in: Proc. ACM MobiCom'14, Maui, HI, Sept 2014, pp. 237–248.

CHAPTER THREE

Designing a cooperative hierarchical model of interdiction median problem with protection and its solution approach: A case study of health-care network

Raheleh Khanduzi[a] and Abdolmotalleb Rastegar[b]

[a]Department of Mathematics and Statistics, Gonbad Kavous University, Gonbad Kavous, Golestan, Iran
[b]Department of Cartography, Golestan University, Gorgan, Golestan, Iran

1. Introduction and background

Facility protection decisions play a crucial and vital role in the strategic planning of client assignment for a wide range of service systems (e.g., health care, manufacturing, communication, airline, computer, supply chain, transportation, etc.). This is because unsatisfactorily protected facilities or an inappropriate number of protection operations can considerably increase defensive costs and budgets and reduce client services. The first theoretical research on the interdiction and protection of facilities started in 2007 by Church and Scaparra [1] when they considered how to protect q of the p operating facilities for minimizing losses following the worst case of r of the p facilities due to attacks. Thenceforth, interdiction median problem with protection (IMP) and its applications were developed in various areas of studies and types of mathematical models. In another study, Scaparra and Church [2] proposed a bi-level IMP of two decision makers called the defender (leader) and the attacker (follower). The leader's subproblem consisted of the defender's decision about protection operations to minimize the worst-case losses subject to attacks and the follower's subproblem consisted of the attacker's decision about disruption strategy to maximize the customer service distance.

In most cases, the papers in first in studies were conducted in the interdiction and protection of median network. Especially, there are modeling

https://doi.org/10.1016/B978-0-12-821187-8.00003-4

approaches which were well-known or widely applied in the bi-level optimization area including IMP by considering capacity and budget constraint (see Aksen et al. [3]; Liberatore et al. [4]), stochastic programming (see Liberatore et al. [5]), partial interdiction or protection (see Aksen et al. [6]; Khanduzi et al. [7]), dynamic programming (see Starita and Scaparra [8]; Khanduzi and Maleki [9]), and fuzzy programming (see Maleki and Khanduzi [10]). Additionally, the interdiction-protection problems were studied on different types of service networks. Most popular service networks are: covering interdiction problem with protection (Mahmoodjanloo et al. [11]; Roboredo et al. [12]), hub interdiction and protection problem (Ghaffarinasab and Atayi [13]; Quadros et al. [14]), hierarchical interdiction problem with protection (Aliakbarian et al. [15]; Forghani et al. [16]) and interdiction-protection problem within a supply chain network (Khanduzi et al. [17]; Azad et al. [18]; Zhou and Poh [19]).

One of the important applications of hierarchical networks is the health-care network. This network is composed of three levels of health facilities where some patients require clinical services, some patients require to be referred to hospitals, and some others require to be referred to medical centers. In such a service network, patients first referred to the clinics and after receiving basic health care and diagnostic services they are referred to a hospital or a medical center for outpatient surgery and inpatient services. In the health-care network, unreliable and imprecise facility location-protection decisions have a crucial effect on the network beyond simple operational cost and desired service to patients; for example, hard access to health facilities, long service distance, injury deaths rise, and destruction of health facilities due to sabotage operations. From this viewpoint, location-protection-interdiction modeling for health-care facilities is a valuable issue. Moreover, because of universally extensive directions, such as decreasing health service time, increasing service quality, reducing the service distance between patients and health facilities, and security of health network and facilities and services, the reliable location of health-care facilities and the associated health-care facility protection problems were conspicuously more vital and meaningful to community of operational researchers.

A disaster was a crucial disruption occurring on a health-care network that caused people, health facilities, and services, economic loss which increased the capability of the designers or decision makers to manage using their protection resources. The designers of health-care network suffered the very costs when a disaster hit. Disasters were generally classified into a natural or manmade. A natural disaster was a natural procedure or event that led to the loss of health facilities and services, hurt, economic trouble, or physical damage. Numerous events like earthquakes, landslips, geysers, overflows, cataclysms, and tornados were all natural disasters that killed many people of society and service networks and destroy facilities of networks. So, several authors have paid considerable attention to the robust and reliable design problem of health-care networks for disaster restraint and management. Mete and Zabinsky [20] developed a stochastic programming model for finding the location of medical supplies to manage disaster events under risk of earthquake disruption. This model included the disaster particular information and possible impacts of disasters via scenarios. This work dealt with two strategies such as loading and routing of vehicles for transporting medical supplies with respect to disaster response. Paul and Batta [21] introduced two mixed-integer formulations for locating hospitals and allocating capacities for relief programs of natural disasters. The aim of first model was to determine the optimal location of hospitals and allocation of capacities to minimize the demand-weighted distance between patients and hospitals due to disaster scenarios. The objective of the second model was to reallocate the capacity of hospitals so that the effectiveness of a health network against the disaster phenomena was maximized. Shishebori and Babadi [22] studied a robust mixed-integer programming model for the location of medical service facilities, which simultaneously considered uncertain parameters, network disruptions, and investment budget restriction to fortify and control the network reliability under disruptions. A reliable two-level multi-flow hierarchical location-allocation model for health-care system in disruptive events was presented by Zarrinpoor et al. [23]. The queuing theory was taken into consideration in the mathematical formulation which managed uncertainty of demand and service parameters. To provide the service quality, a minimum boundary for the probability of patients' waiting time was studied. In another study, Zarrinpoor et al. [24] proposed an extension of reliable hierarchical location-allocation problem for a health-care network under natural disasters. They dealt with stochastic programming by considering an ambiguity of demand, service, and accessibility parameters, arranging patients based on the urgency of their demand and pursuing different service orders to serve all patients' demand, priority queue network, and risk of disruptions.

On the other hand, manmade disasters were the result of technical or human hazards. Examples involved panics, fires, undesirable events in industrial and transport systems, oil drops, terrorist attacks. So, the need for protection programs is derived from the requirement of health-care networks to protect the clinics, hospitals, and medical centers against an attacker's sabotage operations. Indeed, the allocation of defensive/offensive resources is a crucial process. The studies on the interdiction/protection problems addressed hierarchical models without real-world case study. Although the significance of hierarchical interdiction/protection models is obvious, and health-care networks dealt with the interdiction/protection operations on clinics, hospitals, and medical centers, but there is no work in the literature that considers these models on a health-care network. Also, the development of health-care networks has made cooperation fundamental to both attacker and defender. Increasing the performance of health-care network needs the decision makers to effectively interact with other people. The distribution of decisions of leaders and followers need to be taken at the health facilities. The exposure of health facilities makes multiple, simultaneous attackers to launch a creditable danger. Motivation for this study was based on this experience that it was often easier for aggressors to cooperate than for protectors to do so. So, we consider a defender to allocate defensive resources for protecting the health facilities and a number of offenses which exploit cooperation to interdict unprotected health facilities.

With respect to the abovementioned subjects, we proposed a hierarchical bi-level model to analyze how doing the complete defense and cooperative offense policies incorporated into a health-care network with capacitated health facilities and offensive/defensive resource constraint. So, we introduced a Stackelberg game between the defender as a leader and the attacker as the follower based on the allocation of defensive resource to each clinic, hospital, or medical center and internal and cooperation probability of any attacker. The patients were assigned to the closest clinics, hospitals, or medical centers concerning initial capacity at the health facilities for the basic health care, diagnostic, outpatient, and inpatient services. Doing so, a mixed-integer quadratic bi-level formulation is proposed. As far as we are aware, this mathematical formulation has not been comprehensively argued in the literature yet. The contributions of this study were in twofold: (1) incorporating a real issue to a hierarchical network that is a safety clinic, hospital, or medical center with full protection and interdiction operations to cooperative ones. Adding resource constraints for protection operations and interdiction budget for the cost of the action of any attacker, which no work has considered them in a hierarchical interdiction/protection problem simultaneously; and (2) hybriding a metaheuristic algorithm, i.e., tabu search algorithm (TS) with a double-layer neural network (DLNN) to solve the proposed bi-level problem in which the solution space of the upper-level subproblem was searched regarding TS principles, and the corresponding lower-level subproblem was solved using DLNN. In regards to solve the problem, the bi-level programming problems were NP-Hard problem [25], which had received considerable attention in the field of operational researches and computer sciences. Therefore, many authors focused on metaheuristic algorithms to solve the bi-level interdiction/protection problems and their related hybrid types [26, 27]. Since the attacker's subproblem is a mixed-integer quadratic problem (MIQP), exact algorithms solve large-scaled problems with excessive execution times. So, a DLNN is utilized to solve large-scale instances of the problem. From the viewpoint of Yaakob et al. [28], DLNN is an efficient approach to solve MIQPs based on computational times and accuracy. On the other hand, TS was a metaheuristic algorithm mimicking people thinking procedure and was basically a type of global search method. TS had advantages of simplicity, changeability, flexibility, quickness, precision, stability, and robustness. The neighborhood searching capability and qualification of TS was comparatively robust and vigorous. A semi-deterministic nature was another advantage of TS as it performed both as a local and global search approach. This chapter combined the advantages of TS and DLNN that the computational speed and accuracy were enhanced. Also, the approach robustness was satisfactory and converged globally to an optimum solution.

The remainder of this chapter is organized as follows. Section 2 provides a mathematical formulation of the developed model. Section 3 proposes a hybrid solution approach for the model, which is based on a tabu search algorithm and double-layer neural network. Section 4 describes a practical case study to show the applicability of the new model. Finally, Section 5 concludes this study and outlines our future work.

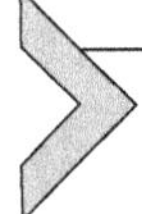

2. Problem description and formulation

2.1 Notation

The sets, parameters, and decision variables used in a cooperative hierarchical and interdiction median problem with protection (CHIMP) are introduced in advance, as follows:

Sets and parameters:

I	Set of clients or patients, $I = \{1, \ldots, n\}$
J_1	Set of health facilities at the first level or clinics,
J_2	Set of health facilities at the second level or hospitals,
J_3	Set of health facilities at the third level or medical centers,
S	Set of services S={basic health care, diagnostic, outpatient, inpatient},
d_{ij}	Shortest distance between patient $i \in I$ and health facility $j \in J_1 \cup J_2 \cup J_3$,
b	Number of protection operations,
c_h	Cost of the action of the hth attacker,
C	Total interdiction budget,
H	Set of attackers to interdict facility $j \in J_1 \cup J_2 \cup J_3$,
q_j	Capacity of health facility $j \in J_1 \cup J_2 \cup J_3$ for all services,
a_i	Demand at patient zone $i \in I$,
σ_h^1	Internal probability of the hth attacker
σ_h^2	Cooperation probability of the hth attacker
α	Percent of the demand that is referred to obtain diagnostic service after getting basic health care,
β	Percent of the demand that is referred to obtain outpatient surgery after getting diagnostic service,
γ	Percent of the demand that is referred to obtain inpatient service after getting diagnostic service,
F_{ij}^1	Set of existing clinics that are farther than clinic $j \in J_1$ from patient $i \in I$, $F_{ij}^1 = \{l \in J_1 \mid l \neq j, d_{il} > d_{ij}\}$
F_{ij}^2	Set of existing hospitals that are farther than hospital $j \in J_2$ from patient $i \in I$, $F_{ij}^2 = \{l \in J_2 \mid l \neq j, d_{il} > d_{ij}\}$
F_{ij}^3	Set of existing medical centers that are farther than medical center $j \in J_3$ from patient $i \in I, F_{ij}^3 = \{l \in J_3 \mid l \neq j, d_{il} > d_{ij}\}$
R_{ij}^1	Set of existing clinics and hospitals (not including clinic $j \in J_1$) that are farther than health facility $j \in J_1, J_2$ from patient $i \in I$, $R_{ij}^1 = \{l \in J_1, J_2 \mid l \neq j, d_{il} > d_{ij}\}$
R_{ij}^2	Set of existing hospitals and medical centers (not including hospital $j \in J_2$) that are farther than health facility $j \in J_2, J_3$ from patient $i \in I$, $R_{ij}^2 = \{l \in J_2, J_3 \mid l \neq j, d_{il} > d_{ij}\}$
T_{jk}^1	Set of existing hospitals (except hospital $k \in J_2$) that are farther than hospital $k \in J_2$ from health facility $j \in J_1 \cup J_2$, $T_{jk}^1 = \{l \in J_2 \mid l \neq k, d'_{jl} > d'_{jk}\}$
T_{jk}^2	Set of existing medical centers (except medical center $k \in J_3$) that are farther than medical center $k \in J_3$ from health facility $j \in J_2 \cup J_3$, $T_{jk}^2 = \{l \in J_3 \mid l \neq k, d''_{jl} > d''_{jk}\}$

Variables:

$$w_j = \begin{cases} 1, & \text{if the facility } j \in J_1 \cup J_2 \cup J_3 \text{ is protected} \\ 0, & \text{otherwise} \end{cases}$$

$$v_{hj} = \begin{cases} 1, & \text{if h}^{th} \text{ attacker is assinged to facility } j \in J_1 \cup J_2 \cup J_3 \\ 0, & \text{otherwise} \end{cases}$$

$$x_{ij}^{1}=\begin{cases}1, & \text{if demand of client } i\in I \text{ for basic health care is assigned to facility } j\in J_1\cup J_2\\ 0, & \text{otherwise}\end{cases}$$

$$x_{ij}^{2}=\begin{cases}1, & \text{if demand of client } i\in I \text{ for diagnostic service is assigned to facility } j\in J_1\cup J_2\\ 0, & \text{otherwise}\end{cases}$$

$$x_{ij}^{3}=\begin{cases}1, & \text{if demand of client } i\in I \text{ for outpatient surgery is assigned to facility } j\in J_2\cup J_3\\ 0, & \text{otherwise}\end{cases}$$

$$x_{ij}^{4}=\begin{cases}1, & \text{if demand of client } i\in I \text{ for inpatient service is assigned to facility } j\in J_3\\ 0, & \text{otherwise}\end{cases}$$

$$y_{jk}^{1}=\begin{cases}1, & \text{if a demand for basic health care is primarily assigned to } j\in J_1\cup J_2 \text{ and then referred to } k\in J_2 \text{ to receive diagnostic service;}\\ 0, & \text{otherwise.}\end{cases}$$

$$y_{jk}^{2}=\begin{cases}1, & \text{if a demand for diagnostic service is primarily assigned to } j\in J_1\cup J_2 \text{ and then referred to } k\in J_2\cup J_3 \text{ to receive outpatient surgery;}\\ 0, & \text{otherwise.}\end{cases}$$

$$y_{jk}^{3}=\begin{cases}1, & \text{if a demand for diagnostic service is primarily assigned to } j\in J_1\cup J_2 \text{ and then referred to } k\in J_3 \text{ to receive inpatient service;}\\ 0, & \text{otherwise.}\end{cases}$$

e_{ij}^{1}	Demand of patient $i \in I$ for basic health care provided by health facility $j \in J_1 \cup J_2$,
e_{ij}^{2}	Demand of patient $i \in I$ for diagnostic service provided by health facility $j \in J_1 \cup J_2$,
e_{ij}^{3}	Demand of patient $i \in I$ for outpatient surgery provided by health facility $j \in J_2 \cup J_3$,
e_{ij}^{4}	Demand of patient $i \in I$ for inpatient service provided by health facility $j \in J_3$,
f_{jk}^{1}	Demand for diagnostic service referred from health facility $j \in J_1 \cup J_2$ to hospital $k \in J_2$,
f_{jk}^{2}	Demand for outpatient surgery referred from health facility $j \in J_1 \cup J_2$ to health facility $k \in J_2 \cup J_3$,
f_{jk}^{3}	Demand for inpatient service referred from health facility $j \in J_1 \cup J_2$ to medical center $k \in J_3$.

Auxiliary variables:

λ_j	Probability that health facility $j \in J_1 \cup J_2 \cup J_3$ is interdicted
μ_{hj}	Probability that health facility $j \in J_1 \cup J_2 \cup J_3$ is interdicted based on the action of attacker h

2.2 New mathematical model

$$\min_{\boldsymbol{w}} Z(\boldsymbol{w}) \tag{1}$$

s.t.

$$\sum_{j \in J_1 \cup J_2 \cup J_3} w_j = b \tag{2}$$

$$w_j \in \{0, 1\} \ , \ \forall j \in J_1 \cup J_2 \cup J_3 \tag{3}$$

where $\boldsymbol{w}$ solve:

$$Z(\boldsymbol{w}) = \max_{v} \sum_{i \in I} (a_i d_{ii_1} (1 - \lambda_{i_1})) + \sum_{u=2}^{|J_1 \cup J_2 \cup J_3|} \left(a_i d_{ii_u} (1 - \lambda_{i_u}) \prod_{q=1}^{u-1} \lambda_{i_q} \right) + a_i \prod_{q=1}^{|J_1 \cup J_2 \cup J_3|} \lambda_{i_q}) \tag{4}$$

s.t.

$$\lambda_j = 1 - \prod_{h=1}^{|H|} \left(1 - \mu_{hj} v_{hj} \right), \forall j \in J_1 \cup J_2 \cup J_3 \tag{5}$$

$$\mu_{hj} = \sigma_h^1 \left(\max_{g \in H} \left(\sigma_g^2 v_{gj} \right) \right), \forall h \in H, \forall j \in J_1 \cup J_2 \cup J_3 \tag{6}$$

$$\sum_{j \in J_1 \cup J_2 \cup J_3} v_{hj} \leq 1, \forall h \in H \tag{7}$$

$$\sum_{j \in J_1 \cup J_2 \cup J_3} \sum_{h \in H} c_h v_{hj} \leq C \tag{8}$$

$$v_{hj} \leq 1 - w_j \ , \ \forall j \in J_1 \cup J_2 \cup J_3, \forall h \in H \tag{9}$$

$$\sum_{j \in J_1 \cup J_2 \cup J_3} \sum_{s \in S} e_{ij}^s = a_i, \forall i \in I \tag{10}$$

$$\sum_{k\in J_2} f_{jk}^1 = \alpha\left(\sum_{i\in I} e_{ij}^1\right),\ \forall j\in J_1\cup J_2 \tag{11}$$

$$\sum_{k\in J_3} f_{jk}^2 = \beta\left(\sum_{i\in I} e_{ij}^2\right),\ \forall j\in J_1\cup J_2 \tag{12}$$

$$\sum_{k\in J_3} f_{jk}^3 = \gamma\left(\sum_{i\in I} e_{ij}^2\right),\ \forall j\in J_1\cup J_2 \tag{13}$$

$$\sum_{i\in I}\sum_{s\in S} e_{ij}^s \le \sum_{i\in I}\sum_{s\in S}\sum_{h\in H} q_j x_{ij}^s\left(1-v_{hj}\right),\ \forall j\in J_1\cup J_2\cup J_3 \tag{14}$$

$$\sum_{i\in J_2\cup J_3}\sum_{t=1}^{3} f_{jk}^t \le \sum_{k\in J_2\cup J_3}\sum_{t=1}^{3}\sum_{h\in H} q_j y_{jk}^t\left(1-v_{hj}\right),\ \forall j\in J_1\cup J_2\cup J_3 \tag{15}$$

$$\sum_{p\in R_{ij}^1} x_{ip}^1 + x_{ip}^2 \le \sum_{h\in H} v_{hj},\ \forall i\in I,\ \forall j\in J_1 \tag{16}$$

$$\sum_{p\in F_{ij}^1} x_{ip}^1 + \sum_{p\in F_{ij}^2} x_{ip}^1 \le \sum_{h\in H} v_{hj},\ \forall i\in I,\ \forall j\in J_2 \tag{17}$$

$$\sum_{p\in F_{ij}^1} x_{ip}^2 + \sum_{p\in F_{ij}^2} x_{ip}^2 \le \sum_{h\in H} v_{hj},\ \forall i\in I,\ \forall j\in J_2 \tag{18}$$

$$\sum_{p\in F_{ij}^2} x_{ip}^3 + \sum_{p\in F_{ij}^3} x_{ip}^3 \le \sum_{h\in H} v_{hj},\ \forall i\in I,\ \forall j\in J_3 \tag{19}$$

$$\sum_{p\in F_{ij}^3} x_{ip}^4 \le \sum_{h\in H} v_{hj},\ \forall i\in I,\ \forall j\in J_3 \tag{20}$$

$$\sum_{p\in R_{ij}^2} x_{ip}^2 + x_{ip}^3 \le \sum_{h\in H} v_{hj},\ \forall i\in I,\ \forall j\in J_2 \tag{21}$$

$$\sum_{p\in T_{jk}^1} y_{jp}^1 \le \sum_{h\in H} v_{hk},\ \forall k\in J_2,\ \forall j\in J_1\cup J_2\cup J_3 \tag{22}$$

$$\sum_{p\in T_{jk}^2} y_{jp}^2 \le \sum_{h\in H} v_{hk},\ \forall k\in J_2\cup J_3,\ \forall j\in J_1\cup J_2\cup J_3 \tag{23}$$

$$\sum_{p\in T_{jk}^2} y_{jp}^3 \le \sum_{h\in H} v_{hk},\ \forall k\in J_3,\ \forall j\in J_1\cup J_2\cup J_3 \tag{24}$$

$$v_{hj},\ x_{ij}^s,\ y_{jk}^1,\ y_{jk}^2,\ y_{jk}^3 \in\{0,\ 1\},\ e_{ij}^s, f_{jk}^1, f_{jk}^2, f_{jk}^3 \ge 0, 0\le \mu_{hj},\ \lambda_j \le 1,\ \forall i\in I,\ \forall k, j\in J_1\cup J_2\cup J_3,\ \forall s\in S,\ \forall h\in H \tag{25}$$

In the proposed bi-level model, Eqs. (1)–(3) show the upper-level problem, whereas Eqs. (4)–(25) are equal to the formulation of the lower-level problem. Eqs. (1) and (4) represent the objective functions of the defender and attacker, respectively. The objective function is profit obtained by planning the attack scenario $\boldsymbol{v}$, i.e., the expected patient service distance after interdiction operation. Constraint (2) demonstrates the resource constraint on the protection operation of clinics, hospitals, and medical centers. In the lower-level subproblem, the probability of clinic, hospital, or medical center $j\in J_1\cup J_2\cup J_3$ being interdicted, shown by λ_j, is computed by Eq. (5), in which μ_{hj} demonstrates the success probability that attacker h disrupts facility $j\in J_1\cup J_2\cup J_3$. Also, μ_{hj} is computed by Eq. (6), which includes the internal probability of an attacker h, i.e., σ_h^1, and the cooperation probability of all attackers, i.e.,σ_h^2. Constraint (7) guarantees that each attack can disturb no more than one clinic, hospital, or medical center. Restriction (8) is the resource constraint on the attacks. The restriction set in Eq. (9) describes a legitimate relationship between the follower's subproblem and the leader's subproblem. It shows that a clinic, hospital, or medical center fortified in the leader's subproblem cannot be ruined in the follower's subproblem. Constraints (10) insure that the basic health care, diagnostic, outpatient, and inpatient needs are served completely for all patients by a clinic, hospital, or medical center. Eq. (11) demonstrates that a particular known percent of patients who receive basic health care are referred to get diagnostic services. Eqs. (12) and (13) represent that a given percent of patients who receive diagnostic service is referred to achieve outpatient surgery and inpatient services, respectively. Restrictions (14) confirm that no demand is served with a destroyed clinic, hospital, and medical center or with a health facility is not considered for basic health care, diagnostic, outpatient or inpatient service. In addition, the set of constraints in Eq. (14) specify the remaining capacity at each clinic, hospital, and medical center to meet the demand of patients, i.e., if health facility j is ruined, then its capacity will be lost. Constraints (15) provide that no demand referred from a health facility to another health facility is allocated to a ruined clinic, hospital, and medical center or to a health facility is not referred to obtain another health-care service.

Constraints (16)–(24) are restrictions on the assignment of patients to closest health facilities. These restrictions warrant each patient to be assigned to the closest remaining clinic, hospital, or medical center. Concerning these restrictions, if the closest clinic, hospital, or medical center to each patient is destroyed then the demand of patient is assigned to the next closest unharmed clinic, hospital, or medical center. Specifically, when a clinic is attacked, its patients are referred to the closest clinic or hospital [see constraint (16)]. Moreover, when a hospital is attacked then its patients are reassigned to the closest hospital for basic health care and diagnostic service [see constraints (17) and (18)], and the others are referred to the closest medical center for outpatient and inpatient service [see constraints (19) and (20)]. Also, when a hospital is attacked, its patients are reallocated to the closest hospital, or medical center for diagnostic and outpatient services [see constraint (21)]. Additionally, if there is a relation between a hospital (medical center) and a disrupted hospital (medical center) before an attack, its patients have to be referred to the next closest hospital (medical center) after an attack [see constraints (22)–(24)]. In the end, the continuous and binary restrictions on the variables in the decision-making process are displayed in constraints (3) and (25).

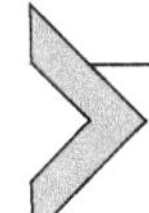

3. Two hybrid algorithms-based tabu search and global approaches

In this section, we develop two hybrid tabu search algorithms to solve the CHIMP. We first describe an exact solution approach based on CPLEX solver that can be executed to evaluate the quality of the solutions of small problem instances generated by the TS [29]. Then, a double-layer neural network (DLNN) is introduced to solve the medium and large problem instances in less time where the solution region of the defender's subproblem was searched concerning TS principles, and the corresponding attacker's subproblem is solved by the DLNN. TS was introduced to solve network programming problems by Glover [30] and it was in accordance with critical oscillation procedure which was generally determined by a structure for memory to control its non-monotonic behavior and to obtain relevant functions of conducting the search. TS was a valid and powerful approach that had been advantageously executed to a number of network programming problems. It utilized a process called a move to specify the neighborhood of any certain solution. TS can be described as an iterative method that evaluated a set of solutions by repeatedly generating moves from one solution to another. TS was defined by the capability to avoid from local minimum and the cycles. This objective was obtained by using a limited-size list of tabu moves, achieved from the current information of the search.

The two essential parts of the TS are the tabu list conditions and the general principles of aspiration. Tabu lists are conducted by including moves in the order in which they are created. If a new typical feature is added to the tabu list, the oldest one will arise from the tabu list. The suitable selection of the tabu list size is critical to the advantageous outcome of the TS algorithm and it depends on the particular problem. Aspiration criteria can abrogate tabu conditions. It implies that the aspiration criteria, when established, can recrudesce a taboo move. In the proposed TS approaches for CHIMP, the neighborhood of a current solution involves all structures on protection operations of clinics, hospitals, and medical centers. Here, we describe in more details the components of the TS to solve CHIMP.

3.1 Initial solution generation

A solution is created in the form of an $1 \times |J_1 \cup J_2 \cup J_3|$ array. This row is the defensive row and shows the number of protection operations on clinics, hospitals, and medical centers.

3.2 Neighborhood structures

A single operator called Swap is utilized to create neighbors. The Swap operator is utilized to vary one of the protected health facilities in the solutions. In this scheme, a protection node and a non-protection node are randomly selected. Then, if the chosen non-protection node is not in the tabu list, the chosen protection node turns into a non-protection node and the chosen non-protection node turns into a protection node.

3.3 Parameter selection for the TS algorithm

The TS method requires two input parameters, i.e., TL_{size} and T_{max}. TL_{size} demonstrates the size of the tabu list, and T_{max} represents the number of iterations without any further significant improvement in the best possible solution. This approach continues until it does not change the best possible solution in the last T_{max} iterations.

3.4 Hybrid tabu search

The new formulation is based on a mixed-integer quadratic bi-level programming model for whose solution, a hybrid TS is developed. In order to obtain a feasible solution of this bi-level optimization problem, we propose two hybrid TS which employ two global search approach as a subroutine: an exact algorithm based on CPLEX solver and a DLNN to solve a mixed-integer quadratic optimization problem where the lower objective function is quadratic in the continuous and in the integer variables. The developed TS-based CPLEX (TS-CPLEX) and TS-based DLNN (TS-DLNN) implement a search in the upper-level subproblem over the protection variables w_j of the clinic, hospitals, and medical centers. TS is an iterative approach that manages the local search to keep it from being trapped in a local minimum.

This is achieved in the TS-CPLEX and TS-DLNN by forbidding the levels of the health-care network from allocating the defensive resources that lead to return to previous solutions. The objective value of the leader can only be achieved by detecting the optimal solution of the follower based on a set of protected clinics, hospitals, and medical centers. This is obtained by solving the follower's subproblem to find the optimal values of v_{hj} (assignment variables for attackers), x^s_{ij} (assignment variables for the demand of patients to health facilities), y^1_{jk}, y^2_{jk}, y^3_{jk} (referral variables for health services), e^s_{ij} (amount of patients' demand for health services by facilities), and $f^1_{jk}, f^2_{jk}, f^3_{jk}$ (amount of referral demands from a health facility to another facility). In this chapter, we designed two hybrid approaches called a TS-CPLEX and TS-DLNN to solve the CHIMP and survey its effectiveness by solving a real case study on a health-care network. In the following, we describe in more details the components of the TS-CPLEX and TS-DLNN for solving the CHIMP.

3.4.1 TS algorithm based on CPLEX solver

To solve the CHIMP using the TS-CPLEX, the defender's subproblem is solved with TS algorithm, while the attacker's solution is achieved by CPLEX solver. Therefore, the iterations of the TS-CPLEX approach include the search for the best protection status of clinics, hospitals, and medical centers (w_j) using TS strategies. When the allocation of defensive resources is specified, the lower-level problem can be solved to optimality by the CPLEX regarding the protection plans of the defender. It gets a plan for assigning the attackers to health facilities and demand of patients to health facilities, refereeing health services and demands from a health facility to another facility and a plan for the amount of patients' demand for health services by facilities, which are the best decisions of the attacker. Then, the solution to the lower-level problem is applied to obtain the objective value of the defender. Also, we execute a mixed-integer quadratic programming problem (MIQP) solver CPLEX 12.5.1 by using a *cplexmiqp* function in MATLAB to solve the lower-level problem to optimality. The pseudo-code of the TS-CPLEX algorithm is shown in Algorithm 1.

Algorithm 1 TS-CPLEX algorithm for the CHIMP

1: Generate a random initial solution w_j
2: Generate tabu list TL with the size of $|N| = |J_1 \cup J_2 \cup J_3|$
3: Compute $Z(\boldsymbol{w})$ as follows: for this initial solution w_j, the attacker's subproblem is solved to optimality using the MIQP solver of CPLEX and v_{hj}, x_{ij}^{s}, y_{jk}^{1}, y_{jk}^{2}, y_{jk}^{3}, e_{ij}^{s}, f_{jk}^{1}, f_{jk}^{2}, f_{jk}^{3} based on w_j shows the optimal solution of follower's subproblem, then the objective function of the problem is computed
4: Set $T_{count} \leftarrow 0$, $Z_{best} \leftarrow Z(\boldsymbol{w})$, $w_{best} \leftarrow w_j^{'}$
5: **while** $T_{count} \prec T_{\max}$ **do**
6: Get a random non-protected based on w_j with $\rho \leftarrow [1, |J_1 \cup J_2 \cup J_3| - b]$
7: **if** $\rho \notin TL$ **then**
8: Generate a new solution $w_j^{'}$ based on w_j with ρ using the "Swap" operator
9: Compute $Z(\boldsymbol{w}^{'})$
10: Update TL
11: **if** $Z(\boldsymbol{w}^{'}) > Z_{best}$ **then**
12: Set $w_{best} \leftarrow w_j^{'}$, $Z_{best} \leftarrow Z(\boldsymbol{w}^{'})$
13: Set $T_{count} \leftarrow 0$
14: **end if**
15: Set $w_j \leftarrow w_j^{'}$, $Z(\boldsymbol{w}) \leftarrow Z(\boldsymbol{w}^{'})$
16: **else**
17: Generate a new solution $w_j^{'}$ based on w_j with ρ using the "Swap" operator
18: Compute $Z(\boldsymbol{w}^{'})$
19: **if** $Z(\boldsymbol{w}^{'}) > Z_{best}$ **then**
20: Set $w_{best} \leftarrow w_j^{'}$, $Z_{best} \leftarrow Z(\boldsymbol{w}^{'})$
21: Set Update the TL
22: **end if**
23: Set $w_j \leftarrow w_j^{'}$, $Z(\boldsymbol{w}) \leftarrow Z(\boldsymbol{w}^{'})$
24: **end if**
25: Set $T_{count} \leftarrow T_{count} + 1$
26: **end while**
27: **return** w_{best}, Z_{best}

3.4.2 TS algorithm based on DLNN

Since the dimension of the attacker's problem is very high and the number of binary variables is very large, the lower-level problem cannot be solved by CPLEX at short execution time. The global algorithms such as branch-and-bound and branch-and-cut are identified by an exponential complexity of the number of variables and constraints, and therefore their application is limited as the dimension of problem increases. So, a neural network will be executed to solve this problem in this chapter. The proposed core solver depends on the double-layer neural network (DLNN) by Yaakob et al. [28]. The w_j binary variables obtained by the TS algorithm form the input for the DLNN. This subsection addresses the problem of efficient approach to solve the lower-level problem of mixed-integer quadratic programming using a DLNN. The attacker's objective function is quadratic in the continuous and integer variables with the linear constraints. The follower's subproblem includes a maximization objective and the $e_{ij}^{s}, f_{jk}^{1}, f_{jk}^{2}, f_{jk}^{3}$ continuous variables and $v_{hj}, x_{ij}^{s}, y_{jk}^{1}, y_{jk}^{2}, y_{jk}^{3}$ binary variables.

A DLNN which includes a Hopfield network and Boltzmann machine had been modeled to solve the attacker's subproblem. The DLNN can be applied to choose a specified number of the unit from those available. DLNN had an upper layer which was applied to choose specified numbers of units from that layer and a lower layer was applied to determine the optimal unit from the specified number chosen in the upper layer. During the implementation mechanism, the network model removes the units in the lower layer that were not chosen by the upper layer. Then, the lower layer was reconstructed by using the chosen units. The DLNN transform the objective value into two energy functions, i.e., the energy function of the upper layer and the energy function of the lower layer based on the weights of the expected return rates of the layers and the output of the unit of the upper layer. It is shown that the output of the DLNN converged globally to an optimal solution. Also, it had a very small model size because of its double-layer framework. The DLNN was well suited to management and large-sized problems and was amenable to a hybrid optimization technique. The flowchart of the proposed method is described in Fig. 1.

4. Computational results

In this section, we give numerical results obtained with TS-DLNN. We validate TS-DLNN solutions with TS-CPLEX, which are described in Section 3. All coding was done in MATLAB (R2018b) software using

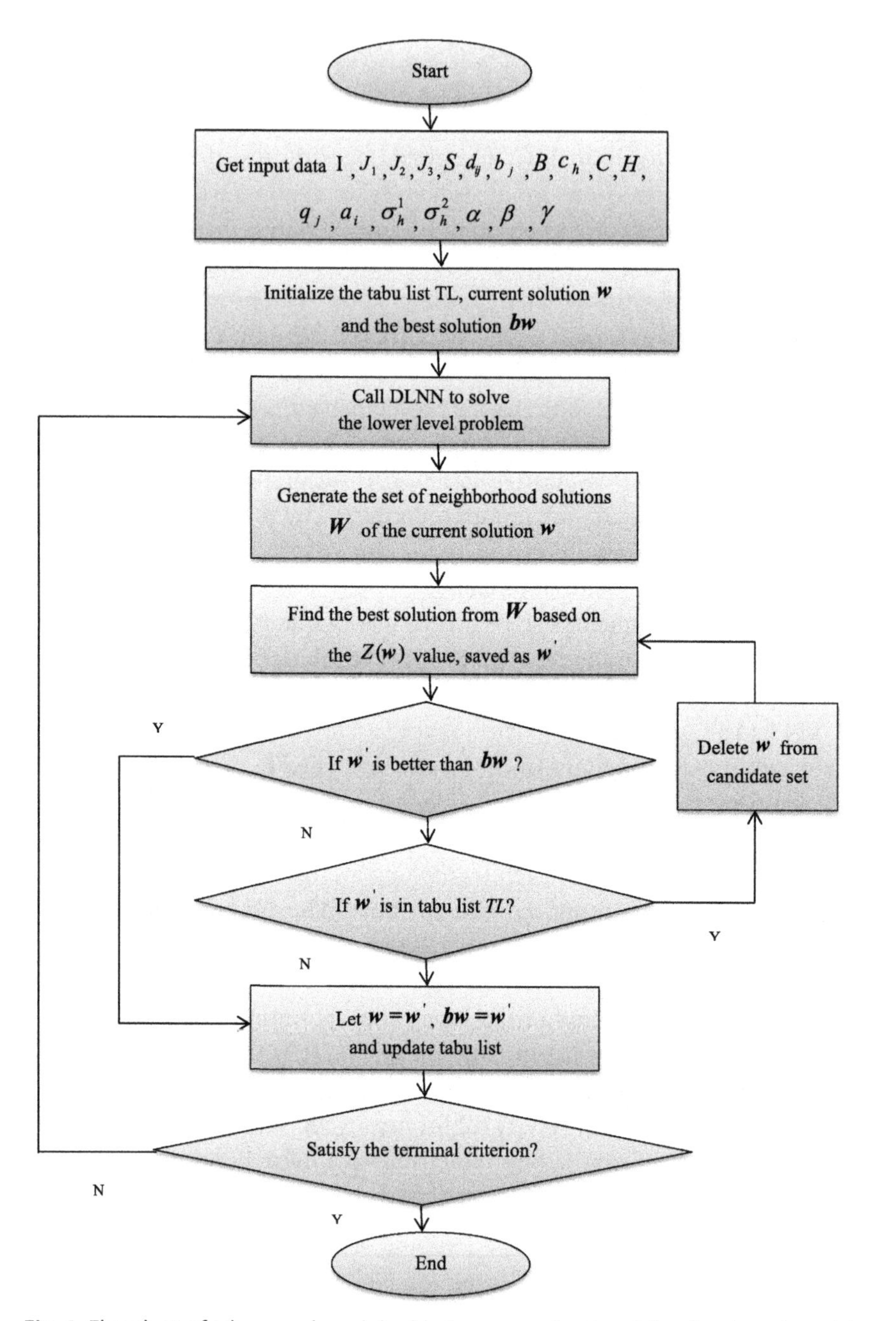

Fig. 1 Flowchart of tabu search and double-layer neural network for developed model.

CPLEX 12.6.0 Optimizer with *cplexmiqp* function. We measured running times on a 64-bit computer, Intel Core i7, 3.3 GHz processor and 4 GB of RAM.

4.1 A real-world case study

We present a realistic case study of the northern province of Iran, Golestan, to demonstrate the applicability of the proposed model. Golestan Province is located in the northeast of Iran and south of the Caspian Sea. Gorgan with a total population of 1.869 million and an area of 20,367 km^2 is the capital city of this province. Golestan Province consists of 14 counties and 29 cities listed in Table 1. Fig. 2 displays the counties of Golestan Province on the geographical map.

The application of the CHIMP is shown as a real-life case study to provide the patients with effective services after attacks to the counties and cities in Golestan Province. The clinics, hospitals, and medical centers in the health-care network of Golestan Province are illustrated in Fig. 3. Table 2 describes the current health-care network in Golestan Province composed of five clinics, nine hospitals, and 16 medical centers. As seen in Table 2, all the cities have clinics, hospitals, or medical centers with different capacity levels. Moreover, Golestan Province is considered a very important province in Iran due to the maritime border and also a border with Turkmenistan. Many Turkmenistani tourists annually refer to hospitals

Table 1 List of counties and cities in Golestan Province.

Number	County name	City name
1	Gorgan	Gorgan, Sarkhon Kalateh, Jelin-e Olya
2	Gonbad-e Kavus	Gonbad-e Kavus, Incheh Borun
3	Torkaman	Bandar Torkaman
4	Aliabad	Aliabad-e Katul, Fazelabad, Sangdevin, Mazraeh
5	Azadshahr	Azadshahr, Neginshahr, Now Deh Khanduz
6	Kordkuy	Kordkuy
7	Kalaleh	Kalaleh, Faraghi
8	AqQala	AqQala, Anbar Olum
9	Minudasht	Minudasht
10	Galikash	Galikash
11	Bandar-e Gaz	Bandar-e Gaz, Now Kandeh
12	Gomishan	Gomishan, Siminshahr
13	Ramian	Ramian, Khan Bebin, Daland, Tatar-e Olya
14	Maraveh Tappeh	Maraveh Tappeh

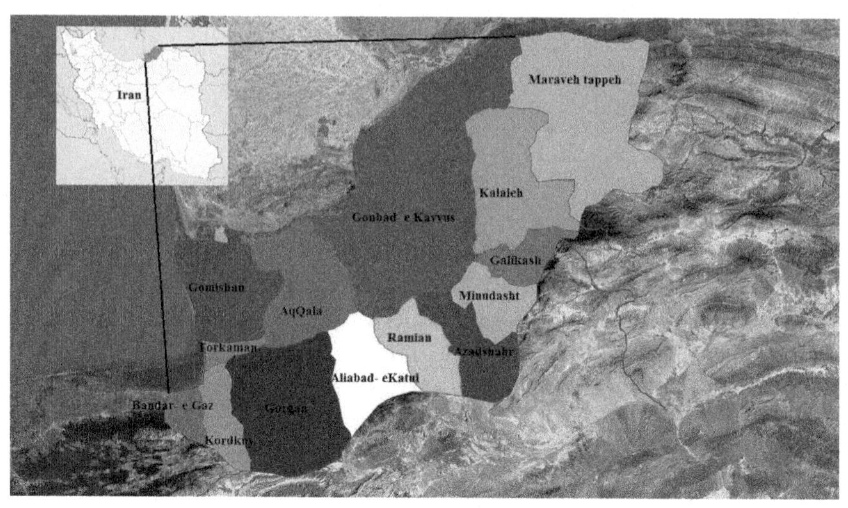

Fig. 2 The geographic map of Golestan Province and its counties.

or medical centers in Golestan Province for surgeries and to treat diseases. Hence, Golestan Province is a critical area in Iran where there is a risk of different disruption operations because of the deliberate sabotage such as disruptive services and delayed action in emergency response networks or attacks. Therefore, the current health-care network may be unable to serve the patients perfectly due to the sabotage operations and destruction of clinics, hospitals, or medical centers.

As clearly seen in Fig. 3, the medical centers providing both outpatient and inpatient services are located in the capital city of Golestan Province (Gorgan), Gonbad-e Kavus, Kordkuy and Bandar-e Gaz, while clinics and hospitals providing basic health care, diagnostic, and outpatient services are scattered throughout the province. In this health-care network, most counties in the province are deprived of medical centers, and because of the failure of these centers, the patients should travel long distances to access outpatient and inpatient services including same-day surgery (or outpatient surgery), laboratory tests, mental health care, X-rays and other radiological operations, medical supplies, preventive and screening operations and definite and biological drugs.

Furthermore, since the existing hospitals do not provide inpatient services, some patients served at a hospital may refer to a medical center for inpatient services. For example, patients should be served by only one hospital in Azadshahr on node 20 that provides basic health care, diagnostic, and

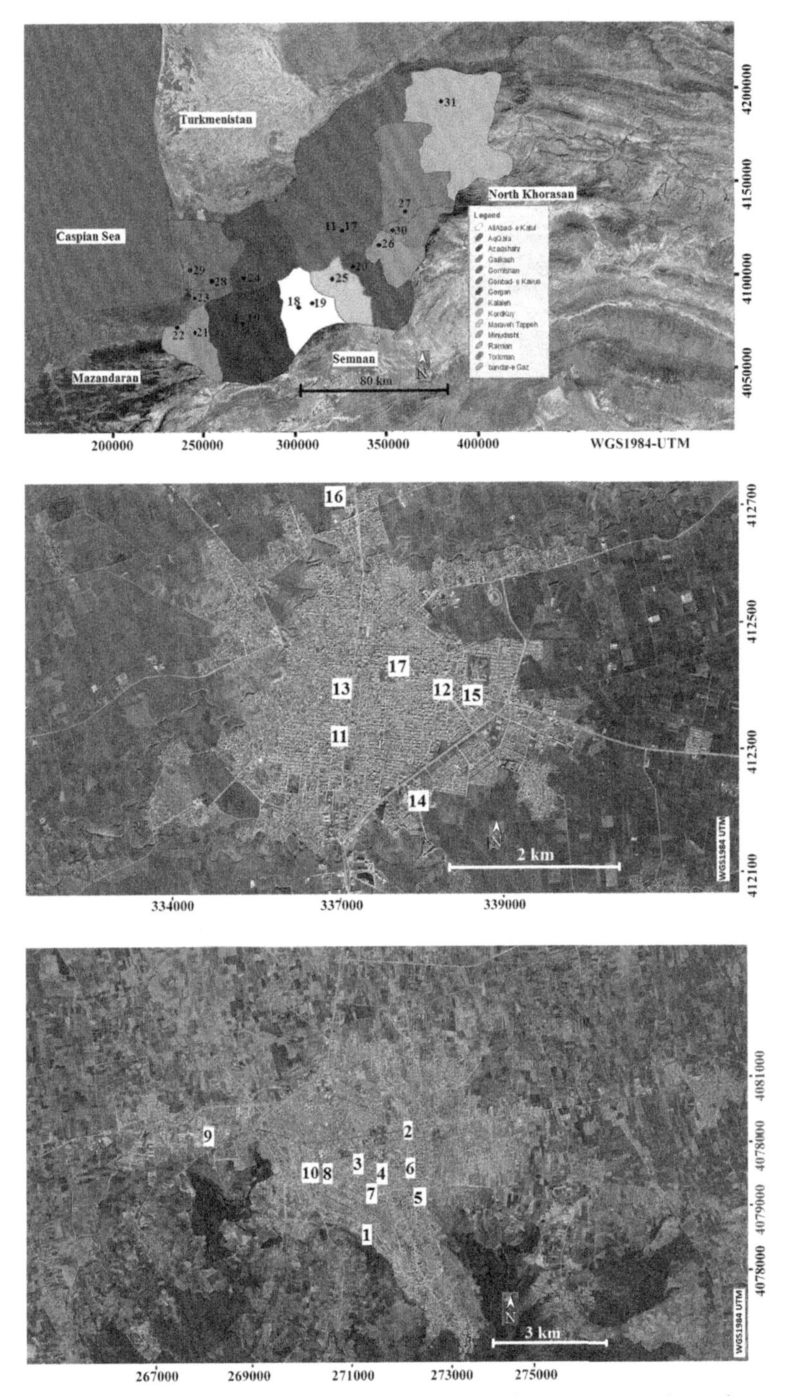

Fig. 3 The geographical map of Golestan Province displaying existing clinics, hospitals, and medical centers.

Table 2 The location of current clinics, hospitals, and medical centers in Golestan Province.

County name	City name	Facility name	Facility type	Node number
Gorgan	Gorgan	Sayyad Shirazi	Medical Center	1
	Gorgan	Dazyani	Hospital	2
	Gorgan	5th Azar	Medical Center	3
	Gorgan	Army 560	Medical Center	4
	Gorgan	Masoud	Medical Center	5
	Gorgan	Taleghani	Medical Center	6
	Gorgan	Dr Mousavi	Medical Center	7
	Gorgan	Falsafi	Medical Center	8
	Gorgan	Hakim Jorjani	Medical Center	9
	Gorgan	Shafa Heart	Medical Center	10
Gonbad-e Kavus https://en.wikipedia.org/wiki/Gonbad-e_Kavus_County	Gonbad-e Kavus	Shohada	Hospital	11
	Gonbad-e Kavus	Beski	Medical Center	12
	Gonbad-e Kavus	Borzouyeh	Medical Center	13
	Gonbad-e Kavus	Khatamolanbia	Medical Center	14
	Gonbad-e Kavus	Shahid Motahari	Medical Center	15
	Gonbad-e Kavus	PayambarAzam	Medical Center	16
	Gonbad-e Kavus	Taleghani	Hospital	17
Aliabad	Aliabad	Baghiatolah Azam	Hospital	18
	Aliabad	Ghaem	Clinic	19
Azadshahr	Azadshahr	Hazrat Masoumeh	Hospital	20
Kordkuy	Kordkuy	Amiralmomenin	Medical Center	21
Bandar-e Gaz	Bandar-e Gaz	Shohada	Medical Center	22
Torkaman	Torkaman	*Imam* Khomeini	Hospital	23
AqQala	AqQala	Al Jalil	Hospital	24
Ramian	Khanbebin	*Imam* Reza	Hospital	25
Minudasht	Minudasht	Fatimatuzzahra	Hospital	26
Kalaleh	Kalaleh	Rasul Akram	Hospital	27
Gomishan	Siminshahr	Siminshahr	Clinic	28
	Gomishan	Health center	Clinic	29
Galikash	Galikash	Shohada 12 dey	Clinic	30
Maraveh Tappeh	Maraveh Tappeh	Loghman Hakim	Clinic	31

outpatient services. So, a patient served at this hospital in Azadshahr must travel about 74, 18, 105, or 121 km to receive inpatient services at medical centers, respectively, in Gorgan, Gonbad-e Kavus, Kordkuy, or Bandar-e Gaz. In some cases, because of the damage or loss of clinics, hospitals and medical centers, patients must travel long distances to receive services from further clinics, hospitals, or medical centers. This may, in turn, cause an increase in the mortality rate of patients. Regarding the above arguments, it is critical to protect the health-care network with a defensive budget and limited capacity to serve patients and improve their access to clinics, hospitals, and medical centers after the sabotage operations.

To examine the proposed model, each city center is considered a client (patient) node with different demands based on the basic health care and diagnostic services, outpatient surgery, and inpatient services. Google Maps are able to compute the travel cost according to the actual road distance between the cities and the clinics, hospitals, or medical centers. To warrant proper and timely health-care services, the minimum acceptable patients' distance is determined regarding the acceptable distance for different routes between the cities and the clinics, hospitals, or medical centers in Golestan Province.

The capacity of health-care facilities would change by the location of the clinics, hospitals, and medical centers, the supplies and service team including doctors, specialists, nurses and personnel, and type of services provided by a clinic, hospital, or medical center. The demand percentage to satisfy diagnostic services after receiving basic health-care services (α), the demand percentage to satisfy outpatient surgery after receiving diagnostic services (β) and the demand percentage to satisfy inpatient services after receiving diagnostic services (γ) are determined according to some historical data and the experts' opinions. We survey the effect of different parameters on the performance of health-care system. The number of attackers ($|H|$), the number of protection operations (b), the total interdiction budget (C), cost of the action of the hth attacker (c_h), internal probability of the hth attacker (σ_h^1), and cooperation probability of the hth attacker (σ_h^2) are considered in a new mathematical model. The value of these parameters is reported in Table 3.

Where U[0, 270] shows a random integer number between and inclusive of 0 and 270, and total interdiction budget (C) is checked in three levels regarding the total cumulative costs: (1) 30% the sum of attacker costs (Low), (2) 50% the sum of attacker costs (Mid), and (3) 70% the sum of attacker costs (High).

Table 3 The characteristics of test instances of the problem.

Parameter	Value
b	$\{10,15,20\}$
$\|H\|$	$\{3,5,7\}$
c_h	$3000 + 100 \times U[0,270]$
C	Low, Mid, High
α	$\{5\%, 10\%\}$
β	$\{5\%, 10\%\}$
γ	$\{1\%, 2\%\}$
σ_h^1	$\frac{h}{\|H\|+1}$
σ_h^2	$\frac{\|H\|-h+1}{\|H\|+1}$

4.2 Performance evaluation of proposed algorithm

Tables 4–11 show the results of applying two methods TS-CPLEX and TS-DLNN on solving the problem instances with changing α, β, γ, b, $|H|$, and C. The first four columns of tables describe the characteristics of problems, and the fifth columns to eighth columns indicate the performance of the proposed algorithms based on precision and computational time. In Tables 4–11, the fifth and seventh columns concern the average of objective functions obtained by the TS-CPLEX and TS-DLNNFBC in 10 replications, respectively. This criterion confirms important and meaningful efficiency of the TS-DLNN on problem instances based on numerical precision. On the basis of the leader's objective values, the TS-DLNN accurately obtains better solutions and is preferable and remarkable to the TS-CPLEX on all the 216 test problems. The TS-DLNN method indicates an obvious and precise improvement on the TS-CPLEX solution. As reported in Tables 4–11, the TS-DLNN algorithm behaves better than TS-CPLEX approach. Such a considerable and enormous difference can be robust, demonstrates the remarkable performance of the neural network, and validates the utilization of TS-CPLEX algorithm for the developed cooperative interdiction-protection problem on the health-care network. As shown in more detail in Tables 4–11, the performance of proposed TS-DLNN makes better solutions than TS-CPLEX when the size of our instances increases. This flow can be observed by tracking the results of two approaches in Tables 4–11 from top to bottom. Therefore, we can conclude that TS-DLNN obtains reliable results for the large-sized instances.

Table 4 Numerical results for the different parameters and $\alpha = \beta = 5\ \%\ , \gamma = 1\%$.

			TS-CPLEX		TS-DLNN	
b	$\|H\|$	C	Average objective	Average time	Average objective	Average t ime
10	3	Low	50,511.28	215.2	50,463.82	102.6
		Mid	51,025.65	260.5	50,976.65	130.5
		High	51,541.01	300.3	51,487.52	152.8
10	5	Low	51,122.47	221.5	51,070.41	115.4
		Mid	51,641.05	280.7	51,589.63	142.6
		High	52,160.52	315.2	52,108.49	162.5
10	7	Low	52,287.08	232.8	52,237.62	127.2
		Mid	52,821.63	290.9	52,768.55	150.5
		High	53,352.19	335.1	53,296.38	180.4
15	3	Low	49,041.55	761.7	48,992.55	382.8
		Mid	49,412.98	1044.1	49,362.97	521.5
		High	49,932.34	1437.3	49,883.41	717.6
15	5	Low	49,631.89	802.8	49,581.44	416.9
		Mid	50,006.28	1101.4	49,956.31	564.8
		High	50,533.47	1500.7	50,482.97	774.2
15	7	Low	50,765.77	841.9	50,714.63	451.4
		Mid	51,146.18	1143.6	51,096.08	621.8
		High	51,688.43	1562.4	51,636.81	839.5
20	3	Low	47,388.69	2315.8	47,341.35	1148.7
		Mid	47,849.57	2864.6	47,801.75	1425.3
		High	48,480.05	3421.2	48,433.25	1702.1
20	5	Low	47,958.32	2422.8	47,910.33	1254.8
		Mid	48,424.72	3002.9	48,376.35	1562.6
		High	49,061.82	3571.3	49,016.89	1859.5
20	7	Low	49,054.45	2518.1	49,004.38	1361.2
		Mid	49,531.48	3130.2	49,482.03	1690.8
		High	50,186.23	3730.4	50,137.17	2014.3

From these tables, it is evident that the integration of neural network and TS improve computational precision.

On the other hand, Tables 4–11 report the results of the developed TS-CPLEX and TS-DLNN algorithms for all 216 available instances of the health-care network of Golestan province concerning the running time. Especially with regard to the execution times, the hybrid TS-DLNN method is the fastest algorithm where its average computational time is around 898 s against the TS-CPLEX algorithm which requires on average 1722 s. As reported by Tables 4–11, hybrid TS-DLNN approach is superior to TS-CPLEX method based on the execution time. The average values of running time show that the TS-DLNN algorithm behaves better than the

Table 5 Numerical results for the different parameters and $\alpha = 10\,\%$, $\beta = 5\,\%$, $\gamma = 1\%$.

			TS-CPLEX		TS-DLNN	
b	$\|H\|$	C	Average objective	Average time	Average o bjective	Average time
10	3	Low	53,718.86	230.8	53,665.23	118.1
		Mid	54,266.01	286.4	54,211.86	148.2
		High	54,940.71	348.9	54,885.85	169.5
10	5	Low	54,362.56	240.5	54,310.57	121.8
		Mid	54,918.28	302.9	54,863.84	160.7
		High	55,601.09	364.7	55,543.62	191.4
10	7	Low	55,603.08	248.6	55,551.78	131.5
		Mid	56,173.46	313.4	56,117.77	171.1
		High	56,871.87	381.8	56,815.06	214.5
15	3	Low	52,145.21	878.5	52,093.15	436.3
		Mid	52,737.33	1176.3	52,684.65	590.5
		High	53,166.28	1580.3	53,110.75	795.5
15	5	Low	52,772.03	921.8	52,719.36	471.5
		Mid	53,371.23	1230.2	53,317.49	637.1
		High	53,803.32	1648.6	53,749.32	855.8
15	7	Low	53,975.12	953.4	53,924.34	521.6
		Mid	54,591.05	1290.2	54,534.45	701.5
		High	55,033.01	1724.6	54,978.08	927.5
20	3	Low	50,164.28	2529.7	50,111.75	1259.7
		Mid	50,464.46	3169.2	50,416.05	1579.1
		High	51,036.06	3682.3	50,985.07	1846.8
20	5	Low	51,073.07	2650.5	51,021.05	1375.2
		Mid	51,646.51	3326.7	51,597.96	1730.1
		High	52,119.58	3850.5	52,065.51	2004.2
20	7	Low	52,240.36	2764.4	52,188.17	1500.6
		Mid	52,828.97	3458.2	52,771.29	1868.3
		High	53,310.79	4018.4	53,257.53	2171.5

TS-CPLEX in order that the average execution times on 216 instances are least. Fig. 2 demonstrates the graphical representation of the effect of proposed solution algorithms on the performance of health-care network in terms of running time with changing α, β, γ, b, $|H|$, and C values. As it can be observed in Fig. 2, the computational time under fortification/interdiction operations and α, β, γ values have an increasing trend with respect to defensive/offensive budget and referral services. This is because by increasing the budgets and referrals, the number of interactions among patients and the clinics, hospitals, and medical centers grows considerably. Fig. 2 also shows that through executing the hybrid TS-DLNN method, the CPU time

Table 6 Numerical results for the different parameters and $\alpha = 5\ \%, \beta = 10\ \%, \gamma = 1\%$.

			TS-CPLEX		TS-DLNN	
b	$\lvert H \rvert$	***C***	**Average objective**	**Average time**	**Average objective**	**Average time**
10	3	Low	54,478.24	241.2	54,423.82	120.6
		Mid	55,001.97	293.1	54,945.02	146.5
		High	55,643.89	363.2	55,590.39	181.6
10	5	Low	55,133.07	252.9	55,077.98	131.5
		Mid	55,663.09	306.4	55,605.24	159.9
		High	56,314.75	369.3	56,258.41	190.1
10	7	Low	56,393.15	263.6	56,336.85	142.5
		Mid	56,935.29	319.8	56,878.41	173.2
		High	57,601.84	396.4	57,544.04	214.6
15	3	Low	52,744.13	904.5	52,687.45	452.5
		Mid	53,400.62	1202.8	53,347.25	606.3
		High	53,811.85	1682.8	53,758.05	841.9
15	5	Low	53,373.09	945.7	53,322.19	483.5
		Mid	54,042.49	1257.9	53,984.05	661.6
		High	54,458.62	1759.7	54,404.76	917.8
15	7	Low	54,593.02	987.1	54,541.17	534.5
		Mid	55,277.65	1323.2	55,222.72	716.6
		High	55,703.29	1825.6	55,646.46	984.3
20	3	Low	51,152.54	2581.7	51,104.14	1290.5
		Mid	51,491.46	3201.3	51,440.25	1600.6
		High	52,277.95	3681.9	52,225.72	1840.7
20	5	Low	51,770.39	2698.5	51,718.89	1408.2
		Mid	52,110.43	3346.8	52,057.41	1746.1
		High	52,903.33	3848.5	52,853.12	2008.4
20	7	Low	52,951.62	2806.4	52,900.71	1525.3
		Mid	53,301.39	3482.3	53,248.88	1881.6
		High	54,110.52	4006.8	54,061.45	2165.4

is improved. This means that the proposed hybrid algorithm needed less running time than TS-CPLEX algorithm. So, the hybrid TS-DLNN algorithm outperforms the TS-CPLEX for the CHIMP. According to Tables 4–11 and Fig. 4, the proposed hybrid approach achieved the best solutions over all instances of the CHIMP and its execution time becomes larger as the problem size increases.

4.3 Sensitivity analysis of parameters

In order to study the effect of various parameters in the new bi-level problem and performance of the health-care system, a sensitivity analysis (SA) is

Table 7 Numerical results for the different parameters and $\alpha = \beta = 10\,\%$, $\gamma = 1\%$.

			TS-CPLEX		TS-DLNN	
b	**|*H*|**	***C***	**Average objective**	**Average time**	**Average objective**	**Average time**
10	3	Low	57,058.65	280.4	57,001.65	140.7
		Mid	57,571.78	369.1	57,512.27	184.8
		High	58,250.16	404.9	58,191.97	202.4
10	5	Low	57,744.49	293.7	57,683.83	143.2
		Mid	58,261.88	385.4	58,205.59	201.3
		High	58,950.33	423.3	58,891.44	220.8
10	7	Low	59,064.26	306.4	59,005.28	166.1
		Mid	59,595.43	402.7	59,535.91	218.4
		High	60,291.64	431.2	60,234.42	239.5
15	3	Low	55,218.86	1049.3	55,163.76	514.8
		Mid	56,032.37	1413.9	55,976.43	711.6
		High	56,320.28	1949.3	56,264.02	974.7
15	5	Low	55,882.59	1097.5	55,826.76	572.5
		Mid	56,705.88	1477.5	56,649.23	776.3
		High	56,993.25	2038.1	56,940.31	1063.8
15	7	Low	57,159.86	1145.6	57,102.76	620.2
		Mid	58,001.91	1552.8	57,942.97	841.5
		High	58,299.94	2116.6	58,241.75	1141.9
20	3	Low	53,584.96	3086.5	53,536.42	1543.5
		Mid	54,192.58	3873.3	54,138.45	1936.7
		High	54,728.24	4483.1	54,673.57	2246.5
20	5	Low	54,234.11	3226.5	54,179.93	1673.3
		Mid	54,842.98	4049.5	54,784.19	2112.8
		High	55,386.08	4697.4	55,330.75	2450.2
20	7	Low	55,473.64	3366.6	55,418.23	1823.5
		Mid	56,092.45	4215.5	56,041.41	2288.8
		High	56,651.94	4901.4	56,595.35	2645.1

carried out with respect to different variations in the parameters of the CHIMP. The numerical results are reported in Tables 4–11 and Figs. 5–8. As declared in this section, the 216 CHIMP instances were generated by different levels of α, β, γ, b, $|H|$, and C parameters and solved by TS-CPLEX and TS-DLNN approaches. Different types of SA techniques can be considered based on α, β, γ, b, $|H|$, and C parameters. Therefore, a reasonable and acceptable SA on the TS-CPLEX and TS-DLNN algorithms and critical parameters is performed and addressed in the context of this subsection as follows:

Table 8 Numerical results for the different parameters and $\alpha = \beta = 5\,\%$, $\gamma = 2\%$.

			TS-CPLEX		TS-DLNN	
b	$\lvert H \rvert$	C	Average objective	Average time	Average objective	Average time
10	3	Low	53,112.75	229.7	53,059.65	114.8
		Mid	53,660.37	283.9	53,606.73	141.9
		High	54,230.35	309.3	54,176.17	159.6
10	5	Low	53,751.12	240.5	53,697.42	125.3
		Mid	54,305.39	296.5	54,251.05	154.8
		High	54,881.25	333.8	54,823.37	174.1
10	7	Low	54,979.62	240.6	54,924.69	135.7
		Mid	55,546.46	309.7	55,490.97	167.5
		High	56,136.54	348.3	56,080.46	188.6
15	3	Low	51,461.56	815.8	51,411.15	407.9
		Mid	52,025.09	1113.3	51,973.12	561.7
		High	52,715.33	1520.2	52,662.67	760.2
15	5	Low	52,081.14	852.5	52,029.11	445.2
		Mid	52,650.43	1174.4	52,595.84	612.8
		High	53,344.97	1589.5	53,295.68	829.3
15	7	Low	53,271.47	890.4	53,218.25	482.1
		Mid	53,853.78	1225.6	53,799.98	653.8
		High	54,568.28	1648.4	54,513.76	898.4
20	3	Low	49,845.19	2463.2	49,794.48	1231.5
		Mid	50,450.02	3017.3	50,399.62	1508.8
		High	50,847.39	3631.2	50,796.06	1815.6
20	5	Low	50,445.34	2575.8	50,394.95	1343.5
		Mid	51,052.43	3144.5	51,003.42	1645.9
		High	51,458.58	3796.5	51,407.17	1980.6
20	7	Low	51,595.29	2687.4	51,546.74	1445.4
		Mid	52,223.34	3281.6	52,171.12	1783.1
		High	52,634.68	3951.2	52,582.19	2135.7

4.3.1 The impact of b and H on the performance of health-care system

The defensive resources and number of attackers for protecting and interdicting clinics, hospitals, and medical centers are two essential parameters of the CHIMP model that can severely impact the performance of health-care system on various levels. Figs. 5 and 6 illustrate the flow of a change in the system planner's objective function regarding the health services provided for three levels of b (number of health facilities that can be protected) and $|H|$ (number of attackers). As is shown in Fig. 5, it can be received that increasing the protection operations lead to reducing the total

Table 9 Numerical results for the different parameters and $\alpha = 10\ \%, \beta = 5\ \%, \gamma = 2\%$.

			TS-CPLEX		TS-DLNN	
b	$\|H\|$	C	Average objective	Average time	Average objective	Average time
10	3	Low	56,362.41	253.3	56,306.12	126.6
		Mid	56,883.65	311.5	56,826.82	155.9
		High	57,464.84	374.7	57,407.43	187.3
10	5	Low	57,032.88	254.5	56,982.89	138.1
		Mid	57,567.39	326.5	57,509.84	170.1
		High	58,155.53	391.8	58,094.43	204.42
10	7	Low	58,341.54	276.3	58,285.25	149.6
		Mid	58,883.11	340.2	58,824.28	184.2
		High	59,484.69	408.4	59,425.27	221.4
15	3	Low	54,558.72	933.6	54,504.22	466.7
		Mid	55,290.18	1244.9	55,234.87	627.4
		High	55,802.27	1685.2	55,746.52	832.7
15	5	Low	55,214.52	975.9	55,155.36	509.1
		Mid	55,951.69	1312.5	55,898.79	684.5
		High	56,473.01	1762.3	56,416.59	919.3
15	7	Low	56,472.46	1008.2	56,420.04	551.5
		Mid	57,233.55	1369.8	57,176.38	741.5
		High	57,763.72	1838.4	57,706.01	995.9
20	3	Low	51,802.27	2663.3	51,746.52	1331.7
		Mid	52,815.11	3357.5	52,762.35	1668.8
		High	53,494.01	3888.6	53,440.57	1944.3
20	5	Low	53,443.95	2774.5	53,394.55	1452.7
		Mid	54,137.01	3510.3	54,082.93	1831.5
		High	54,626.38	4064.7	54,571.81	2120.7
20	7	Low	54,671.56	2905.6	54,616.94	1573.8
		Mid	55,373.33	3653.1	55,319.01	1974.1
		High	55,874.88	4231.5	55,819.06	2297.4

service distance between clinics, hospitals and medical centers, and patient zone. This declares that the maximum level of defensive resources can play a vital role in the performance of the health-care system. In this case, health facilities can be made more reliable to the attacks and the highest level of basic health care, diagnostic, outpatient, and inpatient service can be provided from the nearest uninterdicted clinic, hospital, or medical center. As can be seen in Fig. 6, the maximum of the defender's post-attack objective of the case study is attained by the maximum amount of interdiction budget allocated to an attacker. On the other hand, the maximum number of attackers had increasingly targeted the clinics, hospitals, and medical

Table 10 Numerical results for the different parameters and $\alpha = 5\,\%$, $\beta = 10\,\%$, $\gamma = 2\%$.

			TS-CPLEX		TS-DLNN	
b	**|*H*|**	***C***	**Average objective**	**Average time**	**Average objective**	**Average time**
10	3	Low	57,684.85	252.6	57,627.25	126.2
		Mid	58,280.27	308.1	58,222.05	154.5
		High	58,995.88	382.3	58,936.95	191.5
10	5	Low	58,378.22	264.4	58,319.24	137.7
		Mid	58,980.82	312.5	58,921.87	168.6
		High	59,701.01	389.3	59,645.37	208.8
10	7	Low	59,712.47	275.2	59,652.82	149.4
		Mid	60,328.82	336.1	60,268.55	182.5
		High	61,069.59	416.6	61,004.58	225.5
15	3	Low	56,002.54	959.7	55,947.68	479.9
		Mid	56,591.61	1285.2	56,535.07	642.5
		High	57,239.33	1767.3	57,182.32	878.6
15	5	Low	56,676.71	1003.6	56,620.09	523.2
		Mid	57,274.84	1343.3	57,214.62	700.9
		High	57,927.52	1848.7	57,867.65	969.4
15	7	Low	57,972.07	1047.4	57,914.15	567.2
		Mid	58,580.87	1401.8	58,522.28	759.3
		High	59,251.47	1938.9	59,192.22	1050.2
20	3	Low	54,202.12	2752.6	54,147.97	1376.4
		Mid	54,705.52	3411.2	54,652.87	1705.6
		High	55,381.77	3922.7	55,326.45	1951.3
20	5	Low	54,853.63	2867.1	54,796.83	1501.1
		Mid	55,365.16	3566.8	55,309.86	1860.7
		High	56,047.46	4101.5	55,991.47	2139.6
20	7	Low	56,103.32	3002.8	56,051.27	1626.2
		Mid	56,630.49	3711.4	56,573.92	2005.7
		High	57,328.44	4270.2	57,271.17	2317.9

centers and therefore the patient service distance increased. Indeed, TS-CPLEX and TS-DLNN interdict more and more health facilities as $|H|$ goes from three to seven. This flow is maintained when C increases from low to high, but also, in that case study the average of objective function increases although the number of protected health facilities goes down.

4.3.2 The impact of α,β on the performance of health-care system

We illustrated the $Z(\boldsymbol{w})$ values of the best TS-DLNN solutions obtained from the different levels of the parameters α and β in Fig. 7. The lines corresponding to $\alpha = \beta = 5\%$, $\alpha = 10\,\%$, $\beta = 5\%$, $\alpha = 5\,\%$, $\beta = 10\%$,

Table 11 Numerical results for the different parameters and $\alpha = \beta = 10\%$, $\gamma = 2\%$.

			TS-CPLEX		TS-DLNN	
b	**|*H*|**	***C***	**Average objective**	**Average time**	**Average objective**	**Average time**
10	3	Low	59,664.65	297.2	59,605.05	148.6
		Mid	60,279.36	387.9	60,219.15	195.5
		High	60,946.86	420.7	60,885.97	213.7
10	5	Low	60,381.82	310.3	60,321.57	162.1
		Mid	61,003.92	404.6	60,942.98	211.4
		High	61,672.44	447.5	61,617.82	233.2
10	7	Low	61,761.86	324.4	61,700.16	175.3
		Mid	62,398.18	422.2	62,335.85	224.7
		High	63,089.14	466.4	63,026.11	252.6
15	3	Low	57,790.03	1108.1	57,732.32	559.4
		Mid	58,262.32	1516.4	58,204.12	758.1
		High	59,057.14	2063.9	58,996.15	1031.9
15	5	Low	58,482.66	1169.6	58,426.24	610.2
		Mid	58,962.64	1585.1	58,903.73	825.4
		High	59,767.01	2157.7	59,707.36	1125.8
15	7	Low	59,821.35	1220.2	59,761.59	661.5
		Mid	60,310.24	1644.8	60,249.97	895.6
		High	61,133.01	2251.6	61,071.93	1219.5
20	3	Low	55,982.75	3281.8	55,922.82	1640.4
		Mid	56,552.94	4093.7	56,496.45	2036.9
		High	57,253.67	4795.6	57,200.75	2397.7
20	5	Low	56,655.66	3430.2	56,594.06	1780.6
		Mid	57,232.71	4269.5	57,175.53	2233.2
		High	57,945.91	5013.4	57,883.02	2615.6
20	7	Low	57,950.54	3569.6	57,892.65	1938.2
		Mid	58,540.78	4466.4	58,482.32	2409.5
		High	59,270.28	5221.2	59,210.07	2833.4

and $\alpha = \beta = 10\%$ reveal that as α and β increase $Z(\boldsymbol{w})$ gains larger values. This is a result of the demands of patients referring to a farther clinic or hospital for diagnostic service and outpatient surgery. When α and β are low (5%), the defender can afford to satisfy all patient demands with the nearest health facilities, and the average $Z(\boldsymbol{w})$ decreases. Thus, the increase in the patients' traveling cost is based on the extra distance to serve patients from clinics and hospitals. Actually, the expected distance between the patient zone and a clinic or hospital ascend with the concurrent increase of the parameters α and β.

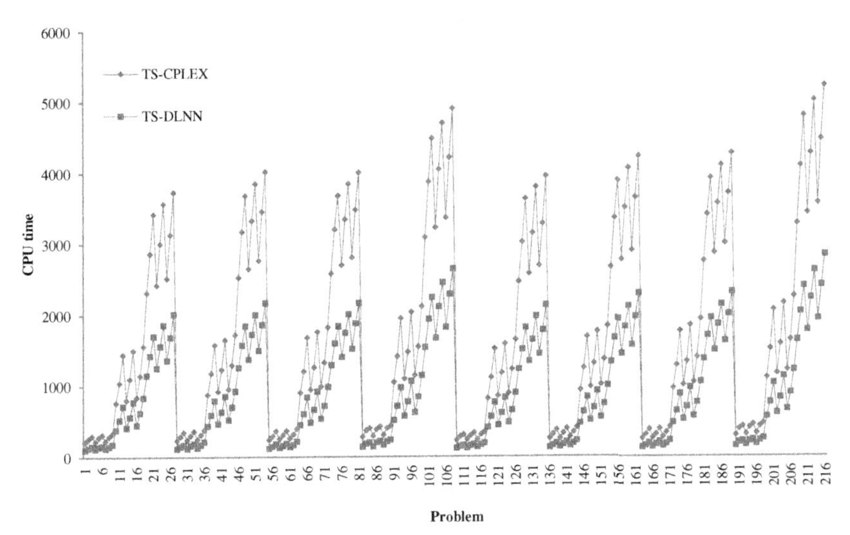

Fig. 4 Comparison of TS-CPLEX and TS-DLNN based on CPU time.

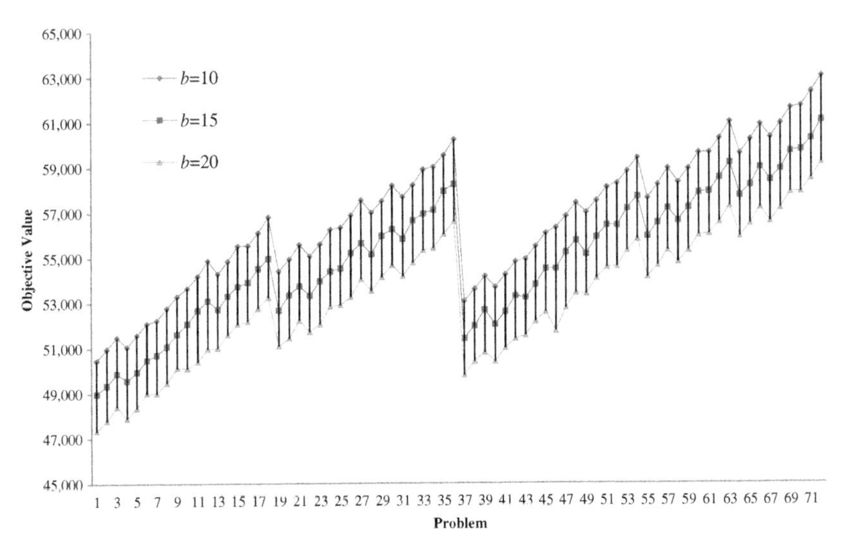

Fig. 5 Impact of protection operations on the performance of health-care system.

As you can perceive from Fig. 7, the increase in the β parameter leads to more patient service distance than the α parameter. For the reason that the diagnostic service is obtained by both clinic and hospital, the patient can receive diagnostic service after receiving basic health care from a similar clinic or hospital. We can deduce that a larger α value would cause less

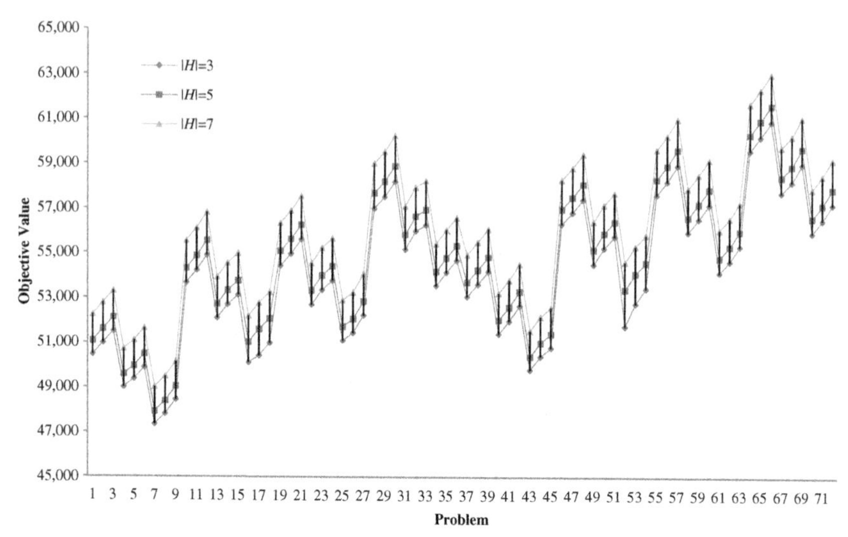

Fig. 6 Impact of number of attackers on the performance of health-care system.

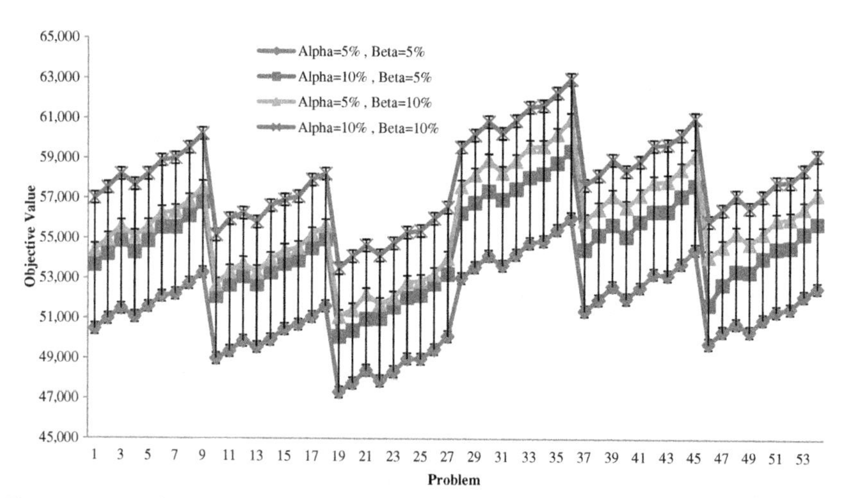

Fig. 7 Impact of α, β on the performance of health-care system.

increase in traveling distance between patients and health facilities. Also, the clinics do not provide the outpatient services and the patients refer to hospitals after receiving diagnostic service. So, the patient service distance ascends as a result of an increase in the unmet demand of the clinics.

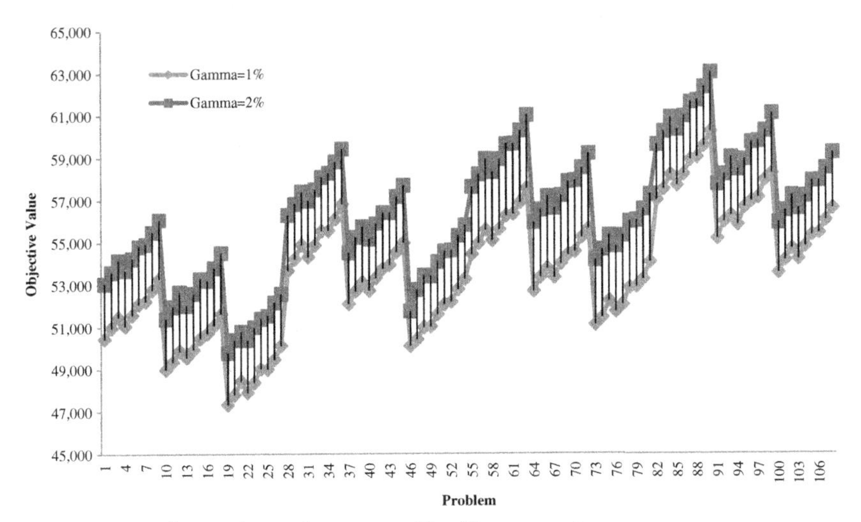

Fig. 8 Impact of γ on the performance of health-care system.

4.3.3 The impact of γ on the performance of health-care system

As the threat level of inpatient service is elevated in the health-care network with γ, Fig. 8 illustrates the service-flow between patients and medical centers. The increase from $\gamma = 1\%$ to $\gamma = 2\%$ causes extra referrals to medical centers and additional demand for inpatient health services. Fig. 8 exhibits how the defender's objective function changes for $\gamma = 1\%$ and $\gamma = 2\%$. When γ increases from $\gamma = 1\%$ to $\gamma = 2\%$, the patient service distance considerably ascends due to extra distances between patients and medical centers to obtain inpatient services. Since hospitals do not provide the inpatient services, a fraction of patients assigned to a hospital may be referred to a medical center for more outpatient or inpatient services. In Golestan province, medical centers were only organized in the counties of Gorgan, Gonbad-e Kavus, Kordkuy and Bandar-e Gaz and patients can get the inpatient services concerning the capacity of existing medical centers. Thereupon, the increasing pattern of the patient service distance is indicated in Fig. 8.

5. Conclusions and future work

In this chapter, we presented a new model of the interdiction median problem with protection on a health-care network called cooperative hierarchical and interdiction median problem with protection (CHIMP). CHIMP was formulated as a bi-level programming or static Stackelberg game between a defender (leader) and several attackers (follower). The leader's decision was to allocate defensive resources to health facilities so that the probability of successfully disturbing a clinic, hospital, and medical center was minimized. Following this protection operation, the attackers searched clinics, hospitals, and medical centers to interdict in a cooperative manner resulting in the capacity loss or elimination of basic health care, diagnostic, outpatient, or inpatient service with a limited interdiction budget.

Since the proposed model was NP-hard, we developed and compared two hybrid metaheuristic approaches, namely, tabu search with CPLEX (TS-CPLEX) and tabu search with a double-layer neural network (TS-DLNN), to solve the CHIMP. DLNN was adopted to obtain high quality of optimal solution for the attacker subproblem. Case study evaluated the impacts of the number of protection operations, number of attackers, total interdiction budget and percent of the demand, that is referred to get diagnostic service after getting basic health care, outpatient surgery after getting diagnostic service, and inpatient service after getting diagnostic service on the solutions that were attained. Our numerical results demonstrated the superior and acceptable efficiency of TS-DLNN concerning both computational time and solution accuracy in comparison with TS-CPLEX, which were validated by the 216 available instances of the practical health-care network.

In future work, we can formulate a hierarchical interdiction/protection model in using the cooperative scenarios among defenders to design the best-case protections and further improve the efficiency of the health-care network. We can also study the use of other metaheuristics, which may lead to generate better solutions based on running time and accuracy. Finally, we are interested in modeling a dynamic version of the hierarchical interdiction/protection problem as well as stochastic or fuzzy models on the health-care network.

Acknowledgments

The first author has been supported by the financial support of the research-technology management of Gonbad Kavous University (Grant no. 6/604). The work also has been supported by the research council of Golestan University for the second author.

References

[1] R.L. Church, M.P. Scaparra, Protecting critical assets: the r-interdiction median problem with fortification, Geogr. Anal. 39 (2) (2007) 129–146.
[2] M.P. Scaparra, R.L. Church, A bilevel mixed-integer program for critical infrastructure protection planning, Comput. Oper. Res. 35 (6) (2008) 1905–1923.

[3] D. Aksen, N. Piyade, N. Aras, The budget constrained r-interdiction median problem with capacity expansion, CEJOR 18 (3) (2010) 269–291.
[4] F. Liberatore, M.P. Scaparra, M.S. Daskin, Hedging against disruptions with ripple effects in location analysis, Omega 40 (1) (2012) 21–30.
[5] F. Liberatore, M.P. Scaparra, M.S. Daskin, Analysis of facility protection strategies against an uncertain number of attacks: the stochastic R-interdiction median problem with fortification, Comput. Oper. Res. 38 (1) (2011) 357–366.
[6] D. Aksen, S.Ş. Akca, N. Aras, A bilevel partial interdiction problem with capacitated facilities and demand outsourcing, Comput. Oper. Res. 41 (2014) 346–358.
[7] R. Khanduzi, H.R. Maleki, R. Akbari, Two novel combined approaches based on TLBO and PSO for a partial interdiction/fortification problem using capacitated facilities and budget constraint, Soft. Comput. 22 (17) (2018) 5901–5919.
[8] S. Starita, M.P. Scaparra, Optimizing dynamic investment decisions for railway systems protection, Eur. J. Oper. Res. 248 (2) (2016) 543–557.
[9] R. Khanduzi, H.R. Maleki, A novel bilevel model and solution algorithms for multi-period interdiction problem with fortification, Appl. Intell. 48 (9) (2018) 2770–2791.
[10] H.R. Maleki, R. Khanduzi, Fuzzy interdiction/fortification location problems on p-median systems, J. Intell. Fuzzy Syst. 30 (3) (2016) 1283–1292.
[11] M. Mahmoodjanloo, S.P. Parvasi, R. Ramezanian, A tri-level covering fortification model for facility protection against disturbance in r-interdiction median problem, Comput. Ind. Eng. 102 (2016) 219–232.
[12] M.C. Roboredo, L. Aizemberg, A.A. Pessoa, An exact approach for the r-interdiction covering problem with fortification, CEJOR 27 (1) (2019) 111–131.
[13] N. Ghaffarinasab, R. Atayi, An implicit enumeration algorithm for the hub interdiction median problem with fortification, Eur. J. Oper. Res. 267 (1) (2018) 23–39.
[14] H. Quadros, M.C. Roboredo, A.A. Pessoa, A branch-and-cut algorithm for the multiple allocation r-hub interdiction median problem with fortification, Expert Syst. Appl. 110 (2018) 311–322.
[15] N. Aliakbarian, F. Dehghanian, M. Salari, A bi-level programming model for protection of hierarchical facilities under imminent attacks, Comput. Oper. Res. 64 (2015) 210–224.
[16] A. Forghani, F. Dehghanian, M. Salari, Y. Ghiami, A bi-level model and solution methods for partial interdiction problem on capacitated hierarchical facilities, Comput. Oper. Res. 114 (2020) 104831.
[17] R. Khanduzi, A.K. Sangaiah, A fast genetic algorithm for a critical protection problem in biomedical supply chain networks, Appl. Soft Comput. 75 (2019) 162–179.
[18] N. Azad, G.K. Saharidis, H. Davoudpour, H. Malekly, S.A. Yektamaram, Strategies for protecting supply chain networks against facility and transportation disruptions: an improved Benders decomposition approach, Ann. Oper. Res. 210 (1) (2013) 125–163.
[19] C. Zhou, K.L. Poh, Optimization of protections against supply chain risks due to intentional acts, in: 17th International IEEE Conference on Intelligent Transportation Systems (ITSC), IEEE, 2014, October, pp. 232–237.
[20] H.O. Mete, Z.B. Zabinsky, Stochastic optimization of medical supply location and distribution in disaster management, Int. J. Prod. Econ. 126 (1) (2010) 76–84.
[21] J.A. Paul, R. Batta, Models for hospital location and capacity allocation for an area prone to natural disasters, Int. J. Oper. Res. 3 (5) (2007) 473.
[22] D. Shishebori, A.Y. Babadi, Robust and reliable medical services network design under uncertain environment and system disruptions, Transport Res E-Log 77 (2015) 268–288.
[23] N. Zarrinpoor, M.S. Fallahnezhad, M.S. Pishvaee, Design of a reliable hierarchical location-allocation model under disruptions for health service networks: a two-stage robust approach, Comput. Ind. Eng. 109 (2017) 130–150.

[24] N. Zarrinpoor, M.S. Fallahnezhad, M.S. Pishvaee, The design of a reliable and robust hierarchical health service network using an accelerated Benders decomposition algorithm, Eur. J. Oper. Res. 265 (3) (2018) 1013–1032.
[25] J.T. Moore, J.F. Bard, The mixed integer linear bilevel programming problem, Oper. Res. 38 (5) (1990) 911–921.
[26] D. Aksen, N. Aras, A matheuristic for leader-follower games involving facility location-protection-interdiction decisions, in: Metaheuristics for Bi-Level Optimization, Springer, Berlin, Heidelberg, 2013, pp. 115–151.
[27] D. Aksen, N. Aras, N. Piyade, A bilevel p-median model for the planning and protection of critical facilities, J. Heuristics 19 (2) (2013) 373–398.
[28] S.B. Yaakob, M.Z. Hasan, A. Ahmed, A double layer neural network for solving mixed-integer quadratic optimization problems, in: AIP Conference Proceedings, vol. 1775 No. 1, AIP Publishing LLC, 2016, p. 030082.
[29] R. Khanduzi, A.K. Sangaiah, Tabu search based on exact approach for protecting hubs against jamming attacks, Comput. Electr. Eng. 79 (2019) 106459.
[30] F. Glover, Future paths for integer programming and links to artificial intelligence, Comput. Oper. Res. 13 (5) (1986) 533–549.

CHAPTER FOUR

Parallel machine learning and deep learning approaches for internet of medical things (IoMT)

S. Sridhar Raj[a] and M. Madiajagan[b]
[a]Pondicherry University, Pondicherry, India
[b]Vellore Institute of Technology, Vellore, India

1. Introduction

The problem of heavy computational processing of deep learning algorithms is resolved by graphical processing units (GPUs). This can improve the performance of the deep learning algorithms by 10 times than the usual execution. In today's world, GPUs are found widely used in the research community in the process of developing parallel architectures for deep learning methods [1–4].

The basic of deep learning algorithms works based on the back-propagation mechanism, which is a kind of feedback mechanism. The second execution has to wait for the output of first one. Thus, parallelism is a big issue in resolving these problems. The bandwidth should be high with a very low latency to attain the best performance out of the GPUs. Network communication generally constitutes the major bottleneck in deep learning parallelism [5–7].

Various methods to improve the performance of the parallelism in deep architectures are namely, the computation can be performed while waiting for the successive communication to avoid the queue and complete the communications in time. The number of parameters can be reduced before transmission by sending only the required parameters.

The physicians and technicians in the hospital are always looking forward to utilize a better medical equipment in order to serve the purpose efficiently. This improvement can be made in the equipments with constant analysis and monitoring of the technologies evolving in the real-world environment. One of the major source of doing this is the internet of things (IoT) techniques, which utilized sensors for data privacy. The sensor data

Intelligent IoT Systems in Personalized Health Care
https://doi.org/10.1016/B978-0-12-821187-8.00004-6

can be used to analyze the blood pressure, heart rate, and ultrasound with high clarity in the output data [8–11].

Some of the benefits in using the medical equipments with possibilities of connected sensors and equipment are [12–16]:

- Full-time monitoring is possible by both the patient and physicians to keep track of their health and to take care of the technical faults, respectively.
- Usage of IoT sensors guarantees the machine tolerance in terms of duration, ability, and usage.
- Balanced scheduling can also be used to stabilize the machines running for a longtime and maintain a perfect balance in the load.
- During lack of resources or over usage, the system can claim resources from the cloud to perform resource and service reservations.
- Equipment managers should be capable of modifying and upgrading interns of efficiency management during failure.

In the further chapters, the review of the IoMT-based methods and applications is discussed.

2. Review of IoMT and deep learning methods

IoMT-based services and platforms are used for monitoring, chronic and fitness programs, which is an efficient and accurate of managing battery life cycle and energy of the resource-constrained tiny wearable devices. The real-time medical data have to be collected and analyzed in the cloud.

k-Means clustering under differential privacy-related methods are proposed to address the privacy budget and decide the centroids for initialization. The number of iterations can be decided based on the methods, which denote the fixed and unfixed iterations with an improved *k*-means algorithm. The mean square error is evaluated between the noisy and true centroids during the selection process to decide on the allocation of budget and iterations. Instead of opting a method for allocations, random selection, equal division of datasets are also performed to compensate the ambiguity in creating the datasets.

DPLK algorithms were proposed to improve the initial centroid selection for *k*-means algorithm by dividing each subset with the original dataset. Distributed environment has been incorporated to reduce the time complexity of the cluster analysis computational speed. MapReduce can be used to perform distributed operations by evolving the parallel *k*-means algorithm, which is more powerful than the sequential programming. Periodic updation of nearest centroids is also concentrated.

Disease diagnosis and analysis has achieved a higher level of IoMT development for disease prediction. As the IoMT technologies increase in the research community, the data generated by the IoMT devices also increase proportionally. To resolve this, big data analysis techniques have to be used to analyze the datasets. Clustering is one way of analyzing huge amount of data without any class label monitoring. In case of issues, there is high possibility of the data to leak, which contains privacy information. Even though certain privacy-preserving methods like *k*-anonymity, diversity, and differential privacy exist, there is still a lot of loop holes to be covered with privacy-preserving algorithms. The trade-off between providing a good accuracy and preserving privacy has to be digested to obtain the outcome as per the situation [17–22].

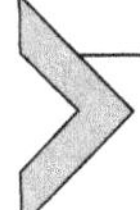

3. Parallel machine learning and deep learning techniques using *k*-means Hadoop frameworks

In this chapter, we discuss about the following techniques adopted for parallel machine learning techniques. They are

- Role of GPUs is to handle the parallel executions, which contributes to faster execution of the deep learning algorithms.
- Hadoop framework for incorporating parallel deep learning in the distributed environment with the IoMT devices.

3.1 Hadoop and parallel computing

A rough clustering is performed with a partial set of the datasets to stabilize the selection of the initial centroids with differentially private *k*-means algorithms. Apache Hadoop is an open-source software framework used to develop applications for processing of data, which are executed in a distributed environment.

Hadoop Distributed File System (HDFS) holds the files in a distributed manner. Mainly, Java is used as a computational high-level language to perform operations in Hadoop. Similar to data residing in a local file system of a personal computer system, in Hadoop, data reside in a distributed file system which is called as a HDFS. The processing model is based on "data locality" concept, wherein computational logic is sent to cluster nodes (server) containing data. This computational logic is nothing, but a compiled version of a program written in a high-level language such as Java. Such a program, processes data stored in Hadoop HDFS. Other Hadoop-related projects at Apache include are Hive, HBase, Mahout, Sqoop, Flume, and ZooKeeper.

Hadoop as a master-slave layout for storing of data and distributed processing of data is shown in Fig. 1. With HDFS as the base file system, the MapReduce is built over the file system to perform various distributed operations. Flume is used as a log collector across the distributed environment. Hive acts as a Structure Query Language (SQL) interface to tackle between the distributed databases. In order to handle the statistical computations, R language is adopted to perform statistical operations. As machine learning is one of the paradigm which is claiming its benefits in all the domains, it does so in distributed environment also. Mahout is the framework used to perform machine learning algorithms over the Hadoop MapReduce framework in a distributed manner. Pig is used to incorporating scripting languages and allows the user to develop their own scripts to integrate and operate the objects in the distributed environment. Oozle and Sqoop maintain the work flow and data exchange, respectively. HBase

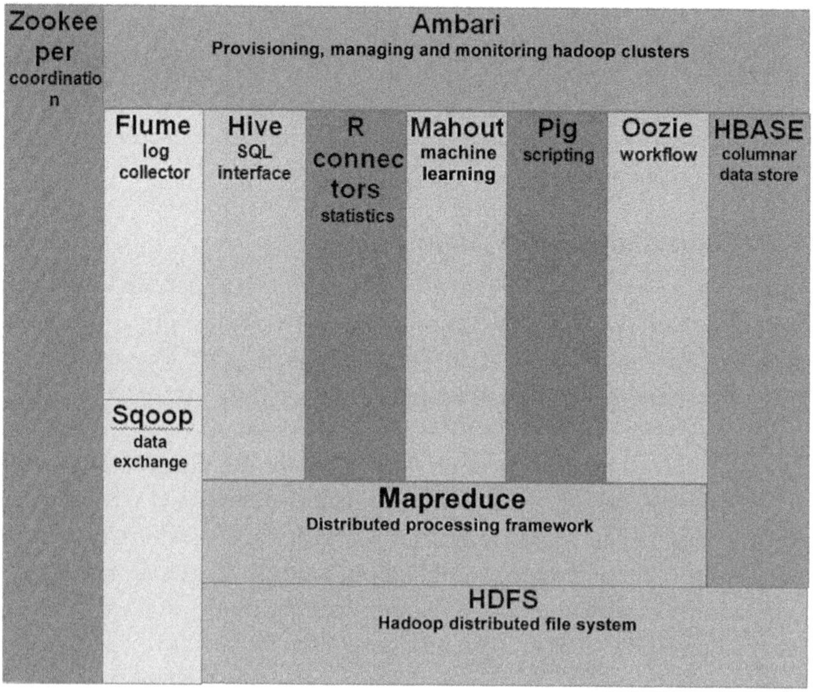

Fig. 1 Hadoop basic architecture for parallelism.

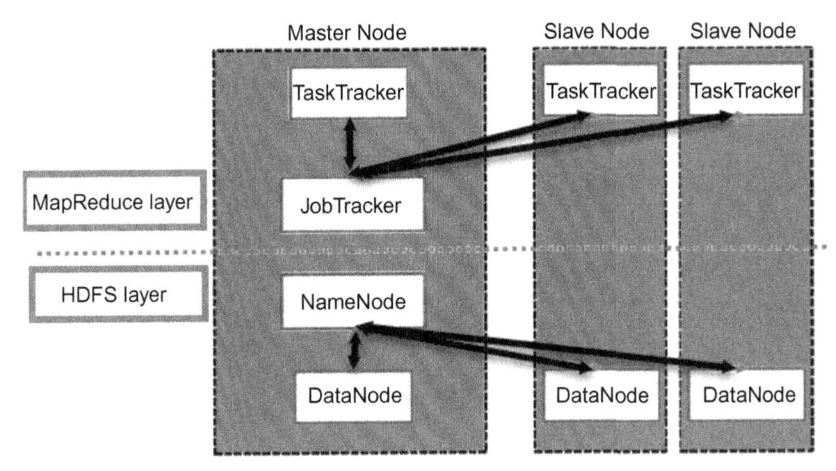

Fig. 2 High-level Hadoop architecture for parallelism.

maintains the columnar data storage space for the data, which is of high priority to perform distributed framework.

As shown in Fig. 2, the namenode denotes files and directory used in the namespace and DataNode shows the management of an HDFS node and enables interaction with the blocks. The master node allows you to conduct parallel processing of data using Hadoop MapReduce. The slave nodes are the additional machines in the Hadoop cluster, which allows you to store data to conduct complex calculations. Moreover, all the slave node comes with Task Tracker and a DataNode. This allows you to synchronize the processes with the NameNode and Job Tracker, respectively [23–26].

3.2 *k*-Means in MapReduce

Privacy-preserving algorithms proposed to handle the variation in the centroids during the execution and safeguard the privacy information in the MapReduce distributed environment. This ensures that a malicious intruder will not be able to acquire information by analyzing the dataset. Since, the records are distributed into multiple clusters and the details of which cluster contains which record is also hidden. This cluster formation in the distributed environment makes the identification of the required record tedious unless the intruder is able to overcome the algorithmic issues.

The number of records in the cluster and its attributes are added together and maintained by the reduces. To make sure that the differential privacy is attained, certain amount of Laplace noise is added. Differentially private *k*-means algorithm consists of a part to deploy into the MapReduce framework, to add noise to the distributed *k*-means algorithm, and determine the condition of the algorithm iterations. Thus, selection of the initial centroids also matters and considered in the algorithm.

Some of the features of Hadoop which can be incorporated in parallelizing the deep learning algorithms for IoMT are

- Hadoop clusters are best suited for unstructured distributed nature of the data. The processing logic embedded consumes less networks bandwidth, which improves the efficiency of Hadoop-based applications.
- Scalability is allowed for adding additional cluster nodes and allows a growth without modifying the application logic.
- Replication of the input data in multiple nodes leads to handle the failure in one node and another cluster node.
- Network topology in Hadoop affects the processing as the cluster size increases along with the network.

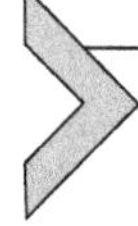

4. Deep learning on internet of medical things (IoMT)

4.1 Deep learning preliminaries

The convolutional neural networks (CNN) is the main revolution in training the image models. The depth of the layers and analysis has made the accurate analysis of the images. This accurate analysis is much needed in the field of medical image analysis, where even a minute details of the image reflects with huge number of information. The main purpose of the CNN used at first is to train the neural networks layers with a set of images with class labels and let the CNN learn the features of each class. The training is carried out based on the fine tuning of the loss function. Once, we achieve the best loss function, the trained model is stored and used for classification accuracy evaluation. To evaluate this model, a new test set is evaluated over the trained model and the rate at which the trained model is predicting the new test cases to its corresponding classes tells the accuracy of the model built shown in Fig. 3.

4.2 IoMT challenge cycle

The ultimate aim is to analyze the data generated by all the IoMT devices irrespective of the environment. Traditional machine learning and deep learning algorithms helps the purpose by performing significant operations to reveal the patterns and underlying relationships to perform prediction.

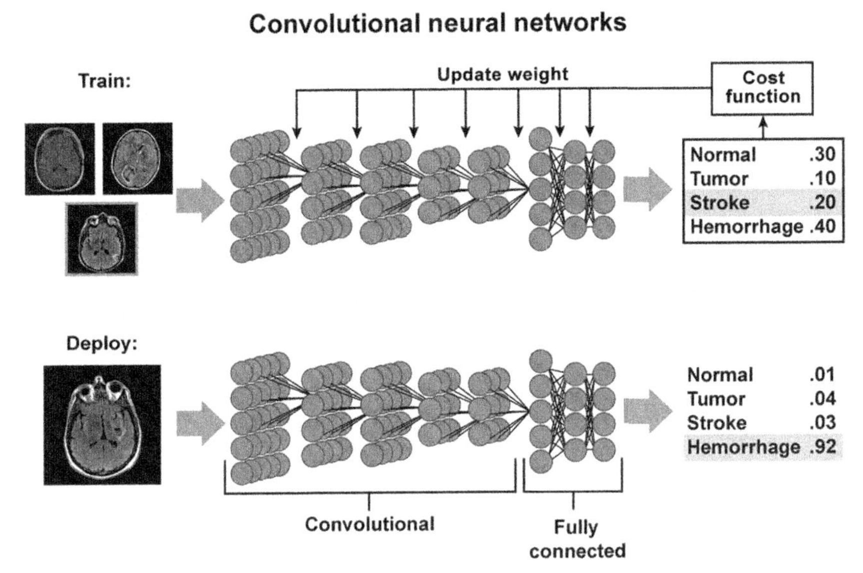

Fig. 3 Convolutional neural networks.

As shown in Fig. 4, the perspective opens up a lot of scope for the companies and researchers to develop the communication between the IoMT components.

4.3 Patient-physician data flow

Fig. 5 shows the overview of the data flow between the patient and the physician. The details of the patient are recorded through the IoMT sensors or devices attached to the patient. After the details are formatted as per the requirement of the physician, the details are communicated to the physician based on the priority of the call to be made.

Fig. 6 shows the detailed steps involved in the communication between the patient and physician.

1. Image digitizer is used to scan the required part to be investigated in the patient using the camera or scanner.
2. Measurement devices include ECG, blood pressure measurement, and oxygen monitor.
3. An interface which displays the recorded content to the patient as well as the physician is displayed.
4. The patient end modem converts the digital data into electrical signal as the draft from the patient's side kept ready for the disposal to the physician.

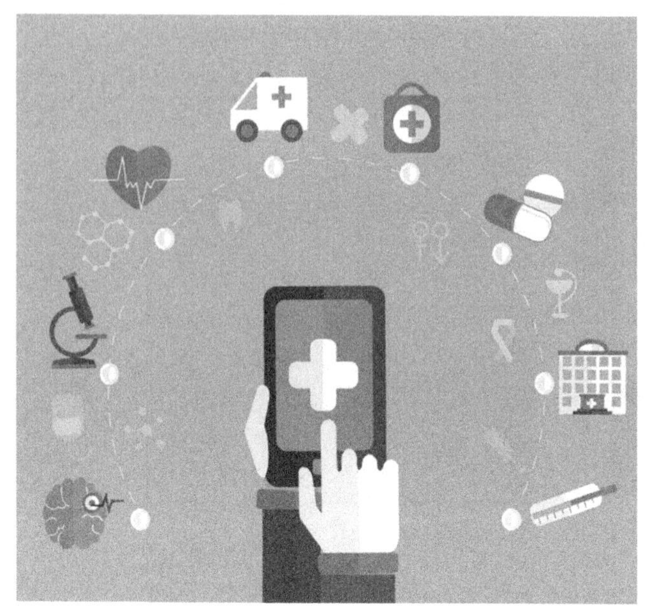

Fig. 4 Challenge IoMT cycle to manage the change.

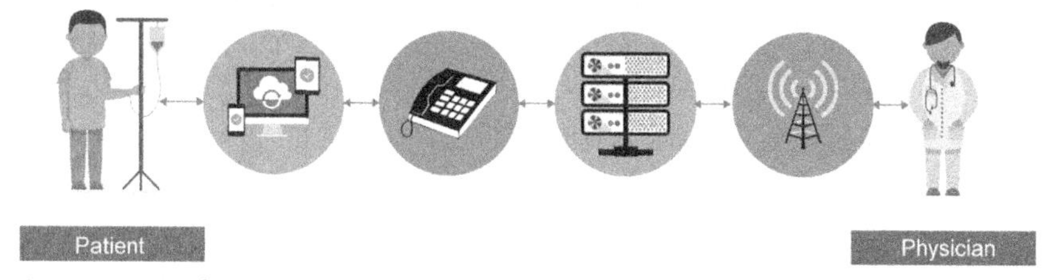

Fig. 5 Patient-physician data flow.

5. The intermediate data analytics tools and storage devices analyze the data and give a decision on the patients records to the physician for further analysis.
6. The data are further interrupted by the physicians to intervene and monitor the visuals.
7. The physician is continuously reported with the real-time data from the network.

4.4 IoMT base cloud architecture and statistics

Fig. 7 shows the overview architecture of the IoMT. The broad categories involve the analytic solutions, device connectivity and data management,

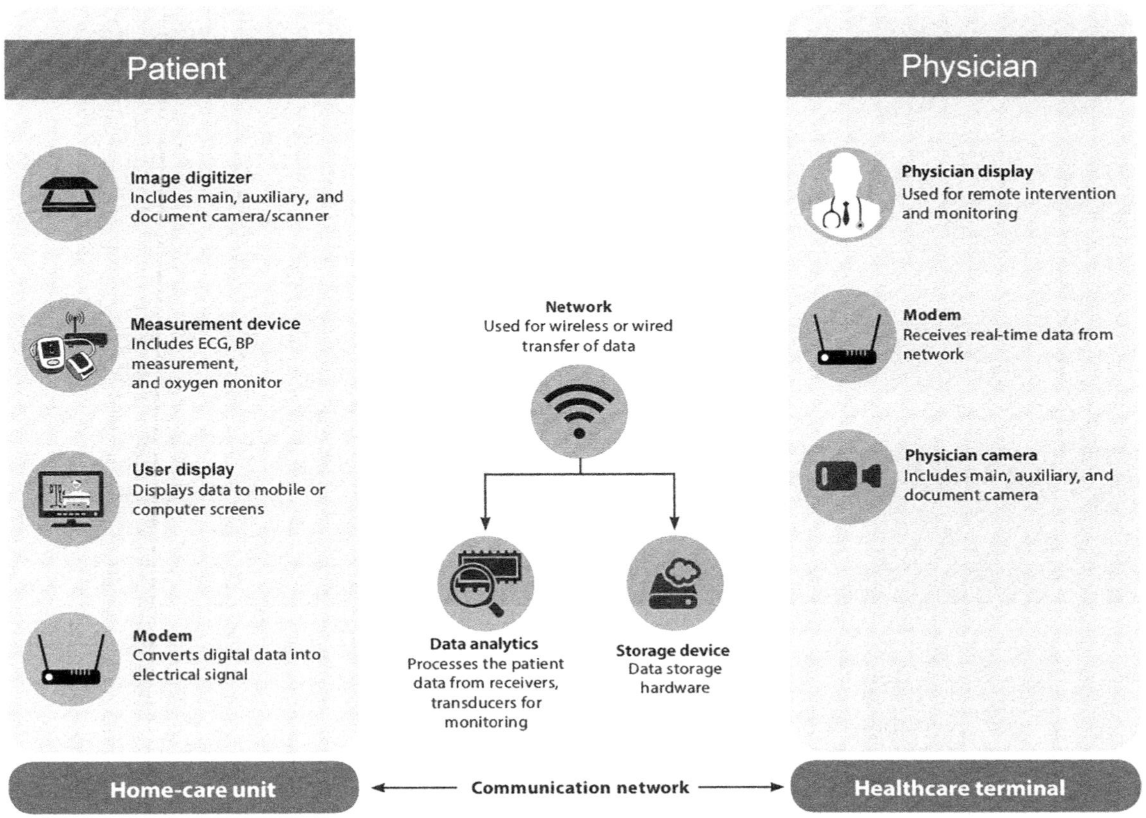

Fig. 6 Detailed flow between the patient and physician.

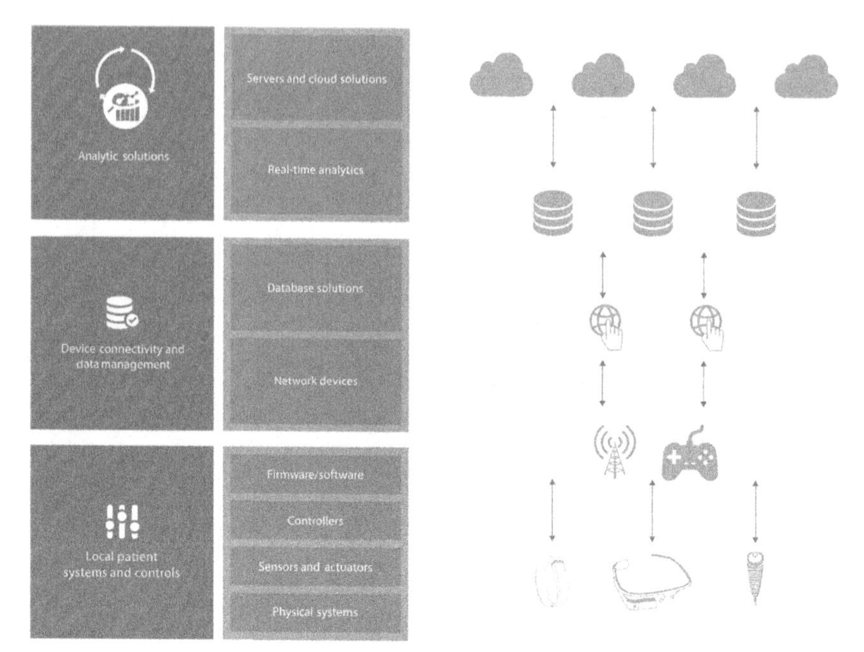

Fig. 7 Overview architecture of IoMT.

and local patient systems and controls. The analytic solutions involve the operations of servers and solutions using cloud. Real-time analytical solutions are also provided to the environment. Database solutions and network connectivity are taken care by the device connectivity and data management module.

The local patient systems and controls contain a firmware with the controllers to manage the sensors and actuators. The data are stored in the cloud and taken to the database with further global searching and sent to the customer. The end user receives the completed products at the end of the process.

Fig. 8 shows the forecast, which has expanded at CAGR of 37% by 2020.

Fig. 9 shows the US market share invested in the conventional medical devices and rapidly adapted due to the widespread clinical acceptance and healthcare policies.

4.5 IoMT system framework levels and mechanisms

IoMT mechanisms based on the biometrics demand a continuous authentication methods to provide reliable security to the devices. It involves four

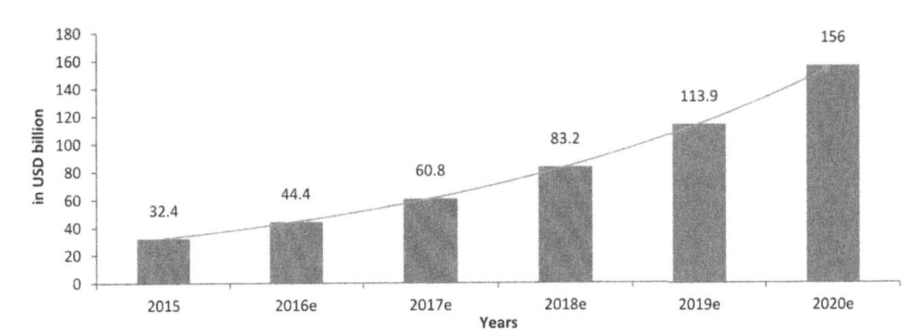

Fig. 8 IoMT market landscape for past 5 years.

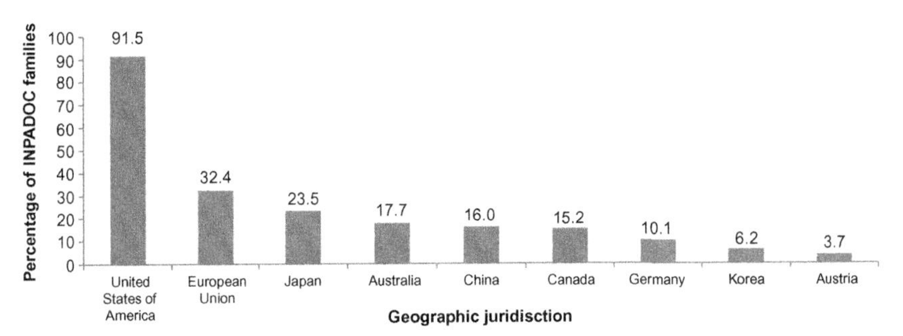

Fig. 9 Patent filing trend in IoMT technology.

important levels namely, device, connectivity, analytical, and data management level:

- Device level consists of IoMT devices, which are used to collect the data from users in the smart environment and makes sure that the collected data are secure. This level also encodes the acquired data to prevent thefts.
- Connectivity level as the name denotes it is capable of transferring the biometric system to the data centers. The communication is established using the 5G, Wi-Fi, Bluetooth, Satellite, and ZigBee. It also possesses security issues when it is linked with the malicious network.
- The most important among the four levels is the analytical level, where the devices are connected through a private weber involving many number of data centers exchanging data at various phases.
- Management level can provide various services by using the appropriate APIs like child care, patient management, etc.

5. Challenges in deep learning-based IoMT

As the IoMT involves two perspectives namely, technical and market-related issues. The challenges in both aspects are listed below:

- IoMT data must be highly secured and prevented from hacking and unauthorized usage of the devices.
- The standards and communication protocols used for maintaining the devices are not streamlined properly.
- Data handling involves lot of errors, which are unnoticed by the system.
- The end-to-end data integrity has to be maintained to avoid wrong interpretation of the information.
- To analyze the IoMT, experts are required to perform the operations.
- The diversity in the devices used for various purpose has to be rectified by providing high interoperability.
- Scalability of the devices, data, and performance maintenance.
- Data overload leading to ambiguity of analysis and mobile handling.

6. Applications of IoMT and deep learning

The major chronic continuous diagnosis and clinical system which denotes the patient and physician, respectively, are discussed below.

6.1 Chronic disease diagnosis

Diseases such as diabetes, cardiac failure, hypertension, blood pressure, electrolyte concentration, etc. are monitored in order to manage the doses for predicting the disease progress. The wearable devices and clinical monitoring setup. The patient side and the physician side devices are shown, respectively.

6.2 Remote-assisted living

Physician acts as the centralized access for all the operations. The network devices are connected to the central point from which the past records and future predictions are obtained to assess the decision on the patients. This remote monitoring is useful in reducing the resource maintenance cost, human resources, etc. Encryption protocols are used to transmit and store critical information ensures the reliability of the solution.

6.3 Additional applications

The home monitoring system which keeps track of the activities of the patient inside a home. The system is connected to the internet to alert the physician when required if the values are abnormal. Step-by-step authentication and alert systems can also built over this to have multiple-step verification before the alert reaches the final destination. Biometric sensors are used to continuously authenticate the person to avoid attacks on the data generated. The authentication would provide a proof of the values are recorded for the right person and avoid any misdirection. The scenario of suggesting a wrong vaccination for a person by observing a different person might occur since all the devices are closely connected with each other.

Sensors connected to the brain cells through neurons. As the brain and deep learning neural networks can be closely related, there are a lot of relations between the brain sensor monitoring and neural networks. The brain cells play a major role in deciding the mental stability of a person, which shows high level of accuracy in what the person's physical and mental status. A sleep monitor which monitors the aforementioned values which also contributes to the patient's health profile. A lot of internal operations takes place only during sleeping and the operations can be monitored clearly to take a proper decision [27, 28].

Wireless capsule endoscopy is a subset of the tele-monitoring and health care for the patient in the medical market, which includes all physiological signals, video, and image signals. Cardiac diseases are treated as emergency cases, where more than one specialist has to monitor a single patient [29, 30].

7. Conclusion and future directions

The chapter discusses about the internet of medical things (IoMT) devices and functionalities. The management of IoMT devices integrated with the other clinical devices. An efficient parallelized *k*-means clustering mechanism is discussed by incorporating the Hadoop MapReduce framework. The architecture of applying deep learning to the IoMT base has been discussed. The technical and market-level challenges of the incorporating IoMT devices are included. In future, this chapter leads to apply deep learning on IoMT applications and modifies base neural network accordingly.

References

[1] F. Al-Turjman, H. Zahmatkesh, L. Mostarda, Quantifying uncertainty in internet of medical things and big-data services using intelligence and deep learning, IEEE Access 7 (2019) 115749–115759.
[2] P.M. Shakeel, S. Baskar, V.R.S. Dhulipala, S. Mishra, M.M. Jaber, Maintaining security and privacy in health care system using learning based deep-Q-networks, J. Med. Syst. 42 (10) (2018) 186.
[3] A. Azmoodeh, A. Dehghantanha, K.-K.R. Choo, Robust malware detection for internet of (battlefield) things devices using deep eigenspace learning, IEEE Trans. Sustain. Comput. 4 (1) (2018) 88–95.
[4] Z. Liu, C. Yao, H. Yu, T. Wu, Deep reinforcement learning with its application for lung cancer detection in medical internet of things, Futur. Gener. Comput. Syst. 97 (2019) 1–9.
[5] M.A. Al-Garadi, A. Mohamed, A. Al-Ali, X. Du, M. Guizani, A survey of machine and deep learning methods for internet of things (IoT) security, arXiv Preprint abs/1807.11023 (2019) 1–42.
[6] U. Iqbal, T.Y. Wah, M.H. Ur Rehman, G. Mujtaba, M. Imran, M. Shoaib, Deep deterministic learning for pattern recognition of different cardiac diseases through the internet of medical things, J. Med. Syst. 42 (12) (2018) 252.
[7] C. Yao, S. Wu, Z. Liu, P. Li, A deep learning model for predicting chemical composition of gallstones with big data in medical internet of things, Futur. Gener. Comput. Syst. 94 (2019) 140–147.
[8] M.-P. Hosseini, D. Pompili, K. Elisevich, H. Soltanian-Zadeh, Optimized deep learning for EEG big data and seizure prediction BCI via internet of things, IEEE Trans. Big Data 3 (4) (2017) 392–404.
[9] P. Sahu, D. Yu, H. Qin, Apply lightweight deep learning on internet of things for low-cost and easy-to-access skin cancer detection, in: Medical Imaging 2018: Imaging Informatics for Healthcare, Research, and Applications, vol. 10579, International Society for Optics and Photonics, 2018, p. 1057912.
[10] J.-G. Lee, S. Jun, Y.-W. Cho, H. Lee, G.B. Kim, J.B. Seo, N. Kim, Deep learning in medical imaging: general overview, Korean J. Radiol. 18 (4) (2017) 570–584.
[11] M.-C. Kwon, M. Ju, S. Choi, Classification of various daily behaviors using deep learning and smart watch, in: 2017 Ninth International Conference on Ubiquitous and Future Networks (ICUFN), IEEE, 2017, pp. 735–740.
[12] C.A. da Costa, C.F. Pasluosta, B. Eskofier, D.B. da Silva, R. da Rosa Righi, Internet of health things: toward intelligent vital signs monitoring in hospital wards, Artif. Intell. Med. 89 (2018) 61–69.
[13] C.F. Pasluosta, H. Gassner, J. Winkler, J. Klucken, B.M. Eskofier, An emerging era in the management of Parkinson's disease: wearable technologies and the internet of things, IEEE J. Biomed. Health Inf. 19 (6) (2015) 1873–1881.
[14] J. Zhu, Y. Song, D. Jiang, H. Song, A new deep-Q-learning-based transmission scheduling mechanism for the cognitive internet of things, IEEE Internet Things J. 5 (4) (2017) 2375–2385.
[15] S. Krech, Medical big data analytics and smart internet of things-enabled mobile-based health monitoring systems, Am. J. Med. Res. 6 (2) (2019) 31–36.
[16] P. Li, Z. Chen, L.T. Yang, Q. Zhang, M.J. Deen, Deep convolutional computation model for feature learning on big data in internet of things, IEEE Trans. Ind Inf. 14 (2) (2017) 790–798.
[17] V. Milutinovic, M. Kotlar, M. Stojanovic, I. Dundic, N. Trifunovic, Z. Babovic, DataFlow systems: from their origins to future applications in data analytics, deep learning, and the internet of things, in: DataFlow Supercomputing EssentialsSpringer, 2017, pp. 127–148.

[18] D.D. Miller, E.W. Brown, Artificial intelligence in medical practice: the question to the answer? Am. J. Med. 131 (2) (2018) 129–133.
[19] P. Xiuqin, Q. Zhang, H. Zhang, S. Li, A fundus retinal vessels segmentation scheme based on the improved deep learning U-Net model, IEEE Access 7 (2019) 122634–122643.
[20] H. Ma, X. Pang, Research and analysis of sport medical data processing algorithms based on deep learning and internet of things, IEEE Access 7 (2019) 118839–118849.
[21] C. Wu, C. Luo, N. Xiong, W. Zhang, T.-H. Kim, A greedy deep learning method for medical disease analysis, IEEE Access 6 (2018) 20021–20030.
[22] M. Sajjad, S. Khan, K. Muhammad, W. Wu, A. Ullah, S.W. Baik, Multi-grade brain tumor classification using deep CNN with extensive data augmentation, J. Comput. Sci. 30 (2019) 174–182.
[23] N. Majumdar, S. Shukla, A. Bhatnagar, Survey on applications of internet of things using machine learning, in: 2019 9th International Conference on Cloud Computing, Data Science & Engineering (Confluence), IEEE, 2019, pp. 562–566.
[24] S. Byrne, Remote medical monitoring and cloud-based internet of things healthcare systems, Am. J. Med. Res. 6 (2) (2019) 19–24.
[25] A. Botta, W. De Donato, V. Persico, A. Pescapé, On the integration of cloud computing and internet of things, in: 2014 International Conference on Future Internet of Things and Cloud, IEEE, 2014, pp. 23–30.
[26] R.F. Molanes, K. Amarasinghe, J. Rodriguez-Andina, M. Manic, Deep learning and reconfigurable platforms in the internet of things: challenges and opportunities in algorithms and hardware, IEEE Ind. Electron. Mag. 12 (2) (2018) 36–49.
[27] J.J. Cimino, Beyond the superhighway: exploiting the internet with medical informatics, J. Am. Med. Inf. Assoc. 4 (4) (1997) 279–284.
[28] J. Qi, P. Yang, G. Min, O. Amft, F. Dong, L. Xu, Advanced internet of things for personalised healthcare systems: a survey, Pervasive Mobile Comput. 41 (2017) 132–149.
[29] A. Morshed, P.P. Jayaraman, T. Sellis, D. Georgakopoulos, M. Villari, R. Ranjan, Deep osmosis: holistic distributed deep learning in osmotic computing, IEEE Cloud Comput. 4 (6) (2017) 22–32.
[30] A.R. Dargazany, P. Stegagno, K. Mankodiya, WearableDL: wearable internet-of-things and deep learning for big data analytics—concept, literature, and future, Mobile Inf. Syst. 2018 (2018) 1–20.

CHAPTER FIVE

Cloud-based IoMT framework for cardiovascular disease prediction and diagnosis in personalized E-health care

Kayode S. Adewole[a], Abimbola G. Akintola[a], Rasheed Gbenga Jimoh[a], Modinat A. Mabayoje[a], Muhammed K. Jimoh[b], Fatima E. Usman-Hamza[a], Abdullateef O. Balogun[a], Arun Kumar Sangaiah[c], and Ahmed O. Ameen[a]

[a]Department of Computer Science, University of Ilorin, Ilorin, Nigeria
[b]Department of Educational Technology, University of Ilorin, Ilorin, Nigeria
[c]School of Computing Science and Engineering, Vellore Institute of Technology (VIT), Vellore, India

1. Introduction

Cardiovascular disease (CVD) is the highest factor that reduced the standard death age between 1990 and 2015 and ischemic heart disease constitute the highest death rate in African and globally being a major cause of sudden deaths [1]. Thus, efficient prediction of such deadly disease will assist greatly in reducing death rate globally. Prediction of the next person to develop CVD is a big research problem among the researchers globally [2]. There are many algorithms for CVD prediction based on factors such as age, drug, and smoking habit but none has yet appeared to give an accurate prediction of who is next to develop the CVD [3].

The increase in chronic diseases and medical conditions can be linked to an increase in population globally [4]. Thus, the explosive growth of diseases and health conditions data in volume, velocity, and diversity called for more intelligent and robust health-care systems to achieve efficient data storage and management capable of fulfilling the needs of the heterogeneous data [4, 5]. Therefore, there is a need for computational intelligence (CI) approaches on medical data to achieve enhanced performance, management and accuracy in analysis, detection, and prediction [6]. Many authors have argued the relevance of the technique of Internet of things (IoT) in handling real-time perception and analysis of health conditions in a ubiquitous manner [4, 5].

Intelligent IoT Systems in Personalized Health Care
https://doi.org/10.1016/B978-0-12-821187-8.00005-8

Internet of medical things (IoMT) is an emerging trend in providing solutions to various health-care problems [7]. IoMT provides an opportunity for technological advances in health care to thrive and also allows the solution to stay with the pace of time [8]. Thus, there is an urgent need for a cloud-based framework for predicting the deadly development of CVD most especially in this era of IoMT. Although some progress has been made in the past two decades on CVD prediction, however, personalized diagnosis of CVD remains a serious challenge [9]. In today's global era, the dynamism in the health-care sector majorly relies on data generated through the adoption of IoMT, however, the efficient usability of such data remains a problem in the medical field [4]. IoMT provides an opportunity for having access to a large volume of medical records which can be harnessed and mined in solving health-related problems not limited to CVD [10].

Bhatia and Sood [5] presented an intelligent framework for effective real-time analysis and prediction of health conditions using IoT technique. According to Gubbi, Buyya, Marusic, and Palaniswami [11], the application of Information and Communication Technology (ICT) in health sectors has greatly improved service delivery in the sector. Thus, the use of ICT in the health sector is sacrosanct. The experienced large volume of data in the health-care systems serves as a major justification for the integration of big data analytics in handling such an increasing volume of data [4]. Bhatia and Sood [5] also attested to the fact that IoT for the health-care system is almost inevitable. The heterogeneous nature of health-care data requires efficient feature extraction method [5]. Hence, a specialized approach like IoMT is desirable to address this peculiarity.

The ever-increasing demands of implementing IoT in the health-care sector has given birth to several frameworks to address real-time prediction, diagnosis and monitoring of the patients. Santos et al. [12] introduced a novel IoT-based system for mobile health (mHealth) and developed a prototype for remote monitoring. Radio frequency identification (RFID) technology, which provides sensing, has been integrated within IoT architecture to monitor user health and trigger remote assistance when required. Santos et al. [13] proposed the use of RFID in the mHealth system that leveraged the strengths of IoT-enabled connected devices. Their study focused on the need to improve health services using IoT and RFID technologies. Yang et al. [14] have developed a prototype of an intelligent home health-care management system for disabled people. A three-tier framework was proposed based on Wireless Sensor Network (WSN) in health-care systems [15].

Mitra et al. [16] proposed a KNOWME wireless body sensor networks (WBSN) platform for end-to-end communications. The authors employed wearable sensors and smartphone to collect and transfer the captured information. This platform comprises four layers targeted at reducing communication complexity. In addition, several cloud-based frameworks have been proposed in the literature. A cloud-centric IoT-based health-care framework has been implemented in the work of Tyagi et al. [17] with various health-care entities such as patients, doctors, and hospitals. Hossain and Muhammad [18] focused on the development of real-time health-care monitoring infrastructure for analyzing a patient's health condition. The authors proposed HealthIIoT based on smartphone technology that is capable of monitoring electrocardiogram (ECG) and other health-care-related data. To address the IoT framework with cloud security for user's privacy protection, watermarking technique has been proposed for the security of data. Seales et al. [19] proposed IoT framework that is based on content-centric cloud architecture. Wan et al. [20] proposed platform production services that are based on IoT and cloud architectures. Hassan et al. [21] introduced an intelligent hybrid context-aware model to monitor patients at home using the hybrid architecture comprising local and cloud components. However, despite the huge number of studies conducted in the literature, addressing the problem of cloud unavailability due to network disconnection to provide reliable health-care monitoring system still remains an open research issue. In addition, data aggregation from heterogeneous IoT data sources is also a major challenge. Therefore, this chapter proposed a cloud-centric framework that addresses the research issues.

The remaining parts of this chapter are organized as follows: Section 2 discusses the fundamental concepts of cloud computing. Section 3 presents an extensive discussion on IoT and IoMT. IoT communication technologies, sensor devices in IoMT, IoT architecture, and health-care framework, as well as cloud-based IoT health-care architectures, were discussed in this section. Section 4 highlights the rise of CVD. Section 5 presents the taxonomy of CI techniques that have been used for CVD prediction and diagnosis. Section 6 briefly discusses the rationale for chosen CI techniques for CVD prediction and diagnosis. Section 7 presents experimental results and evaluation of CVD prediction systems. A cloud-based framework for CVD prediction and diagnosis that addressed some of the research issues is proposed in Section 8. Section 9 focuses on the practical case study of the proposed cloud-based framework. Finally, Section 10 concludes the chapter and discusses future research directions for the realization of efficient health-care systems.

2. Fundamental concepts of cloud computing

The cloud makes it easy to access information from anywhere at any time. Using cloud architecture, a user does not need to worry about physical storage location as the hardware that stores the user's data as well as the software that performs the computation are being managed by the cloud provider. The cloud provider owns and takes care of hardware and software necessary to run your home or business applications. In 1950, the concept of Cloud Computing emerged during the implementation of mainframe computers with accessibility to thin/static clients. This development has made cloud computing to evolve from static clients to dynamic clients and from software to services. The concepts of cloud computing built are subscription-based where one can obtain networked storage space and computer resources. The basic requirement to access the cloud is to have an Internet connection [22, 23].

Cloud computing offers numerous advantages including ease of access to applications, availability of large storage space, on-demand self-service that is cost effective, management of load balancing for reliability, provision of online development and deployment tools, and programming runtime environment. The main disadvantage of cloud computing is that since third party is responsible for the provision of data and infrastructure management in the cloud, it becomes very risky to such service providers with sensitive information. This triggered the need to provide more secured techniques to handle data storage and processing in the cloud.

2.1 Deployment models

There are four deployment models of cloud computing: (1) Public Cloud Model that provides easy access to systems and services to the general public. For instance, Google, Microsoft, and Amazon offer cloud services via the Internet. The public cloud model has the benefits of being cost effective, reliable, flexible and highly scalable, and the main disadvantage is its low security, (2) Private Cloud Model enables access to systems and services within an organization. The private cloud may be managed internally or by a third party but it is operated only within a single organization. The benefits of the private cloud model are better security, control, and efficiency while its disadvantages include restriction in area, limited scalability, and inflexible pricing, (3) Hybrid Cloud Model involves a combination of public and private cloud models where noncritical activities are carried out by means of a public cloud while the critical ones are performed using private cloud. The benefits of such a cloud deployment model include scalability, flexibility, cost effectiveness, and better security. The disadvantages include networking issues, infrastructural dependency, and security compliance, and (4) Community Cloud Model enables access to systems and services by a group of organizations. It shares the infrastructure between different establishments from a specific community. Similarly, it is possible to manage this cloud model internally or through a third party. The benefits of a community cloud model are cost-effectiveness and better security than the public cloud. Community cloud provides the same benefits as that of private cloud at low cost [22].

2.2 Service models

Cloud computing provides different service models as shown in Fig. 1. These include Infrastructure as a Service (IaaS), Platform as a Service (PaaS), and Software as a Service (SaaS) models [24].

IaaS gives access to major resources, for example, physical machines, virtual machines, virtual storage, and so on. Aside from these resources, IaaS additionally provides virtual machine disk storage, load balancers, virtual local area network (VLANs), IP addresses, and software bundles. The advantages of IaaS are not limited to its capability to enable the cloud provider to unreservedly allocate the resources over the Internet in a cost-effective way, better control of computing resources based on administrative access to

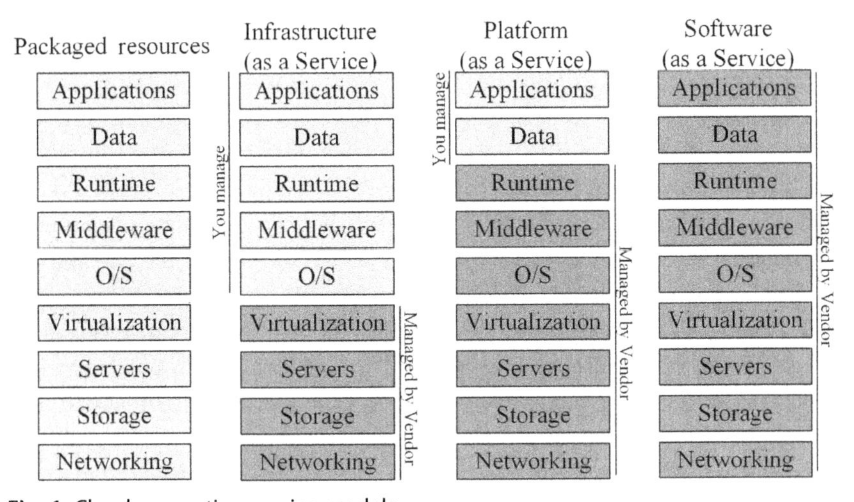

Fig. 1 Cloud computing service models.

virtual machines, adaptable and effective renting of computer hardware, portability, and interoperability with legacy applications. IaaS enables the consumer to access resources in the cloud by issuing administrative commands to the cloud provider to run the virtual machine, start the webserver, store data on cloud's server, and install new applications through the provision of administrative privilege. In IaaS, resources like storage, bandwidth, virtual machines, IP addresses, firewalls, and many more are offered to the consumers on rent. The payment for the usage of such resources is based on the length of time that the consumer spent utilizing those resources. As IaaS provides an opportunity for the consumer to run legacy software, however, this provision exposes the consumers to all the vulnerabilities inherent by this legacy software. Consumers can run virtual machines in three different modes: running, suspended, and off mode. IaaS makes provision for a hypervisor, which is a software layer that supports hardware and enables virtualization and splitting of a physical computer into multiple virtual machines.

PaaS provides a runtime environment and tools for applications development and deployment. It makes provision point-and-click tools, which can be used by a novice in software development to create web applications. For instance, Force.com and Google App Engine are examples of PaaS offering vendors. The benefits of a PaaS service model include lower administrative overhead, the lower total cost of ownership, and the existence of scalable solutions. The main problem of PaaS is lack of portability between PaaS clouds and the PaaS applications, which poses resource constraints on applications. The purpose of PaaS is to provide browser-based development environment allowing developers to create new databases and manipulate application source codes either through point-and-click tools or Application Programming Interface (API). PaaS offers built-in tools for defining workflow and business rules as well as approval processes. It has the capability for built-in security, scalability, and web service interfaces. PaaS is simple to integrate with other applications either on the same or outside the platform.

SaaS model provides software application as a service to the consumers. SaaS enables software that is deployed on a hosted service to be accessible via the Internet. SaaS software includes Customer Relationship Management applications, Billing and Invoicing Systems, Help Desk Applications, and Human Resource applications. Some of the SaaS applications are not customizable such as an Office Suite but SaaS provides API, which allows the developer to develop a customized application. In SaaS, the software can be accessed via the Internet and they are maintained by the software vendors. The software is licensed per subscription or usage-based and it is billed periodically. One of the advantages of SaaS applications is cost effectiveness and since the applications require no maintenance fee at the consumer's end, it is easy to access. In addition, these applications are available on-demand and it is possible to scale them up and down based on need. SaaS applications provide share data model and can be automatically upgraded and updated. This enables multiple users to share a single instance of the application without the need to hardcode the operations of the application for each individual end user. This means that individual end users are working on the same version of the application on the cloud infrastructure. The use of SaaS applications available on the cloud has provided several benefits some of which include efficiency and scalability. The deployment of SaaS application requires little or no installation and thus reduces software complexity at the end user's side. This also reduces the risk of software configuration with very low distribution cost. The SaaS application is license-free with low deployment cost, more portable applications, less vendor lock-in and more scalable solutions. However, there are several issues associated with SaaS and these include browser-based risks and vulnerabilities, network dependence, and lack of portability between SaaS clouds [22, 23].

3. IoT and IoMT

IoT provides worldwide interconnection of items on remote terminals that can be distinctively identified for information gathering and sharing under stable Internet connectivity [11, 25]. In an IoT environment, many of the items around are connected together to make human life easier through radio frequency and sensors. This brings about a huge volume of data that needs to be processed in order to make meaning and decision-making [11, 26]. The occurrence of IoT needs the consideration of the scenario at hand with the users, appropriate software structural design, and analytical tools [11]. IoT can be applied to virtually all aspect of human life today, for example, education, transportation, health-care, personal use, and many more. In the health-care sector, it is referred to as the IoMT. The growth of data in biomedical and health-care communities has increased tremendously. Being able to analyze the data accurately and timely will lead to early detection of diseases, comparison of diagnosis, and adequate care for the patients as well as provision of accurate and reliable results at a reduced cost [26, 27]. Fig. 2 shows some of the domains where IoT has demonstrated successful implementation.

Today, the information needed in the medical field is available in a different format and can be secured using biometrics [28]. This information can

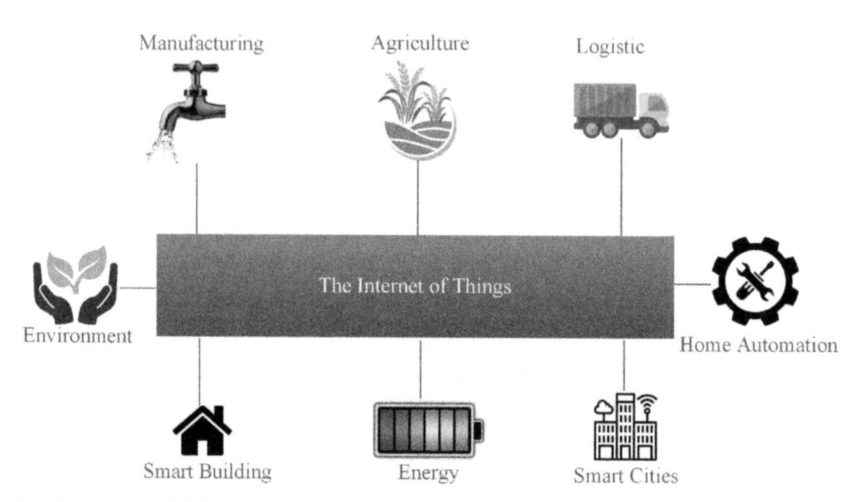

Fig. 2 Internet of Things.

be studied to make a meaningful prediction and diagnosis through available technology. IoMT in IoT makes use of several techniques of artificial intelligence, data mining, and machine learning in classifying accumulated medical information to identify hidden patterns. The patterns are then used in prediction and diagnosis. Some of these classification techniques are discussed extensively in Section 5 of this chapter and they include decision tree, Bayesian network, support vector machine (SVM), k-nearest neighbor (KNN) and artificial neural network (ANN) among others. For instance, the Bayesian network as a well-known model for the management of uncertainty through the representation of probability and graph theory has been studied for CVD [29]. Abdar et al. [30] compared five machine learning approaches for the prediction of heart diseases. In their study, decision tree had the highest accuracy followed by KNN and SVM. The performance was influenced by the nature of data input available, which was either discrete or continuous values. Using some parameters like age, gender, and blood pressure, Fuster-Parra et al. [29] were able to identify the relationship between features with variable interdependencies using Bayesian network for epidemiological studies, in particular for CVD diagnosis. The study was also able to make a new predictive model. Bandyopadhyay et al. [31] developed a model for prediction of the occurrence of a CVD within 5 years using electronic health data. Bayesian network features were combined with inverse probability of censoring weights for censored events data. Chen et al. [27] proposed a new algorithm for multimodal risk prediction using structured and unstructured data with a convolutional neural network.

3.1 IoT communication technologies

One of the major pillars of IoT realization is communication technologies. Several communication technologies have been used to realize the IoT vision. Data obtained from sensors in IoT deployment can be analyzed and displayed on the smartphone and at the same time transmitted to health-care centers for further observation through the use of wireless communication platforms. However, despite the numerous IoT communication technologies, there has not been a one-size-fits-all communication solution. This means that each communication technology has its own strengths and weaknesses in terms of scalability, communication range, cost, and network demands. Thus, the available technologies are best suited for diverse scenarios and environments. Some of the communication technologies available for IoT realization include Low-Power Wide Area Networks, Cellular networks (like 3G, 4G, and 5G), ZigBee, Bluetooth, Wi-Fi, and RFID [32–34]. These existing communication technology platforms provide good Internet connectivity with high speed while moving from place to place, thus, enabling individuals to stay connected with their health-care providers [33]. Behr [32] has conducted an extensive study on the available wireless communication technologies for the realization of IoT.

As stated earlier, each of these communication technologies has strengths and weaknesses. Table 1 highlighted the comparison of some of the available wireless communication technologies for IoT realization. For instance, in the connected health-care sector, cellular and BLE-enables technologies have been highly applicable. BLE-enabled devices are mostly used in conjunction with electronic devices such as smartphones for transferring data to the cloud. Today, BLE has been widely integrated into wearables and fitness devices in medical domain as well as smart home devices. Through this arrangement, data can be efficiently transferred and visualized on handheld devices like smartphones, iPad, and tablets [32].

3.2 Sensor devices in IoMT

In the IoT ecosystem, two things are greatly significant, which are Internet and physical devices such as sensors. Sensing devices have played a major role in IoT realization and several of these devices are available for personalized medical health. In realizing IoT vision, different applications demand varying types of sensors for data collection from the environment. For instance, a wearable sensor attached to the patient's body can collect patient information like body temperature, blood pressor, ECG pulse, heart rate, SPO2

Table 1 Comparison of some IoT communication technologies.

Key IoT Verticals	LPWAN (Star)	Cellular (Star)	Zigbee (Mostly Mesh)	BLE (Star and Mesh)	Wi-Fi (Star and Mesh)	RFID (Point-to-point)
Industrial IoT	✓	•	•			
Smart Meter	✓					
Smart City	✓					
Smart Building	✓		•	•		
Smart Home			✓	✓	✓	
Wearables	•			✓		
Connected Car					•	
Connected Health		✓		✓		
Smart Retail		•		✓	•	✓
Logistics and Asset Tracking	•	✓				✓
Smart Agriculture	✓					

✓ Strongly applicable. • Moderately applicable.

among others. Sensors connectivity and the network is one of the important layers in IoT architecture. This layer forms an important part of the IoT ecosystem, which provides network connectivity to other layers. Sensors of the IoT ecosystem act as the front end. These sensors can be connected either directly or indirectly to IoT networks when the signal conversion and processing have been addressed. However, all sensors are not the same and there are several types that exist in the market nowadays. For instance, researchers have monitored 2–3 days of continuous physiological parameters based on sensors devices. The related parameters collected are used to update the relevant health-care records [33, 35]. Sensors enabled patient's personalized health data and activities to be collected and such data can be transferred to the cloud for further analysis.

The advancement in energy efficient and better computing and communication technologies have globally transformed the telecommunication industry. Additionally, the availability of improved sensors and battery technologies has provided enabled modern handheld devices such as smartphones, iPad, and tablets to seamlessly connect to the Internet for entertainment, gaming, and health and fitness monitoring. Smartphones have witnessed tremendous growth in the telecommunication market over the past few years. Smartphone-based health-care systems have a strong tendency to provide a cost-effective option for long-term health-care management. These systems can allow health-care doctors to monitor the health status of their patients remotely without intruding into their daily activities. Smartphones now have different embedded sensors including an image sensor, accelerometer, ambient light sensor, global positioning system (GPS) sensor, gyroscope, microphone, and fingerprint [36]. These sensors help in measuring several health parameters of patients including their heart rate, heart rate variability, respiratory rate, body glucose, and blood pressure just to mention but a few. This turns communication gadget into a perpetual and long-term health-care monitoring tool. Fig. 3 shows some sensors that have been used for data collection in IoMT. These sensors enable health personnel such as a doctor to monitor patient's activities and health parameters in real time. A comprehensive review of the various smartphone sensors has been provided in the work of Majumder and Deen [36].

For instance, Hassan et al. [21] presented the use of IoT sensors to capture a patient's context-aware data. The authors proposed a hybrid framework integrated with Patient's Local Module (PLM) that ensures convergence between IoMT sensors devices and clouds. The hybrid model transferred a number of computations to the edge of the network in PLM, which include changing of low-level data to a higher level of the construct in order to improve the computations in the cloud part of the hybrid framework. This arrangement provides an opportunity for better storage, processing, and big data analytics, and to develop a system for classifying patient's health condition. Some researchers are of the opinion that the use of embedded sensors can provide better health-care monitoring than wearable sensors [15, 33]. It has been argued that wearable sensors are intrusive to the patients and can lead to extra burden on users when compared with embedded sensors, which are usually unobtrusive and reduce usability issues. From this perspective, Gupta et al. [33] proposed IoT cloud-based architecture that monitors patients during their regular activities at the health-care centers to alert health-care personnel.

3.3 IoT architecture and health-care framework

In the past few years, diverse definitions and architectures of IoT have been proposed by many establishments as listed in the work of Chen et al. [37]. In this study, IoT has been referenced with different terms such as "global

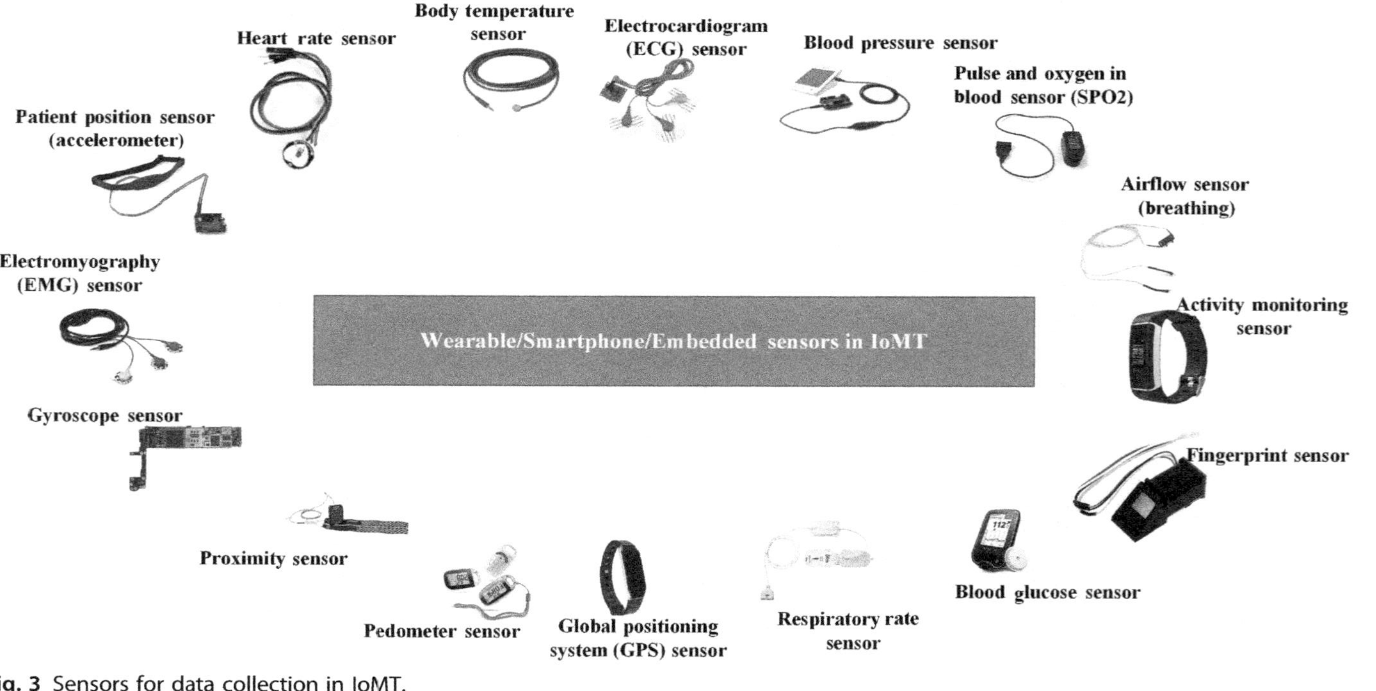

Fig. 3 Sensors for data collection in IoMT.

infrastructure," "global network infrastructure," or "worldwide network." Each of these terms makes provision for physical and virtual objects to connect through the available communication technologies for information sharing. Chen et al. [37] emphasized the need for open and generic architecture for IoT due to the high cost of implementation that may be involved in data integration and private protocols from different organizations [33]. In support of Chen et al. [37] the proposal, Kang and Zhongyi [38] proposed an open and generic architecture for IoT with features like flexibility, scalability, and aligns with the standard protocols. This architecture provides the opportunity of integrating multiple IoT applications. In the same vein, a three-layer generic IoT architecture has been proposed by China communication standard association consisting of sensing, network, and application layers [33]. A four-layer IoT sensing framework has been proposed in the work of Wan et al. [20] comprises sensing and information gathering, information handling and application, information delivery [33]. Wan et al. [20] are also concerned with the privacy and security of the proposed framework. Domingo [39] proposed three-tier IoT architecture for disabled people, which includes layers like perception, network, and application. This architecture has been implemented for several application areas like school, shopping as well as smart homes. Lambda architecture that is based on big data technology using MapReduce framework has been proposed in the work of Han et al. [40]. This architecture emphasizes the significance of the real-time data analytics from big data generated by IoT sensing devices. Lambda architecture has a speed layer that can analyze data generated in real time, addressing the real-world problem of general big data analytics. This architecture merges the results analyzed in a batch layer to produce the final analytical results. Specifically, this architecture was introduced to the developed predictive system for CVD [40].

Due to the ever-increasing demands of implementing IoT in the healthcare sector, several frameworks have been proposed to address real-time prediction, diagnosis, and monitoring of the patients. Santos et al. [12] introduced a novel IoT-based system for mobile health (mHealth) and developed a prototype for remote monitoring [33]. RFID technology, which provides sensing has been integrated within IoT architecture to monitor user health and trigger remote assistance when required. The authors recorded different users' activities related to patients health with the use of different body-centric RFID tags such as the wearable and implantable tags [33]. They discussed the possibility of data gathering from these RFID tags, however, this study failed to present compressive results that demonstrate the various scenarios discussed. Santos et al. [13] proposed the use of RFID in the mHealth system that leveraged the strengths of IoT-enabled connected devices. This study focused on the need to improve health services using IoT and RFID technologies. Yang et al. [14] develop a prototype of an intelligent home health-care management system for disabled people. They implemented the prototype of the proposed system based on WBSN using wireless sensors like pressure sensor, heart rate sensor, and home environment sensing devices. The smartphone was used to implement the prototype for sending and receiving information. Several challenges hampered the feasibility of their proposed system in real-life scenarios. These include the limitation in the transmission speed of the bluetooth device considered, which slow down the operation of the proposed system as well as the inability to address security challenges related to data management. A three-tier framework was proposed based on WSN in health-care systems [15]. The authors discussed several IEEE data communication standards that can provide the capability for information sharing in the IoT environment. These standards include IEEE 802.11-2012 for WLAN, IEEE 802.11-2011 for cognitive radio, IEEE 802.20 for a high-velocity mobile broadband system, IEEE 802.16-2009 for WiMAX, and IEEE 802.15.12005 for Bluetooth [33, 41]. Among these standards, the authors suggested the use of IEEE 802.16 for health-care systems due to its collision-free characteristic as well as high Quality of Service (QoS) in real time. Similarly, security issues relating to data protection within the proposed framework have not been discussed in this study. Mitra et al. [16] proposed a KNOWME WBSN platform for end-to-end communications. The authors employed wearable sensors and smartphone to collect and transfer the captured information. This platform comprises four layers targeted at reducing communication complexity.

3.4 Cloud-based health-care architectures

WSN has the capability to connect different sensing nodes to realize the vision of IoT. This technology has given rise to several health-care-related applications and simplifies data collection and analysis. Advances in the development of sensing devices aided the significant growth of mobile health-care applications with provision to diagnose different diseases. Mobile health care, popularly called mHealth, has only become possible owing to the rapid advancement in sensors technology. According to the International Telecommunication Union (ITU), the number of mobile phone subscribers worldwide has grown to about 5.9 billion [42]. This development offered a great opportunity for mHealth applications as many organizations and researchers now leverage this market to provide quality health-care systems through the deployment of new mHealth applications. A number of mHealth applications are now available in the market for monitoring blood pressure [43], heart rate [44], and physical activity [45] to mention but a few. Smartphones, tablets, and iPads now provide the capability to aid personalized health care where patient health condition can be monitored in real time. Some of these include WIHMD [43], BlueBox [44] and HeartToGo [46]. One of the notable challenges with these applications is the security and privacy of the user's data. In addition, the availability of these applications relied so much on a smartphone, however, one of the limitations of the smartphone is the battery life. Smartphone applications sense incoming data from WSN nodes for activities such as heart rate, ECG pulse, blood sugar levels, and blood pressure, which are analyzed to alert patients in case of an emergency [33].

As an extension to the traditional IoT architecture and health-care frameworks, IoT based frameworks integrated into the cloud provides several benefits over the traditional existing health-care frameworks. Among these benefits include the easy deployment of health-care applications, enhancement of information security, accurate and quick delivery of data, storage availability, energy saving and cost reduction. The authors in Ref. [33] have introduced a cloud-based framework targeted towards improved data security and privacy-preserving. A cloud-centric IoT-based health-care framework has been implemented in the work of Tyagi et al. [17] with various health-care entities such as patients, doctors, and hospitals.

Hossain and Muhammad [18] focused on the development of real-time health-care monitoring infrastructure for analyzing a patient's health condition. The authors proposed HealthIIoT based on smartphone technology that is capable of monitoring ECG and other health-care-related data. This framework is centered around cloud architecture. Addressing the IoT framework with cloud security to address the user's privacy, the watermarking technique has been proposed for the security of data. Hassanalieragh et al. [35] conducted a comprehensive review and introduced guidelines for remote health-care monitoring. The authors proposed an IoT infrastructure with cloud capability. The use of wearable sensors has been explored in their study and the authors concluded that cloud-based data analytics is more efficient compared with cloudlet. Seales et al. [19] proposed IoT framework that is based on content-centric cloud architecture with many benefits. The proposed PHINet cloud architecture record user's data from user body sensors in real time. These data are used to update the database automatically. The authors emphasized that the use of clouds makes data easy to handle for analysis for users personalized health care. Wan et al. [20] proposed platform production services that are based on IoT and cloud architectures. Hassan et al. [21] introduced an intelligent hybrid context-aware model to monitor patients at home using the hybrid architecture comprising local and cloud components. Provision is made in the cloud-based portion of the architecture to store and analyze the big data generated by the ambient-assisted living systems used to monitor patients suffering from chronic diseases from home. The local portion of the model monitors patients when there is Internet failure or any other failure in the cloud system. The proposed model utilizes context-aware approaches by monitoring physiological signals, patient activities, and ambient conditions concurrently to obtain the real-time health status of the patient.

Similarly, Balamurugan et al. [47] presented an enhanced architectural model for context-aware monitoring called BDCaM. This framework uses cloud computing platforms to process every generated context of ambient-assisted living (AAL) systems. Distributed servers in the cloud store and process the context generated to extract required information for decision-making. This cloud-based architecture is centered on a two-step learning methodology where the first phase identifies the correlations between context attributes and the threshold values of vital signs. MapReduce FP-Growth algorithm was used over a long-term context data of a particular patient to generates a set of association rules that are specific to that patient. This ensures the development of personalized health-care system targets toward modern technology. The second phase of the system employed supervised learning over a new large set of context data produced through the association rules generated in the first phase. This methodology enables the system to accurately predict any patient situation. Aliçkovic and Subasi [24] leverage the availability of cloud platform services such as SaaS to develop health-care system capable of predicting diseases. The main objective of their system is to design a mobile health-care system integrated with data mining techniques to provide relatively high-precise analytical results and advanced SaaS cloud computing platform offering low cost and high-quality services. The system is particularly designed to use existing mobile devices such as smartphones, tablets, smartwatches, wearable sensors for monitoring, and delivering health-care services for both patients and health-care givers. However, their proposed framework is yet to be implemented due to the time limit in gathering IoT sensor data for analysis.

4. The rise of cardiovascular diseases

CVD is a class of diseases that involves the heart or blood vessels [48]. CVD includes coronary artery diseases (CAD) such as angina and myocardial infarction (commonly known as a heart attack), heart failure, hypertensive heart disease, rheumatic heart disease, cardiomyopathy, heart arrhythmia, congenital heart disease, valvular heart disease, carditis, aortic aneurysms, peripheral artery disease, thromboembolic disease, and venous thrombosis [48] among others. CVD currently account for nearly half of the non-communicable diseases (NCD) and have overtaken communicable diseases as the world's major disease burden, with CVD remaining the leading global cause of death, accounting for 17.3 million deaths per year. This number is expected to grow to about 23.6 million by 2030 [49]. According to WHO [50], 17.7 million people died from CVD in 2015, representing 31% of all global deaths. Of these deaths, an estimated 7.4 million were due to CAD and 6.7 million were due to stroke. Increasingly, the population affected are those in low- and middle-income countries where 80% of these deaths occur at younger ages than in higher-income countries, and where the human and financial resources to address them are mostly limited [49]. Out of the 17 million premature deaths, under the age of 70, due to NCD in 2015, 82% are in low- and middle-income countries with CVD accounting for 37%. However, due to the rise in the number of diseases, IoMT research has witnessed tremendous growth. In fact, several organizations, like CISCO and General Electric have predicted that IoT-based health-care market in future will become a $117 billion market by the year 2020 [51]. Therefore, the domain of IoMT research is broad and several studies have been conducted in the literature to predict and diagnose CVD.

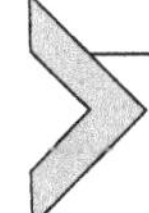

5. Taxonomy of CI techniques for CVD prediction in IoMT systems

CI deals with the design and development of biologically and linguistically motivated computational models. CI covers methods that are primarily based on ANN, Fuzzy logic, evolutionary algorithms, SVM, artificial immune systems (AIS), swarm intelligence, rough sets, and chaotic systems as well as hybrid methods that combine two or more techniques. In addition, CI has also been extended to other computational methods including ambient intelligence, cultural learning, artificial endocrine networks, and social reasoning. CI involves computational models and tools that deal with

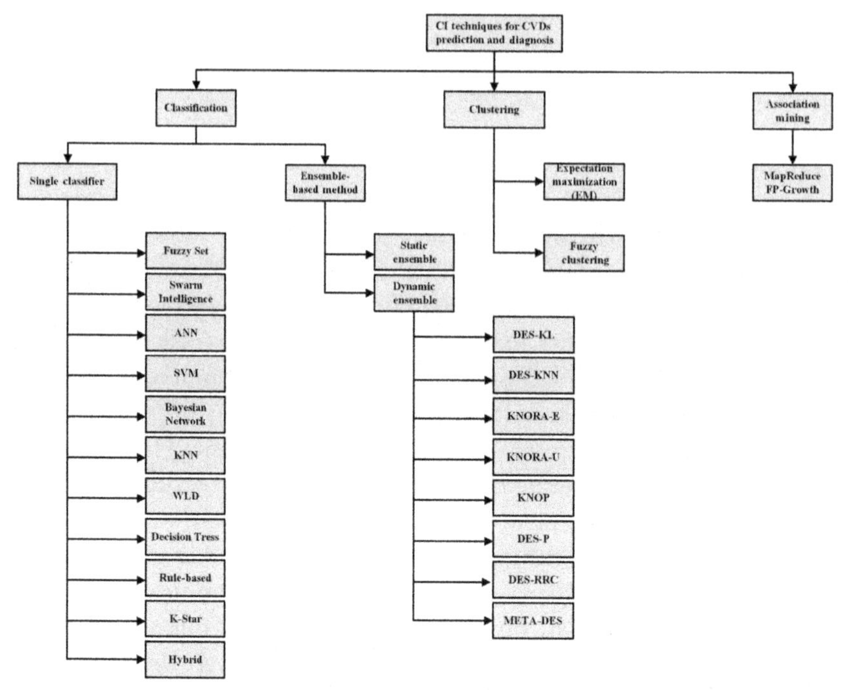

Fig. 4 Taxonomy of CI techniques for CVD prediction and diagnosis.

learning, adaptation, and heuristic optimization to provide solutions to problems that are difficult to solve using traditional computational algorithms. CI techniques have played a major role in developing intelligent health-care systems such as systems for CVD detection and diagnosis for personalized health monitoring. Fig. 4 presents the taxonomy of CI techniques used for CVD detection and diagnosis.

5.1 Classification techniques

The most commonly used CI techniques for CVD prediction and diagnosis are those techniques that belong to classification data mining. In this technique, the target class is known apriori from the data samples collected using IoMT devices. The goal is to predict the health status (i.e., the target class) of patients taken into consideration important features that describe the type of CVD under analysis. In some cases, patient's health condition is diagnosed to provide alert for personalized health-care services. Before proceeding to the CVD prediction and diagnosis stage, the data samples are preprocessed and necessary features extracted for analysis. Several methods have been used during the preprocessing and feature extraction stages. The most important features of QRS wave which provide useful information about patient's heart condition are P wave, QRS complex, T wave, and QRS intervals. Other features such as morphological, statistical, and transform features have also been employed for CVD prediction and diagnosis [52].

For instance, Discrete Cosine Transform (DCT), Discrete Wavelet Transform (DWT), and Empirical Mode Decomposition (EMD) techniques have been widely used for signal decomposition of the collected data into different frequency bands [24, 53]. These techniques can be combined with dimensionality reduction techniques like principal component analysis (PCA), multiscale principal component analysis (MSPCA), and independent component analysis (ICA) among others [24, 54]. Filter techniques are applied to remove noise and to prepare the samples for further processing.

To select the most discriminative features for CVD prediction and diagnosis, a number of feature selection approaches, which include analysis of variance (ANOVA) [53], information gain [55] as well as swarm intelligence optimization algorithm like Whale Optimization Algorithm (WOA) [56] and Grey Wolf Optimization (GWO) [57] have been used. WOA has been combined with Naïve Bayes classification algorithm to produce a robust classification model for CVD prediction and diagnosis.

To predict heart rate variability (HRV) [58] proposed a novel approach to mortality prediction of intensive care unit (ICU) cardiovascular patient using Fuzzy logic technique. The authors observed that in order to effectively predict the patient's future health status, the related physiological effects should be accurately identified. Biological signal processing technique was employed to extract useful information from ECG signals. A two-dimensional return mapping of the HRV signals was constructed and new features obtained from the mapping. The implementation of these methods leads to more favorable treatment of patients and the allocation of appropriate equipment to them as required. Nikolaiev and Timoshenko [59] presented a general architecture of an automatic system for heart pathologies detection for CVD. This system is a set of interconnected software and hardware components that aim to monitor human biological signals, detect abnormalities and pathologies, and predict the appearance of some diseases. It has been pointed out that with the convergence of modern mobile apps, machine learning techniques, and new intelligent sensors, it is possible to greatly improve patient health condition.

ANN based on deep learning framework has been presented in the literature for prediction of cardiovascular risk factors from retinal fundus photographs. The authors employed deep learning architecture to extract new knowledge from retinal fundus images [60]. The images were preprocessed and scale-normalized by detecting the circular mask of the fundus image and resizing the diameter of the fundus to be 587 pixels wide. Images for which the circular mask could not be detected or those that were of poor quality were excluded from the training stage. The optimization algorithm used to train the network weights was a distributed stochastic gradient descent implementation. To speedup the training, batch normalization, as well as pre-initialization using weights from the same network trained to classify objects in the ImageNet dataset was used. The study provided evidence that deep learning may uncover additional signals in retinal images that will allow for better cardiovascular risk stratification. In particular, they could enable cardiovascular assessment at the population level by leveraging the existing infrastructure used to screen for diabetic eye disease. Poplin et al. [61] presented deep learning architecture that is based on restricted Boltzmann machine (RBM) and deep belief networks (DBN) for ECG classification. The study specifically targeted the detection of ventricular and supraventricular heartbeats from ECG data. The results obtained proved that with a suitable choice of parameters, RBM and DBN can achieve high average recognition accuracies of ventricular ectopic beats of 93.63% and 95.57% for supraventricular ectopic beats at a low sampling rate of 114 Hz. The results further showed that deep learning framework can achieve state-of-the-art performance at lower sampling rates and simple features when compared to traditional methods. An integrated decision support system based on ANN and Fuzzy Analytic Hierarchy Process, Fuzzy AHP, for heart failure risk prediction have been studied [62]. In this research, the commonly used heart failure attributes were considered and their contributions were determined by an experienced cardiac clinician. Fuzzy AHP technique was used to compute the global weights for the attributes based on their individual contribution. Then the global weights that represent the contributions of the attributes were applied to train an ANN classifier for the prediction of heart failure risks in patients. The improvement of the heart failure risk prediction in the study was attributed to the kind of heart failure attributes employed as well as the method used for prediction [62].

Cardiac arrhythmia is a type of CVD that seriously affects millions of people around the world and accounts for approximately 80% of the sudden cardiac death [53]. There is life-threatening and non-life-threatening cardiac arrhythmia. The life-threatening arrhythmia threatening the patient's life and immediate treatment is required for prevention. Non-life-threatening arrhythmia imposes a long-term health risk to patients and at the same time, requires special care to avoid further deterioration of heart function. To predict cardiac arrhythmia, Desai et al. [53] proposed a KNN algorithm fed with attributes from six different approaches. In this study, the performance of KNN algorithm was evaluated using features from DCT combined with PCA, DCT combined with ICA, DWT combined with PCA, DWT combined with ICA, EMD combined with PCA, and finally EMD combined with ICA. Each of these phases was subjected to ANOVA feature selection technique to select discriminative features for predicting cardiac arrhythmia. Experimental results show that KNN classifier achieved an accuracy of 99.77% based on DWT combined with ICA method. In predicting cardiac arrhythmia, machine learning algorithms such as SVM, weighted linear discriminant (WLD), decision tree, optimum-path forest (OPF) and JRip algorithms have also been employed [56, 63]. Nair et al. [64] demonstrated the use of Spark big data framework to predict the health status of a patient in real time. Open-source big data engine built around Apache Spark and decision tree algorithm in Spark machine learning library has been utilized to provide scalable analytics platform in the cloud. The authors further demonstrated the applicability of Spark-based machine learning model on streaming big data with a real-time health status prediction use case. Alickovic and Subasi [24] proposed the use of Random Forest, C4.5 and Classification and Regression Trees (CART) algorithms to predict cardiac arrhythmia. The study observed that Random Forest classifier achieved better accuracy of over 99% for ECG signal classification.

Hybrid intelligent systems for CVDs prediction and diagnosis have been studied extensively. For instance, Nilashi et al. [54] combined different methods to produce a knowledge base system for disease prediction. This approach employed noise removal and prediction techniques using PCA, CART, and Fuzzy rule-based techniques. Hybrid real-time remote monitoring (HRRM) framework has been studied in Ref. [56]. The authors proposed IoT cloud architecture that leveraged cloud computing capability and local module to enhance prediction capability. In their study, context-aware data of patients collected through IoMT devices were used to remotely monitor patient health status. The proposed HRRM framework innovates a PLM that do a convergence between IoMT sensors and the clouds. The purpose of PLM is to handle the scenario when there is Internet disconnection to access the cloud portions of the architecture. Furthermore, the authors proposed a cloud classification method for dealing with the imbalanced dataset by minimizing errors especially in the minority class that signifies serious situations. To address class imbalance problem, four techniques were investigated, which are Class Balancer (CB), Random under Sampling (RUS), Synthetic Minority over Sampling (SMOTE), and Random over Sampling (ROS). A hybrid algorithm of Naïve Bayes (NB) and WOA has been introduced for feature subset selection that achieves the highest accuracy. The NB-WOA functions as a safe-failure unit that chooses when to stop the monitoring using HRMM in the case of the failure of dominant sensors.

Aside from the use of single classification algorithms, ensemble-based methods have been applied to predict CVD. For static ensemble, Logitboost, Random Forest, Gradient Boosting Machines have been studied [55, 65]. Motwani et al. [55] proposed information gain for feature selection and boosted ensemble algorithm, Logitboost, for classification to predict coronary heart disease (CHD). They investigated the feasibility and accuracy of machine learning to predict 5-year all-cause mortality (ACM) in patients undergoing coronary computed tomographic angiography (CCTA) and compared the performance to existing clinical or CCTA metrics. Totally 25 clinical and 44 CCTA parameters were evaluated, including segment stenosis score (SSS), segment involvement score (SIS), modified Duke index (DI), number of segments with noncalcified, mixed, or calcified plaques, age, sex, gender, standard cardiovascular risk factors, and Framingham risk score (FRS). The authors observed that the use of ensemble method improved the classification performance. He et al. [63] proposed dynamic ensemble selection (DES) for cardiac arrhythmia detection from IoT-based ECGs. The proposed DES provides accurate detection of cardiac arrhythmia. This model uses a result regulator that employs different features to improve the classification outcome from the DES method. The DES framework comprises five stages: preprocessing, feature extraction, classifier pool training, dynamic selection classification, and result refinement. For noise removal, the authors employed a median filter and DWT from the baseline ECG signals. At the classifier pool training phase, the SMOTE-ENN technique was employed for class imbalance in the training data. To increase the diversity of the classifier pool, six classifiers have been studied, which include multilayers perceptron (MLP), SVM, Linear SVM, Bayesian model with Gaussian kernel, decision tree, and KNN. These classifiers were trained using different training subsets. At the dynamic selection classification stage, eight DES techniques have been used, which are DES-KNN, DES-KL, DES-P, DES-RRC, KNORA-E, KNORA-U, KNOP, META-DES. Finally, the results were refined using a result regulator.

5.2 Clustering techniques

As opposed to classification techniques, clustering or unsupervised techniques attempt to group data into different classes based on their similarity. In this approach, the target class is not provided and the goal of the clustering model is to identify the categories of the records in the dataset. This technique attempts to learn from the data by observing the similarities among instances in the dataset. Clustering approach is very useful in pattern analysis and for grouping medical records into different categories based on the patients' health characteristics. In the domain of health care, Nilashi et al. [54] proposed a hybrid intelligent model that integrates the clustering technique with a classification model to produce a knowledge-based system for disease diagnosis. In their study, the expectation-maximization (EM) clustering algorithm, which is based on a Gaussian mixture model, has been used with PCA. This approach helps the authors to cluster medical data for the experiment into similar groups. The study relied on Fuzzy rule-based method to learn the prediction models from the clusters and PCA has been used for dimensionality reduction to address multicollinearity in the dataset.

5.3 Association mining

Association rule mining is a popular and well-researched method for discovering interesting relations between variables in large databases. It is intended to identify strong rules discovered in databases using different measures of interestingness. Based on the concept of strong rules. Association rule learning, also known as dependency modeling, searches for relationships between features in the dataset. This approach has been studied in the domain like stock analysis to uncover interesting patterns in market basket data. Balamurugan et al. [47] proposed enhancement of architectural model for context-aware monitoring that uses cloud computing platforms. In this study, the generated contexts of AAL systems were sent to the cloud where a number of distributed servers store and process those contexts to extract required information for decision-making. Particularly, the authors proposed two-stage learning where the first stage identifies the correlations between context attributes and the threshold values of vital signs. With the use of MapReduce FP-Growth algorithm over a long-term context data of a particular patient, the system is able to generate a set of association rules that are specific to that patient. The second stage of the system employs a supervised learning technique over large context-aware data using the rules discovered in the first stage. On average, this system produced an accuracy of 76% using the FP-growth association rule learner.

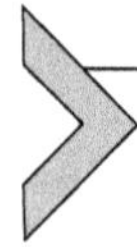

6. The rationale for choosing CI techniques for CVD prediction

There are great continuous efforts toward the improvement of human health services, where IoT applications, as situated within CI techniques, have achieved tremendous success. However, the understanding and application of these underlying mechanisms, IoT application, in relation to health-care practices still need to be further explored. Therefore, this section briefly highlights the rationales behind the adoption of CI techniques for CVD prediction and diagnosis. Behavioral analysis science, aimed at preventing people's health challenges, such as mild cognitive impairment (MCI) and frailty is an important endeavor whose great success has been attributed to CI paradigm [66]. Hence, the use of innovative technologies that are CI inclined enables discreet capturing of personal data for automatic recognition of behavioral changes for effective and efficient health-care diagnosis and monitoring. This enhances behavioral health-care processes and improves the personalized health care of patients especially for CVD prediction and diagnosis. A number of CI techniques have been investigated as discussed in the previous section. There have been success stories from the application CI techniques toward improving models' classification accuracy and subsequently enhancing health-care systems performance in CVD prediction and diagnosis. However, more still need to be done to assess CVD prediction and diagnosis systems based on several evaluation metrics as the use of a single evaluation metric such as model classification accuracy cannot guarantee optimal performance of the systems.

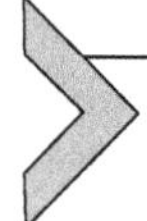

7. Experimental results and evaluation of CVD prediction systems

Several evaluation metrics such as classification accuracy, sensitivity, specificity, and receiver operating characteristic (ROC) have been employed by researchers to evaluate predictive systems for CVD. For instance, in a study conducted by Kim et al. [67], the proposed statistical DBN-based predictive model for CVD risk recorded an accuracy and ROC curve of 83.9% and 79%, respectively. Thus, the proposed DBN performed better than other prediction algorithms such as NB, logistics regression, ANN, SVM, and Random forest. Poplin et al. [61] presented deep learning architecture that is based on RBM and DBN for ECG classification. In their study, DBN was developed to classify ventricular and supraventricular heartbeats from ECG data. The results obtained proved that using a suitable choice of parameters, RBM and DBN demonstrated high average recognition accuracy of ventricular ectopic beats of 93.63% and 95.57% for supraventricular ectopic beats at a low sampling rate of 114 Hz. The study conducted by Wiharto et al. [57] also reported that their proposed system produced an average performance with a sensitivity of 80.1% and specificity of 95% using a k-star algorithm to diagnose the level of CHD. Kim, Lee, and Lee [68] have used decision tree technique and Fuzzy logic to develop predictive models for predicting CHD risk with an accuracy and ROC values of 69.51% and 59.4% respectively. Arabasadi, Alizadehsani, Roshanzamir, Moosaei, and Yarifard [69] applied hybrid neural network-Genetic algorithm on heart disease detection using Z-Alizadeh Sani dataset and achieved accuracy, sensitivity, and specificity of 93.85%, 97%, and 92%, respectively. Samuel et al. [62] have developed a hybrid decision support system based on ANN and Fuzzy AHP for heart failure risk prediction. The accuracy obtained by this model was 91.10%. Table 2 shows a summary of some studies based on the CI techniques employed and their performances.

8. Cloud-based IoMT framework for CVD prediction

CVD prediction and diagnosis is a challenging task and several frameworks have been studied as discussed in the previous sections. One of the major challenges facing the realization of IoMT toward developing intelligent personalized health-care systems is data collection and aggregation from multiple IoMT devices. In most cases, data are often not available in real time and there is the difficulty of comparing or linking a variety of data since IoT devices generate complex and heterogeneous data on clinical examinations, monitoring, and health care. Data aggregation from heterogeneous IoT data sources is really a major issue that needs to be critically addressed. Hence, it would be motivating to investigate for a set of biomarkers, which of the IoT sensors improve the performance of the intelligent systems. It would also be interesting to investigate if there exist any other context data that can boost the model's performance. Furthermore, more studies need to be carried out to ascertain the reliability of attributes selected from each biomarker. Another interesting issue in IoMT realization for CVD prediction

Table 2 Performance of systems for CVDs prediction and diagnosis.

Author	CI Technique	Type of Disease	Accuracy	ROC	Sensitivity	Specificity
Kim et al. [68]	Decision tree and Fuzzy logic	CHD	69.51%	0.594		
Arabasadi et al. [69]	Hybrid Neural Network-Genetic Algorithm	CAD	93.85%		97%	92%
Samuel et al. [62]	Hybrid ANN and Fuzzy AHP	Heart failure prediction	91.10%			
Abdar et al. [30]	C5.0, ANN, SVM, KNN, and Logistic regression	Heart disease	C5.0 (93.02%), KNN (88.37%), SVM (86.05%), ANN (80.23%)			
Kim et al. [67]	Statistical feature selection using the nonparametric Mann-Whitney U-test, chi-square, and DBN-based predictive model	CVD	83.9%	0.790		
Ballinger et al. [70]	Semi-supervised, multi-task LSTM	CVD	Accuracy for diabetes (0.8451), high cholesterol (0.7441), high blood pressure (0.8086), and sleep apnea (0.8298)			
Wiharto et al. [57]	k-Star algorithm	CHD	87.5%		80.1%	95%
Al-Tashi, Rais, and Jadid [71]	Grey Wolf Optimization (GWO) and SVM	CAD	89.83%		93%	91%
Poplin et al. [61]	RBM with DBN	CVD	Accuracies of ventricular ectopic beats (93.63%) and supraventricular ectopic beats (95.57%)			

and diagnosis is the need to develop a framework that can switch effectively with minimal processing time between cloud and local classification models to provide real-time and up-to-date prediction and diagnosis of CVD.

To address the aforementioned research issues, this chapter proposes a cloud-based framework shown in Fig. 5 for CVD prediction and diagnosis. The proposed framework comprises five modules: data collection/network module, cloud-based data repository, and learning module, model update module, a connection detection module, and prediction/diagnosis and alert management module. The first module functions as WBAN consisting of sensors, communication and networking as the data acquisition module. The data from this module are transferred to the cloud data repository for modeling. At the cloud layer, a Cloud-based Personalized Health-care

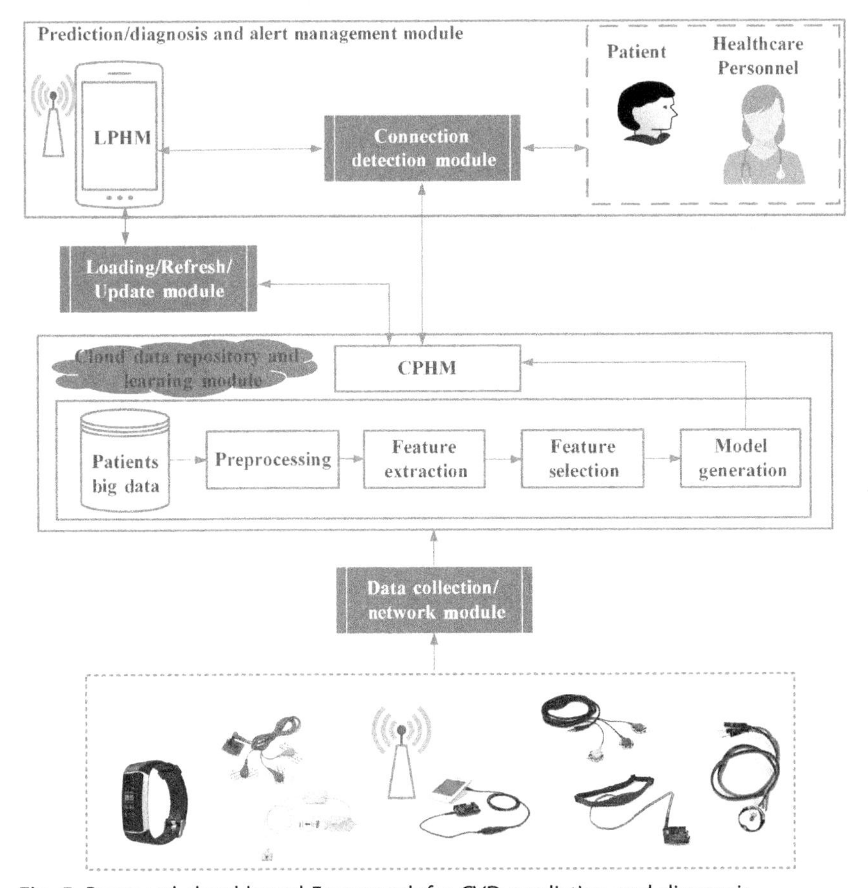

Fig. 5 Proposed cloud-based Framework for CVD prediction and diagnosis.

Model (CPHM), which has been effectively trained and evaluated can be used to predict and diagnose CVD when an Internet connection is available. This framework also provides an update module that automatically updates the Local Personalized Health-care Model (LPHM) on the user's smartphone. The LPHM model functions as the local intelligent model in case of Internet connection failure to the CPHM. To achieve this switching, the framework provides a connection detection module, which automatically detects if the user's smartphone is connected to the Internet or not. This connection ensures scalable and efficient cloud-based architecture that is simple, robust, and accurate for CVD prediction and diagnosis.

9. Practical case of CVD prediction

This section presents the experimental results of the preliminary investigation of the proposed framework for CVD prediction and diagnosis. In this section, five experiments were conducted based on SVM with linear kernel, SVM with the polynomial kernel, SVM with radial basis kernel, Naïve Bayes, and KNN algorithms. The experiments were conducted using the R data analytics tool. The system has a random-access memory of 32 GB and 2.90 GHz Intel Core i9 with both CPU and GPU support. The operating system is Windows 10 Pro 64-bit.

9.1 Dataset description

The CVD dataset used for the experimental analyses was obtained from Kaggle at https://www.kaggle.com/sulianova/cardiovascular-disease-dataset. This dataset contains 70,000 records of patients with 12 attributes including the target class. The target class indicates the presence or absence of CVD. Table 3 summarizes this dataset and Table 4 presents the description of the attributes available in the dataset.

Table 3 Dataset description.

Item	Description
Dataset	Cardiovascular disease dataset
No of attributes	12 plus target
Class/target attribute	0—Absence of CVD, 1—Presence of CVD
No of records with 0	35,021
No of records with 1	34,979
Total records	70,000

Table 4 Description of dataset attributes

Attribute	Nature	Code	Attribute type
Age	Objective	age	int
Height	Objective	height	int
Weight	Objective	weight	float
Gender	Objective	gender	categorical
Systolic blood pressure	Examination	ap_hi	int
Diastolic blood pressure	Examination	ap_lo	int
Cholesterol	Examination	cholesterol	categorical
Glucose	Examination	gluc	categorical
Smoking	Subjective	smoke	binary
Alcohol intake	Subjective	alco	binary
Physical activity	Subjective	active	binary
Presence or absence of cardiovascular disease	Target attribute	cardio	binary

The nature of the attributes is categorized into three: objective, subjective, and examination. The objective attributes are factual information about the patient. Subjective attributes are those provided by the patients. Lastly, the examination attributes are from the results of a medical examination of the patients' health condition. This dataset was recently added to the repository on January 20, 2019 for CVD prediction and diagnosis. This chapter reports the first state-of-the-art experiments based on this newly added CVD dataset particularly with the use of SVM, Naïve Bayes, and KNN algorithms for analyses. The parameters configuration of the algorithm investigated is shown in Table 5.

9.2 Support vector machine

SVM is a statistical supervised learning classification algorithm for data analysis and pattern recognition based on labeled samples. SVM was developed by Vapnik and coworkers [72]. The algorithm can serve the purpose of both

Table 5 Parameters configuration of selected algorithms.

Algorithm	Parameter configuration
SVM + Linear Kernel	$C=1$
SVM + Polynomial Kernel	degree = 3, scale = 0.1, $C=1$
SVM + RBF Kernel	sigma = 0.1286661, $C=1$
NB	fL = 0, adjust = 1, usekernel = TRUE
KNN	$k=9$

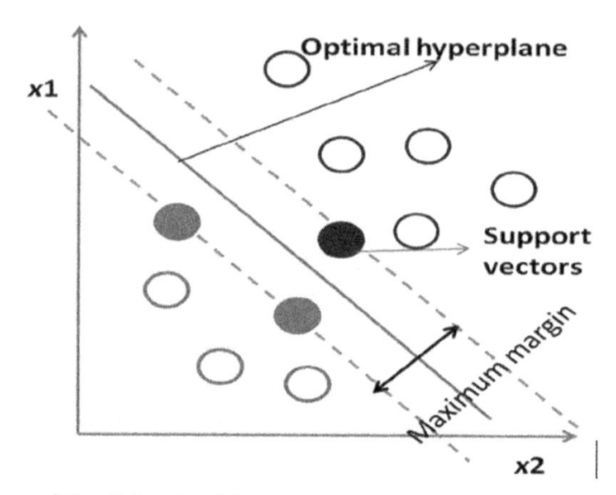

Fig. 6 Operation of the SVM algorithm.

classification and regression tasks. The aim of SVM is to define a hyperplane that separates the boundary between different classes in a dataset. The hyperplane separates the classes by maximizing the margin among the closest points known as support vectors from each class to the hyperplane as shown in Fig. 6. To address the nonlinear separable problem, SVM employs kernel functions to find an optimal separating hyperplane by projecting the training data from low to high dimensional space. SVM uses kernel functions such as linear, radial basis function (RBF), and polynomial kernel.

9.2.1 Experimental results using SVM

The summary of the result of the experiments conducted using linear, polynomial, and RBF kernels is shown in Table 6 based on accuracy, sensitivity and specificity, positive predicted value (PPV), and negative predicted value (NPV). The results that appeared in bold are the promising outcomes as produced by the classifiers. These results revealed that SVM classifier can help in providing real-life prediction of CVD cases based on the evaluation metrics considered. Fig. 7 shows the comparison of the results to visualize the

Table 6 Results of SVM for CVD prediction.

	Threefold cross-validation				
Classifier	**Accuracy**	**Sensitivity**	**Specificity**	**PPV**	**NPV**
SVM + Linear Kernel	0.7259	**0.8163**	0.6355	0.6916	**0.7755**
SVM + Polynomial Kernel	**0.7375**	0.7868	0.6881	**0.7164**	0.7632
SVM + RBF Kernel	0.7291	0.7626	**0.6955**	0.7149	0.7453

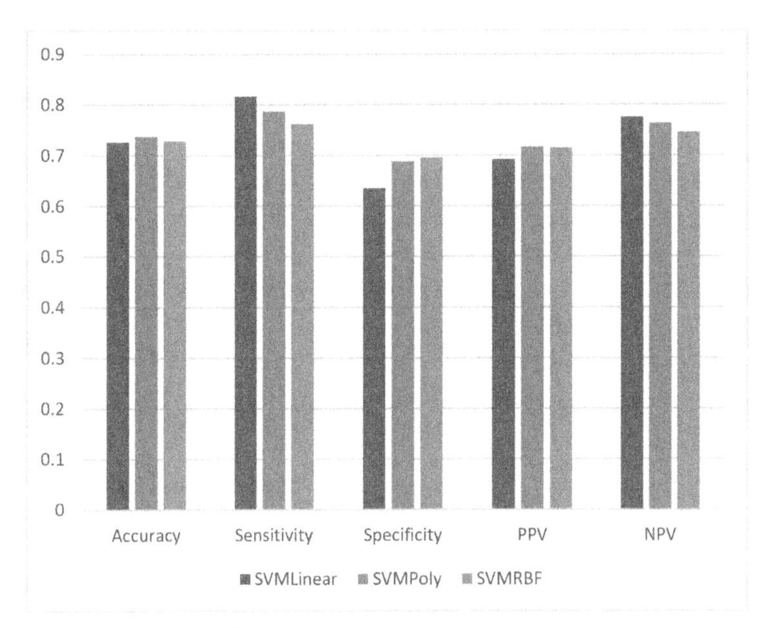

Fig. 7 Results based on SVM.

performance of each of the SVM models. For each experiment, k-fold cross-validation method was used to train the models where the value of k was set to three. In threefold cross-validation, the labeled samples for the training dataset are divided into three subsets of equal size. In each round of the training, one out of three subsets is held as the testing set to validate the classifier, while the remaining two subsets are used to train the classification algorithm. The training and testing datasets were partitioned in the ratio of 70–30 respectively.

The results of the experiments based on the different SVM kernels show that SVM with polynomial kernel performs better than other models in terms of accuracy. This model achieved an average accuracy of 73.75%. The model also demonstrated the best performance based on PPV metric with 71.64%. The second model is the SVM with a linear kernel, which produces the best results based on Sensitivity and NPV. The Sensitivity and NPV of 81.63% and 77.55% were obtained, respectively, for the SVM with a linear kernel.

9.3 Naïve Bayes

Naïve Bayes classifier uses the Bayes' probability theory which assumes that all attributes of a given class in a dataset are independent. This means that

Naïve Bayes work on the assumption that all attributes contribute equally to predicting the target attribute. Naïve Bayes algorithm operates as follows:

Let D represent a training set of tuples with their associated class labels. Assume each tuple is represented by an n-dimensional attribute vector, $X = (x_1, x_2, \ldots, x_n)$, depicting n measurements made on the tuple from n attributes respectively as $A_1, A_2, \ldots, A_n$.

Suppose that there are m classes, $C_1, C_2, \ldots .C_m$. Given a tuple X, the classifier will predict that X belongs to the class having the highest posterior probability, conditioned on X. Formally, Naïve Bayes will predict that tuple X belongs to the class C_i if and only if

$$P(C_i|X) > P(C_j|X) \text{ for } 1 \leq j \leq m, j \neq i \tag{1}$$

The goal is to maximize $P(C_i|X)$.The class C_i for which $P(C_i|X)$ is called the maximum posterior hypothesis. By Bayes' theorem we have

$$P(C_i|X) = \frac{P(X|C_i)P(C_i)}{P(X)} \tag{2}$$

As $P(X)$ is constant for all classes, only $P(X|C_i)P(C_i)$ need to be maximized. If the class prior probabilities are not known, then it is commonly assumed that the classes are equally likely, that is, $P(C_1) = P(C_2) = \ldots = P(C_m)$, and we would, therefore, maximize $P(X|C_i)P(C_i)$. Note that the class prior probabilities may be estimated by $P(C_i) = |C_i, D|/|D|$, where $|C_i, D|$ is the number of training tuples of class C_i in D. In order to predict the class label of X, $P(X|C_i)P(C_i)$ is evaluated for each class C_i. The classifier predicts that the label of tuple X is the class C_i if and only if

$$P(X|C_i)P(C_i) > P(X|C_j)P(C_j) \text{ for } 1 \leq j \leq m, j \neq i \tag{3}$$

In other words, the predicted class label is the class C_i for which $P(X|C_i)$ $P(C_i)$ is maximum.

9.3.1 Experimental results using Naïve Bayes

Fig. 8 presents the classification results using the Naïve Bayes algorithm for CVD prediction. The results show that the Naïve Bayes algorithm achieved accuracy, sensitivity, specificity, PPV, and NPV of 65.28%, 91.07%, 39.45%, 60.10%, and 81.53%, respectively. The accuracy of the Naïve Bayes is lower when compared with SVM models, However, the algorithm produces promising results based on sensitivity metric.

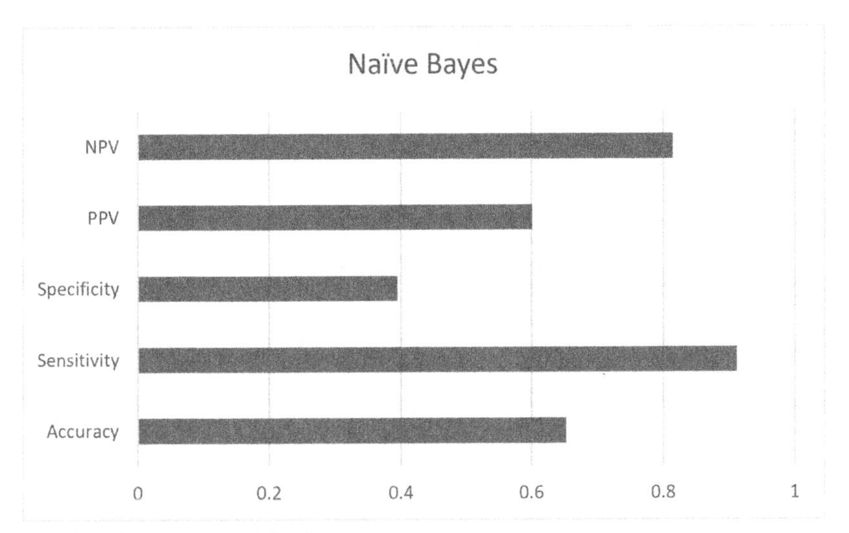

Fig. 8 Results based on Naïve Bayes.

9.4 K-Nearest neighbor

K-Nearest neighbor (KNN) is an instance-based learner, which performs a classification task based on a similarity measure. KNN labeled new instance based on its closest (K) neighboring instances. To get the closest neighboring instance, distance metrics such as Euclidean distance is employed. KNN has three requirements, namely: a set of stored records (training dataset), a distance metric to compute the similarity between records and then the value of parameter K. This section reports the performance of KNN for CVD prediction based on the performance metrics employed for analysis.

9.4.1 Experimental results using KNN

Fig. 9 shows the performance of KNN algorithms for CVD prediction. According to the figure, KNN produces accuracy, sensitivity, specificity, PPV, NPV of 65.41%, 68.11%, 62.70%, 64.64%, and 66.26%, respectively. The results obtained by KNN is low when compared with SVM models. The subsequent section presents a comparison of the different models developed in this chapter.

9.5 Comparison of the models

This section presents the results of the different models developed in this chapter for CVD prediction. As stated earlier, the experiments conducted

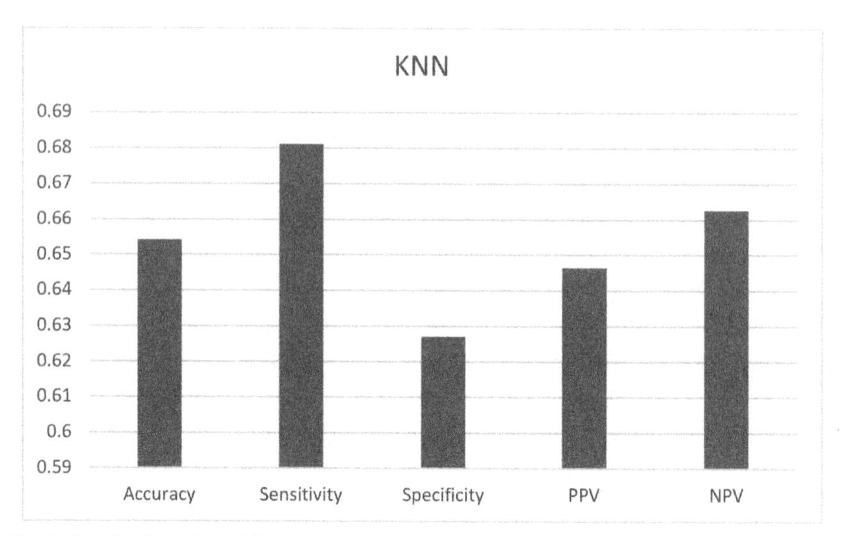

Fig. 9 Results based on KNN.

in this study serve as the preliminary investigation of the practical case of CVD prediction and diagnosis using a newly introduced dataset of patients that are suffering from CVD as well as normal patients. Further experiments need to be conducted to improve the performance of the models. Table 7 highlights the best model based on each performance metrics. Fig. 10 presents the trend in performance of the different models investigated in this study. NB and SVM with polynomial kernel demonstrated promising results when compared with other models investigated based on the values that appeared in bold. This outcome further confirmed the applicability of the selected classifiers for real-life prediction of CVD cases.

Table 7 Models comparison.

	Threefold cross validation				
Classifier	**Accuracy**	**Sensitivity**	**Specificity**	**PPV**	**NPV**
SVM + Linear Kernel	0.7259	0.8163	0.6355	0.6916	0.7755
SVM + Polynomial Kernel	**0.7375**	0.7868	0.6881	**0.7164**	0.7632
SVM + RBF Kernel	0.7291	0.7626	**0.6955**	0.7149	0.7453
Naïve Bayes	0.6528	**0.9107**	0.3945	0.601	**0.8153**
KNN	0.6541	0.6811	0.627	0.6464	0.6626

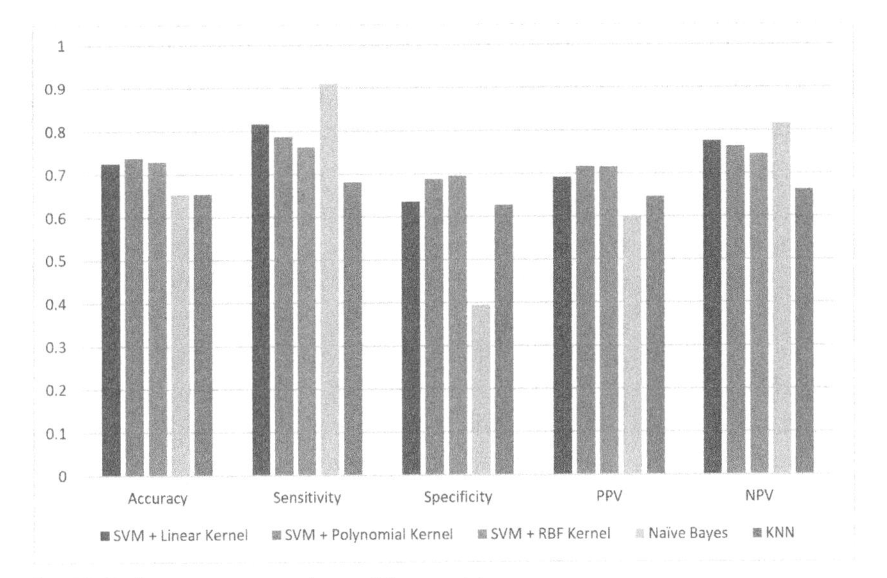

Fig. 10 Performance comparison of the models.

10. Conclusion and future research direction

In this chapter, the fundamental concepts of cloud computing and IoT applications within the context of health-care data management have been elaborately addressed. The use of IoT applications in health-care practices, technically related to the IoMT, is a new and fast-growing trend in the health-related fields. From the wide range of research domains relating to the deployment of IoMT technologies and cloud-based applications in several medical case studies, it is established that IoMT and cloud computing have provided a wide range of new frontiers for the generation, development, analysis, and management of huge volume of data frequency, especially for the online and other health-care environments. In this regard, cloud-based computing technology, as a popular approach in the management of a large volume of health-care data has provided ease of practice from the personalized e-healthcare perspective. A cloud-based framework for CVD prediction and diagnosis has been proposed in this chapter to address some of the research issues highlighted in the chapter. The framework comprising both the cloud architecture and IoMT for personalized e-healthcare has been implemented for prediction and early diagnosis of CVD based on a preliminary investigation using a newly introduced CVD dataset. By the use of cloud-based IoMT framework, necessary patient-related data, including the frequently changing health parameters, are collected from sensors and analyzed. Findings show that the cloud-based IoMT framework developed is contributing to a notable reduction in the spread of chronic diseases like CVD.

It is worth noting for the purpose of future research that the study in the cloud-based IoMT framework for CVD prediction and diagnosis in personalized e-healthcare has recorded progress, but a number of research questions and technical issues need to be addressed. Several challenges with regards to information security, mobility control, and applications in the course of applications design processes are identified in the previous studies. Therefore, the need to focus on such challenges, especially as they relate to IoMT and disease prediction. In this vein, an intelligent security model should be extensively studied to mitigate the identified risks. While various IoMT applications are abundantly available on the Internet, many shortcomings are attributed to them, including lack of security and privacy, reliability, efficiency, and acceptability. Addressing the problem of load balancing and information distribution throughout the cloud servers are of great concern for future research. More efficient security algorithms need to be implemented such as DNA encryption, fully homomorphic encryption and Bcrypt on the cloud just to mention a few.

The previous research that focused on security has introduced RSA at private cloud and AES at public cloud. An existing study claimed that RSA is better in terms of security features. This position could be further strengthened by extending the proposed framework to include more health parameters and activities. This is because data security and privacy are quite important concerns for IoMT-based health-care monitoring system. Mainly because the recorded health-care data in the cloud may possibly suffer various kinds of security threats, such as linking attacks and unauthorized access. In a similar perspective, machine learning methods need to be further integrated to address issues relating to heterogeneous and constantly evolving sensory inputs. There is a need to investigate more CI techniques for health-care systems especially for the detection and diagnosis of CVD. A related issue in this regard that must be addressed is the constraints connected to the data of patients suffering from chronic disease. These data are often inaccessible and, in most cases, not available in real time.

In addition, comparing or linking of varieties of data from multiple sensor devices often pose difficulties because devices generate complex and heterogeneous data on clinical examinations, monitoring, and health care.

Hence, it would be relevant to determine which of the IoMT sensors data are better in performance for a set of biomarkers in order to get a better diagnosis result. More work needs to be done to check the consistency of attributes chosen for each sensor device. A more dynamic health-care monitoring system can be developed by this approach. Many efforts are required to develop reliable and accurate machine learning algorithms by the iteration of a couple of models, unnecessary data elimination, and removal of noise components. Potential bias and overall performance are needed to be assessed in future research as soon as the models are developed.

References

[1] G.A. Roth, C. Johnson, A. Abajobir, F. Abd-Allah, S.F. Abera, G. Abyu, ... K. Alam, Global, regional, and national burden of cardiovascular diseases for 10 causes, 1990 to 2015, J. Am. Coll. Cardiol. 70 (1) (2017) 1–25.
[2] E. Björnson, J. Borén, A. Mardinoglu, Personalized cardiovascular disease prediction and treatment—a review of existing strategies and novel systems medicine tools, Front. Physiol. 7 (2016) 2.
[3] L. Yasnitsky, A. Dumler, A. Poleshchuk, C. Bogdanov, F. Cherepanov, Artificial neural networks for obtaining new medical knowledge: diagnostics and prediction of cardiovascular disease progression, Biol. Med. (Aligarh) 7 (2) (2015).
[4] A. Alamri, Big data with integrated cloud computing for prediction of health conditions, in: Paper presented at the 2019 International Conference on Platform Technology and Service (PlatCon), 2019.
[5] M. Bhatia, S.K. Sood, A comprehensive health assessment framework to facilitate IoT-assisted smart workouts: a predictive healthcare perspective, Comput. Ind. 92 (2017) 50–66.
[6] A. Kalantari, A. Kamsin, S. Shamshirband, A. Gani, H. Alinejad-Rokny, A.T. Chronopoulos, Computational intelligence approaches for classification of medical data: state-of-the-art, future challenges and research directions, Neurocomputing 276 (2018) 2–22.
[7] J.J. Rodrigues, D.B.D.R. Segundo, H.A. Junqueira, M.H. Sabino, R.M. Prince, J. Al-Muhtadi, V.H.C. De Albuquerque, Enabling technologies for the internet of health things, IEEE Access 6 (2018) 13129–13141.
[8] E.R. Dorsey, E.J. Topol, State of telehealth, N. Engl. J. Med. 375 (2) (2016) 154–161.
[9] G.A. Mensah, G.S. Wei, P.D. Sorlie, L.J. Fine, Y. Rosenberg, P.G. Kaufmann, ... M. M. Engelgau, Decline in cardiovascular mortality: possible causes and implications, Circ. Res. 120 (2) (2017) 366–380.
[10] H.M. Krumholz, Big data and new knowledge in medicine: the thinking, training, and tools needed for a learning health system, Health Aff. 33 (7) (2014) 1163–1170.
[11] J. Gubbi, R. Buyya, S. Marusic, M. Palaniswami, Internet of Things (IoT): a vision, architectural elements, and future directions, Futur. Gener. Comput. Syst. 29 (7) (2013) 1645–1660.
[12] J. Santos, J.J. Rodrigues, B.M. Silva, J. Casal, K. Saleem, V. Denisov, An IoT-based mobile gateway for intelligent personal assistants on mobile health environments, J. Netw. Comput. Appl. 71 (2016) 194–204.
[13] A. Santos, J. Macedo, A. Costa, M.J. Nicolau, Internet of things and smart objects for M-health monitoring and control, Procedia Technol. 16 (2014) 1351–1360.

[14] L. Yang, Y. Ge, W. Li, W. Rao, W. Shen, A home mobile healthcare system for wheelchair users, in: Paper Presented at the Proceedings of the 2014 IEEE 18th International Conference on Computer Supported Cooperative Work in Design (CSCWD), 2014.
[15] Y. Zhang, L. Sun, H. Song, X. Cao, Ubiquitous WSN for healthcare: recent advances and future prospects, IEEE Internet Things J. 1 (4) (2014) 311–318.
[16] U. Mitra, B.A. Emken, S. Lee, M. Li, V. Rozgic, G. Thatte, … S. Narayanan, KNOWME: a case study in wireless body area sensor network design, IEEE Commun. Mag. 50 (5) (2012) 116–125.
[17] S. Tyagi, A. Agarwal, P. Maheshwari, A conceptual framework for IoT-based healthcare system using cloud computing, in: Paper Presented at the 2016 6th International Conference-Cloud System and Big Data Engineering (Confluence), 2016.
[18] M.S. Hossain, G. Muhammad, Cloud-assisted industrial internet of things (iiot)–enabled framework for health monitoring, Comput. Netw. 101 (2016) 192–202.
[19] C. Seales, T. Do, E. Belyi, S. Kumar, PHINet: a plug-n-play content-centric testbed framework for health-internet of things, in: Paper presented at the 2015 IEEE International Conference on Mobile Services, 2015.
[20] J. Wan, C. Zou, K. Zhou, R. Lu, D. Li, IoT sensing framework with inter-cloud computing capability in vehicular networking, Electron. Commer. Res. 14 (3) (2014) 389–416.
[21] M.K. Hassan, A.I. El Desouky, S.M. Elghamrawy, A.M. Sarhan, Intelligent hybrid remote patient-monitoring model with cloud-based framework for knowledge discovery, Comput. Electr. Eng. 70 (2018) 1034–1048.
[22] G. Lewis, Basics About Cloud Computing, Software Engineering Institute Carniege Mellon University, Pittsburgh, 2010.
[23] P. Mell, The NIST Definition of Cloud Computing, Retrieved from: https://nvlpubs.nist.gov/nistpubs/Legacy/SP/nistspecialpublication800-145.pdf, 2011.
[24] E. Alickovic, A. Subasi, Medical decision support system for diagnosis of heart arrhythmia using DWT and random forests classifier, J. Med. Syst. 40 (4) (2016) 108.
[25] S.K. Polu, S.K. Polu, Design of an IoT based heart attack detection system, Int. J. 5 (2019) 53–57.
[26] G.J. Joyia, R.M. Liaqat, A. Farooq, S. Rehman, Internet of medical things (IOMT): applications, benefits and future challenges in healthcare domain. J. Commun. (2017)https://doi.org/10.12720/jcm.12.4.240-247.
[27] M. Chen, Y. Hao, K. Hwang, L. Wang, L. Wang, Disease prediction by machine learning over big data from healthcare communities, IEEE Access 5 (2017) 8869–8879.
[28] Y. Xin, L. Kong, Z. Liu, C. Wang, H. Zhu, M. Gao, … X. Xu, Multimodal feature-level fusion for biometrics identification system on IoMT platform, IEEE Access 6 (2018) 21418–21426.
[29] P. Fuster-Parra, P. Tauler, M. Bennasar-Veny, A. Ligęza, A. Lopez-Gonzalez, A. Aguilo, Bayesian network modeling: a case study of an epidemiologic system analysis of cardiovascular risk, Comput. Methods Prog. Biomed. 126 (2016) 128–142.
[30] M. Abdar, S.R.N. Kalhori, T. Sutikno, I.M.I. Subroto, G. Arji, Comparing performance of data mining algorithms in prediction heart diseases, Int. J. Electric. Comput. Eng. 5 (6) (2015) 2088–8708.
[31] S. Bandyopadhyay, J. Wolfson, D.M. Vock, G. Vazquez-Benitez, G. Adomavicius, M. Elidrisi, … P.J. O'Connor, Data mining for censored time-to-event data: a Bayesian network model for predicting cardiovascular risk from electronic health record data, Data Min. Knowl. Disc. 29 (4) (2015) 1033–1069.
[32] A. Behr, Best Uses of Wireless IoT Communication Technology, Retrieved from: https://industrytoday.com/article/best-uses-of-wireless-iot-communication-technology/, 2018.

[33] P.K. Gupta, B.T. Maharaj, R. Malekian, A novel and secure IoT based cloud centric architecture to perform predictive analysis of users activities in sustainable health centres, Multimed. Tools Appl. 76 (18) (2017) 18489–18512.
[34] A.D. Usman, A.R. Wada, S.M. Sani, Compact single feed millimeter wave antenna for 5G wireless mobile application using electromagnetic bandgap, Int. J. Inform. Process. Commun. 6 (2) (2018) 213–219.
[35] M. Hassanalieragh, A. Page, T. Soyata, G. Sharma, M. Aktas, G. Mateos, … S. Andreescu, Health monitoring and management using Internet-of-Things (IoT) sensing with cloud-based processing: opportunities and challenges, in: Paper presented at the 2015 IEEE International Conference on Services Computing, 2015.
[36] S. Majumder, M.J. Deen, Smartphone sensors for health monitoring and diagnosis, Sensors 19 (9) (2019) 2164.
[37] S. Chen, H. Xu, D. Liu, B. Hu, H. Wang, A vision of IoT: applications, challenges, and opportunities with china perspective, IEEE Internet Things J. 1 (4) (2014) 349–359.
[38] Y. Kang, Z. Zhongyi, Summarize on Internet of Things and exploration into technical system framework, in: Paper Presented at the 2012 IEEE Symposium on Robotics and Applications (ISRA), 2012.
[39] M.C. Domingo, An overview of the Internet of Things for people with disabilities, J. Netw. Comput. Appl. 35 (2) (2012) 584–596.
[40] S. Han, K. Kim, E. Cha, K. Kim, H. Shon, System framework for cardiovascular disease prediction based on big data technology, Symmetry 9 (12) (2017) 293.
[41] V.N. Adama, I.S. Shehu, S.A. Adepoju, R.G. Jimoh, Towards designing mobile banking user interfaces for novice users, in: Paper presented at the International Conference of Design, User Experience, and Usability, 2017.
[42] ITU, One Third of the World's Population Is Online, Retrieved from: www.itu.int/ITU-D/ict/facts/2011/material/ICTFactsFigures2011.pdf, 2011.
[43] J.M. Kang, T. Yoo, H.C. Kim, A wrist-worn integrated health monitoring instrument with a tele-reporting device for telemedicine and telecare, IEEE Trans. Instrum. Meas. 55 (5) (2006) 1655–1661.
[44] L. Pollonini, N.O. Rajan, S. Xu, S. Madala, C.C. Dacso, A novel handheld device for use in remote patient monitoring of heart failure patients—design and preliminary validation on healthy subjects, J. Med. Syst. 36 (2) (2012) 653–659.
[45] J.R. Kwapisz, G.M. Weiss, S.A. Moore, Activity recognition using cell phone accelerometers, ACM SigKDD Explor. Newsl. 12 (2) (2011) 74–82.
[46] J.J. Oresko, Z. Jin, J. Cheng, S. Huang, Y. Sun, H. Duschl, A.C. Cheng, A wearable smartphone-based platform for real-time cardiovascular disease detection via electrocardiogram processing, IEEE Trans. Inf. Technol. Biomed. 14 (3) (2010) 734–740.
[47] S. Balamurugan, R. Shermy, V. Prabhakaran, R.G.K. Shanker, Internet of Ambience: An IoT Based Context Aware Monitoring Strategy for Ambient Assisted Living, (2016).
[48] S. Mendis, P. Puska, B. Norrving, World Health Organization, Global Atlas on Cardiovascular Disease Prevention and Control, World Health Organization, Geneva, 2011.
[49] L.J. Laslett, P. Alagona, B.A. Clark, J.P. Drozda, F. Saldivar, S.R. Wilson, … M. Hart, The worldwide environment of cardiovascular disease: prevalence, diagnosis, therapy, and policy issues: a report from the American College of Cardiology, J. Am. Coll. Cardiol. 60 (25 Supplement) (2012) S1–S49.
[50] WHO, Cardiovascular Diseases, Retrieved from: http://www.who.int/news-room/fact-sheets/detail/cardiovascular-diseases-(cvds, 2018.
[51] T.J. McCue, $117 Billion Market For Internet of Things In Healthcare By 2020, Retrieved from: https://www.forbes.com/sites/tjmccue/2015/04/22/117-billion-market-for-internet-of-things-in-healthcare-by-2020/#1c74466a69d9, 2015.

[52] R. Begg, D.T. Lai, M. Palaniswami, Computational Intelligence in Biomedical Engineering, CRC Press, 2007.
[53] U. Desai, R.J. Martis, C. Gurudas Nayak, G. Seshikala, K. Sarika, K.R. Shetty, Decision support system for arrhythmia beats using ECG signals with DCT, DWT and EMD methods: a comparative study, J. Mech. Med. Biol. 16 (01) (2016) 1640012.
[54] M. Nilashi, O. Bin Ibrahim, H. Ahmadi, L. Shahmoradi, An analytical method for diseases prediction using machine learning techniques, Comput. Chem. Eng. 106 (2017) 212–223.
[55] M. Motwani, D. Dey, D.S. Berman, G. Germano, S. Achenbach, M.H. Al-Mallah, … T.Q. Callister, Machine learning for prediction of all-cause mortality in patients with suspected coronary artery disease: a 5-year multicentre prospective registry analysis, Eur. Heart J. 38 (7) (2016) 500–507.
[56] M.K. Hassan, A.I. El Desouky, S.M. Elghamrawy, A.M. Sarhan, A Hybrid Real-time remote monitoring framework with NB-WOA algorithm for patients with chronic diseases, Futur. Gener. Comput. Syst. 93 (2019) 77–95.
[57] W. Wiharto, H. Kusnanto, H. Herianto, Intelligence system for diagnosis level of coronary heart disease with K-star algorithm, Healthc. Inform. Res. 22 (1) (2016) 30–38.
[58] M.K. Moridani, S.K. Setarehdan, A.M. Nasrabadi, E. Hajinasrollah, A novel approach to mortality prediction of ICU cardiovascular patient based on fuzzy logic method, Biomed. Signal Process. Control 45 (2018) 160–173.
[59] S. Nikolaiev, Y. Timoshenko, Reinvention of the cardiovascular diseases prevention and prediction due to ubiquitous convergence of mobile apps and machine learning, in: Paper presented at the 2015 Information Technologies in Innovation Business Conference (ITIB), 2015.
[60] S.M. Mathews, C. Kambhamettu, K.E. Barner, A novel application of deep learning for single-lead ECG classification, Comput. Biol. Med. 99 (2018) 53–62.
[61] R. Poplin, A.V. Varadarajan, K. Blumer, Y. Liu, M.V. McConnell, G.S. Corrado, … D.R. Webster, Prediction of cardiovascular risk factors from retinal fundus photographs via deep learning, Nat. Biomed. Eng. 2 (3) (2018) 158.
[62] O.W. Samuel, G.M. Asogbon, A.K. Sangaiah, P. Fang, G. Li, An integrated decision support system based on ANN and Fuzzy_AHP for heart failure risk prediction, Expert Syst. Appl. 68 (2017) 163–172.
[63] J. He, J. Rong, L. Sun, H. Wang, Y. Zhang, J. Ma, D-ECG: a dynamic framework for cardiac arrhythmia detection from IoT-based ECGs, in: Paper Presented at the International Conference on Web Information Systems Engineering, 2018.
[64] L.R. Nair, S.D. Shetty, S.D. Shetty, Applying spark based machine learning model on streaming big data for health status prediction, Comput. Electr. Eng. 65 (2018) 393–399.
[65] S.F. Weng, J. Reps, J. Kai, J.M. Garibaldi, N. Qureshi, Can machine-learning improve cardiovascular risk prediction using routine clinical data? PLoS One 12 (4) (2017) e0174944.
[66] L. Mainetti, L. Patrono, P. Rametta, Capturing behavioral changes of elderly people through unobtruisive sensing technologies, in: Paper Presented at the 2016 24th International Conference on Software, Telecommunications and Computer Networks (SoftCOM), 2016.
[67] J. Kim, U. Kang, Y. Lee, Statistics and deep belief network-based cardiovascular risk prediction, Healthc. Inform. Res. 23 (3) (2017) 169–175.
[68] J. Kim, J. Lee, Y. Lee, Data-mining-based coronary heart disease risk prediction model using fuzzy logic and decision tree, Healthc. Inform. Res. 21 (3) (2015) 167–174.
[69] Z. Arabasadi, R. Alizadehsani, M. Roshanzamir, H. Moosaei, A.A. Yarifard, Computer aided decision making for heart disease detection using hybrid neural network-Genetic algorithm, Comput. Methods Prog. Biomed. 141 (2017) 19–26.

[70] B. Ballinger, J. Hsieh, A. Singh, N. Sohoni, J. Wang, G.H. Tison, …J.E. Olgin, DeepHeart: semi-supervised sequence learning for cardiovascular risk prediction, in: Paper presented at the Thirty-Second AAAI Conference on Artificial Intelligence, 2018.
[71] Q. Al-Tashi, H. Rais, S. Jadid, Feature selection method based on grey wolf optimization for coronary artery disease classification, in: Paper presented at the International Conference of Reliable Information and Communication Technology, 2018.
[72] A.J. Smola, B. Schölkopf, A tutorial on support vector regression, Stat. Comput. 14 (3) (2004) 199–222.

CHAPTER SIX

A study on security privacy issues and solutions in internet of medical things—A review

G. Sripriyanka and Anand Mahendran
School of Computer Science and Engineering, Vellore Institute of Technology, Vellore, Tamil Nadu, India

1. Introduction

Internet of things (IoT) is defined as the interconnection and communication of devices, objects, sensors, etc. without human involvement. Gartner predicts nearly 25 billion (IoT) devices and sensors will be connected by the year 2020 [1]. The data is collected from various devices for transformation. Smart technology provides various services to the user by using sensor information. Nowadays we have "anyone, anytime, anywhere connected with anything" remotely by using the Internet. IoT is using different types of objects such as mobile phones, sensors, RFID, actuators, etc. for interaction. Current smart technology can increase the economy of the world and improve the quality of human life. Cisco predicted that approximately the emergence of new devices can be invented a lot in 2020 [2]. The Internet of things (IoT) was introduced by Kevin Ashton in 1999 in the RFID journal. A computer can track, count everything, and understand the situation without human dependency. The scope of this technology is to connect numerous devices via the Internet and control from anywhere and this can make our life smarter. There are various necessities for IoMT applications. Several wireless applications are supported in this IoMT smart technology for service offering to the user. A hybrid network consisting of multiple wireless technologies will be used in the future of the IoMT infrastructure. Subsequently, it is important to characterize the corresponding necessities of rising IoMT applications with the goal. That we can decide the ideal remote advances can satisfy the future IoMT vision [3]. Many applications were developed in this system, including smart home, smart city, smart health care, smart industry, smart wearables, smart grid, smart agriculture, etc. as shown in Fig. 1 [4].

Intelligent IoT Systems in Personalized Health Care
https://doi.org/10.1016/B978-0-12-821187-8.00006-X

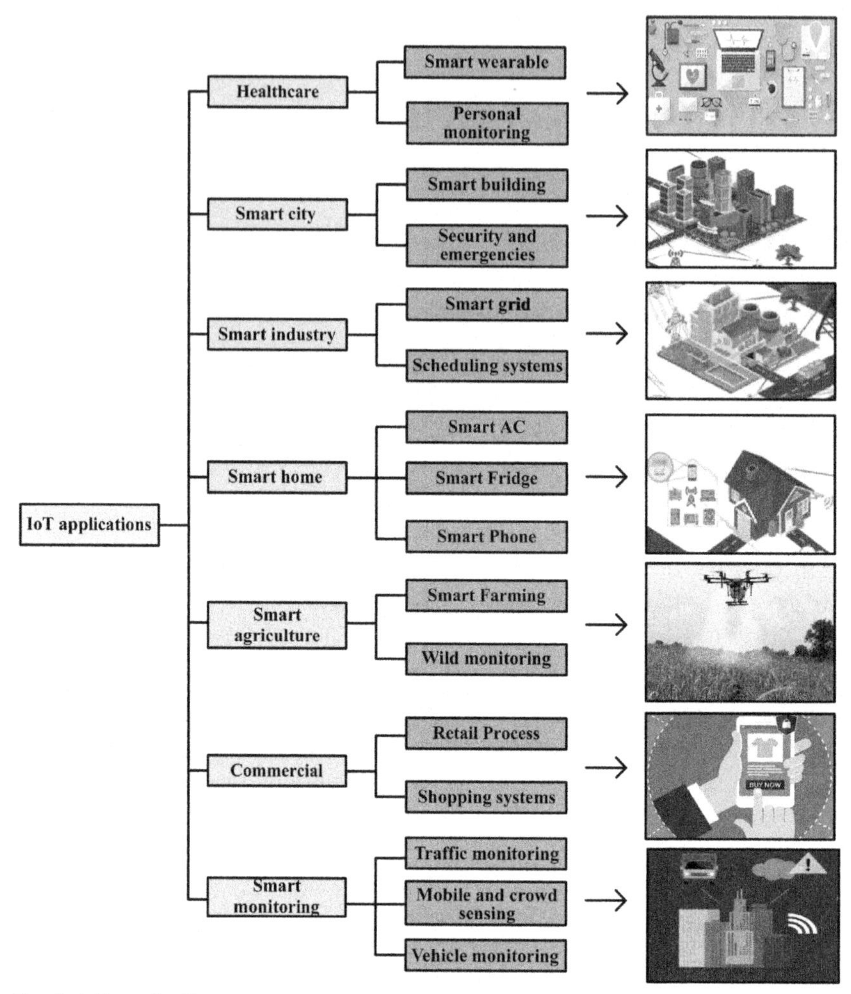

Fig. 1 IoT applications.

The collection of interconnected devices and applications is linked over in the network, which deals with the connected medical things in smart health-care technology. IoMT is one of the most innovative technologies in the IoT environment. According to the statistical report, 87% of health-care establishments are connected to smart technology in 2019 [5]. It increases productivity and reduces the cost of medical expenses. Medical clinics and committed facilities are utilizing associated articles to improve the conveyance of medicare. They can follow medications and perform mechanized electronic graphing. Specialists can follow patient visits and access services even from remote areas [6]. The IoMT patient monitoring approach is used to recognize the patient's location in an emergency. IoMT has been utilized to create pill bottles that track medication adherence. IoMT innovation is legitimately advertised to shoppers for self-checking and gathering biometric data. Example: A smart thermometer which peruses temperature utilizing temperature sensors in cell phones or different gadgets.

These days, there are gadgets in the showcase that can get an electrocardiogram (ECG) at home. Such gadgets help track and gather persistent data directly from their homes and aid telemedicine administration [7, 8]. Examples include observing the remote framework to identify and anticipate incessant ailments including diabetes determination or unhappiness in today's life in a private situation of the patient. In this work, because of the related expenses of reviving or supplanting the batteries of the wearable gadgets, the issue of vitality restrictions is considered. In this paper, we discuss and highlight battery-powered well-being wearable gadgets to monitoring the patients. By sending the asymmetry of the system assets, the vitality proficiency of the wearable equipment is correspondingly improved [9]. The reviewed framework is dependent on Bluetooth low power and is a segment of a private stage that is furnished with camcorders on the body and environmental sensors for recognizing to acknowledge quick insight and settling on choice by methods for AI systems. To assess the proposed system, prototype wearable equipment and three prototype collector units were applied. The enhancement concept of this paper explains the reliability of this health-care technology [4].

IoT is the model that has the competency to modernize on our regular life, in divisions of reaching from our households, entertainment, transportation, well-being, manufacturing to communication with the government sector. This smart technology is offering are indisputable with more advantages and luxury; in any case, these may accompany a huge danger of individual character and information security. A few inquiries have led to locating a superior method to kill security chances and limit the impact on clients' protection prerequisites [10]. The development sector offers an IoMT application proposal with incredible benefit for humans' lifetime, though they can complement enormous charge thinking about the personality's protection and safety assurance [5]. The IoMT developers have not implemented a strong security practice on gadgets, so safekeeping authorities warn that uncontrolled gadgets can pose tremendous risks when they are linked to Internet communication [11]. In such a case with fast and specialized conditions with more security, protection applications have been developed and sold for users with recent technology. But still, people face challenges in lack of updates and technology. IoMT technology describes the programming types of equipment with fluctuating memory size, transmission capacity, and handling power. What we need is start to finish information insurance all through the IoMT design and application terminals. The security challenge's existing approaches are not offering effective solutions to IoMT applications. It has become indispensable for IoMT to have amended security instruments in request to keep pace with this quick-moving world and give solid items and administration to every one of its clients [12].

Computational Intelligence (CI) in current technology is a fundamental reference source that gives pertinent hypothetical structures and the most recent exact look into discoveries in the region of computational insight and the IoT. Computational Intelligence is the hypothesis, structure, application, and improvement of naturally and etymologically propelled computational ideal models. Generally, the three primary mainstays of CI have been neural networks, fuzzy systems, and evolutionary computation. In any case, in time numerous nature propelled figuring standards have developed. Along these lines, CI is an advancing field and at present notwithstanding the three fundamental constituents, it envelops figuring ideal models like surrounding knowledge, counterfeit life, social learning, fake endocrine systems, social thinking, and fake hormone systems. CI assumes a significant job in creating fruitful keen frameworks, including games and intellectual formative frameworks. In the course of the most recent couple of years, there have been blasts of research on deep learning, specifically profound convolutional neural systems. These days, profound learning has become the central strategy for computerized reasoning. Truth be told, the absolute best artificial intelligence (AI) frameworks depend on this system [13, 14].

Machine learning (ML) is a field of essence that may insinuate personal computers gaining from past comprehension (historical data) to upgrade future execution. The primary focal point of this field is the capacity of machines to utilize modified learning systems. Learning implies comprehension or considering changes concerning past experiences. ML is the main thrust for man-made consciousness (artificial intelligence). The preferred position of ML frameworks is that it utilizes intelligent models, heuristic learning, information acquisitions, and choice trees for a major organization. Thus, it gives controllability, perceptibility, and unflinching quality to the framework under an organization. On later occasions, these concept specialists moved their emphasis more to calculations which are practical, dependable, and computationally less mind-boggling in learning methods [15].

Internet of medical thing (IoMT) devices change the lives of people where security assumes a significant job in communication technology. Customary security arrangements cannot give total answers for security dangers in remote health care. IoMT gadgets are associated with a system that contains constrained memory and insignificant registration of assets. The continuous information transmission through medical gadgets can be messed with the gadgets unusually. Bioinspired procedures are utilized to take care of the security issues of designing methods. These procedures are proficient, successful, and progressively affordable when contrasted with other existing arrangements. Discovery and counteraction of digital assaults in IoMT is a major test. For instance, health-care surveillance cameras and other web switches do not contain any patches to introduce frequently. Another security issue is how to introduce patches for billions of gadgets. To determine every one of these issues, this review proposes an enhanced fake resistant-based plan strategy to give novel security arrangements in health care. The proposed structure system will recommend the best approach to accomplish and fulfill the security prerequisites by IoMT. The contribution information of each piece of the IoMT framework can be gathered and examined to decide typical examples of collaboration, subsequently recognizing malevolent conduct at beginning times. Also, ML strategies could be significant in foreseeing new assaults, which are regularly changes of past assaults, since they can cleverly anticipate future obscure assaults by gaining from existing models. Subsequently, IoMT frameworks must have a change from only encouraging secure correspondence among gadgets to security-based knowledge empowered by learning strategies for viable and secure frameworks [16].

2. Internet of things applications

In this section, we discuss the wide variety of applications developed in IoT for the customer's usage. These IoT applications are highly reliable, save time and cost, and are well organized as shown in Fig. 1.

2.1 Smart city

Smart city, similarly recommends, it's a huge development and extents a wide-ranging collection of consumption in current environments, from aquatic circulation to ecofriendly observing and from traffic controlling to wastage management. The enthusiasm behind why it is therefore prevailing is that it tries to evacuate the uneasiness and problems with people who sleep in urban communities. IoT arrangements offered within the shrewd town half tackle completely different city-related problems, including traffic, decreasing air and commotion contamination, and making urban communities safer. Smart city activities need to be monitored for providing various solutions to the user. IoT creates massive datasets that must be broken down and handled to execute smart city administrations. Enormous information stages, some portions of the city information and communication technology (ICT) framework, need to sort, investigate, and process the information assembled from the IoT. City administration coordinates ICT answers to communicate open administrations, simultaneously captivating networks in neighborhood administration, therefore advancing participation. For example, the Greater London Authority activity, where the city lobby is utilizing an open basic stage to impart information to neighborhood networks.

2.2 Smart home

Smart home is the first and foremost effective application in the IoT infrastructure that plays an important role in home security and remote technology and also provides leading significant IoT applications on all channels. The number of people scanning for good homes expands every month by around 60,000 people. Another intriguing issue is that the information on smart homes for IoT investigation incorporates 256 organizations and new businesses. A lot of organizations are presently effectively engaged with keen homes, even as comparative applications within the ground.

2.3 Health care

IoT applications will mostly rework responsive therapeutic-based structures into proactive successfulness frameworks. The assets that moderate and flow therapeutic analysis utilizes would like basic real information. It for the foremost half that utilizes the remaining information, controlled things, and volunteers for meditative assessment. IoT opens approaches to associate the degree of the ocean of great info through investigation, constant field info, and testing. The net of things also improves the current gadgets in power, accuracy, and accessibility. IoT centers on creating frameworks rather than simple hardware.

IoT empowered care contrivance works are operating properly during this era. IoT has different applications in social welfare which range from remote checking gear to progress and also the keen sensors to hardware coordination. It will probably improve; however, doctors convey care and, what is more, keep patients secure and sound. Human services IoT technology will be allowing the well-being users to spend extra time for interacting with their medical care physicians, which might support quiet commitment and fulfillment. From individual well-being sensors to careful robots, IoT in social welfare carries new instruments fresh from the foremost recent innovation within the biological system that aids in improved human services in health care. IoT upsets social welfare and provides skillful pocket-accommodating answers for each patient and human services [17]. Care devices are connected with smart sensors to collect the data from a patient and send that into the cloud through the connected gateway. Now the data is stored in public networks for next-level action.

2.4 Smart grid

In harmonizing IoT and Hertz, the smart grid-iron is established for replacing the traditional power system and it offers more efficient and reliable power service to the customers. This smart grid technology is developed to increase the usage of distributed power properties and rechargeable vehicles through energy distributed generators. These are introduced to reducing the CO_2 discharge and increasing the energy storage competency. And it maintains the proper collaborations between the service providers and the consumers with bidirectional network communications and smart metering systems. With these methods, the brilliant framework can accomplish extraordinary dependability, effectiveness, well-being, and intuitiveness. By incorporating IoT, countless savvy meters can be sent to houses and structures associated with brilliant matrix correspondence systems. Brilliant meters can screen vitality age, stockpiling, and utilization and can collaborate with utility suppliers to report vitality request data of clients and get constant power valuing for clients. With the guide of haze/edge processing foundation, the huge measure of information gathered from savvy meters can be put away and prepared with the goal that the compelling activities of the keen matrix can be bolstered [18, 19].

3. Internet of medical things

IoT offers innovative services to the user in all domains with some disadvantages. One of the most important IoT smart technologies in the human infrastructure is remote health-care monitoring and provides medical services efficiently. This application is used to create a good bond between the patient and the health-care environment conveniently. Certainly, the IoMT joins the dependability and security of predictable medicinal gadgets and the effect, curiosity, and versatility abilities of conventional IoT. It can tackle the issue of maturing also ceaseless sicknesses by having the option to deal with various gadgets conveyed for various patients, although being sufficiently exclusive to manage an assortment of ailments calling for exceptionally heterogeneous checking and activation prerequisites. Also, IoMT similarly answers extra challenges, for example, the patient's portability (i.e., the unavoidable checking of patients in their everyday lives, in a restriction to telemedicine frameworks, which are intensely centered on in-home). An increasingly coordinated and IoT-empowered e-health approach demonstrates the fundamental needs of every person in these zones [8, 20].

In view of the difficult idea of these concerns, new innovative solutions for requesting social health-care frameworks in developed nations are altering how we convey human services. The expansion of individualized computing gadgets, alongside gains in computational force now in these gadgets, is empowering the improvement of health care and offering responses to address the requirements of both our maturing populace and patients with constant illnesses. IoMT is communication between numerous individual restorative gadgets as well as between gadgets and human services suppliers, for example, emergency clinics, medicinal scientists, or privately owned businesses. The coming of the IoMT is primarily brought about by the increment being used, and the advancement of associated and disseminated restorative devices is bringing both talented impending applications and various difficulties. Since individual recovering devices regularly come as wearable appliances, we will concentrate on the combination of wearable medicinal gadgets to the IoMT [21]. These results are additionally basic in numerous IoT use cases in medicinal services, yet they are not generally accomplished. Besides, there are such a significant number of ways to deal with the digitization of medicinal services records that by and by an IoT organization needs to consider these distinctions on the off chance that it is connected with an individual patient. Not all well-being information from associated gadgets eventually arrives in the Electronic Medical Report condition. There are a lot of other data frameworks and frameworks of knowledge, contingent upon sort of information, gadget, degree, and reason. Also, there is a move toward real-time health systems (RTHS), which goes past and incorporates mindfulness and constant information capacities in an IoT and associated/wearable gadget viewpoint. Frameworks are a piece of the more extensive setting and procedures inside this RTHS frameworks approach [22].

The IoT is portrayed as a system of physical gadgets that utilizes availability to empower the trading of information. These gadgets are not the perplexing mechanical headways. They do, in any case, streamline forms and empower social insurance laborers to finish assignments in an opportune way. Organizations that represent considerable authority in human services or innovation will in general intensely put resources into IoT. At present, most technological gadgets accompany a huge type of network, such as wearables also, for example, biosensors to X-beam machines with Wi-Fi or connected with Bluetooth for remote treatment. IoMT can interconnect with more medicinal gadgets to give basic information of patient that help well-being professionals also play beyond their employments. IoT in medicinal services encourages unremarkable but significant assignments to improve understanding results and takes a portion of the weight off well-being specialists. For example, remote patient checking, treatment progress perception, and the lodging of immunizations are for the most part capacities of medicinal gadgets with coordinated IoT [23].

IoT in expansive terms is a mixture of interconnected gadgets and presentations which are associated through on the web PC systems shown in Table 1 [24]. IoMT or net of restorative things could be a subdivision that manages interconnected medicative gadgets/gear and health care related to care IT. Medical gadgets supplied with Wi-Fi or nevertheless close to field correspondence modernization change machine-to-machine communication that is the premise of IoMT. IoMT is another approach for modified consumer regions to driven applications in particular and IoT technology for human services in the smart era, checking with its particular requirements and challenges. Instances of "shrewd things" in the IoMT would incorporate sensor wearables (for example, for Parkinson's illness, various scleroses), sensors, and gadgets for diabetes, pulse, and electrocardiogram (ECG), and shrewd things for insulin and inhalers. The creators recognized the necessity of interoperability as a significant test for medical things. This significant issue manages the requirement for restrictive conventions from

Table 1 Advance medical things in healthcare.

S.no	Advance medical things in health care	Description
1.	Clinical efficiency	Medical clinics and devoted centers are utilizing associated articles to improve the conveyance of Medicare. They can follow medications and perform robotized electronic graphing. Specialists can follow the persistent visits and access EMR even they are from remote areas. The IoT sensors are utilized for the remote area to following of patients and medical equipment in constant
2.	Consumer/home monitoring	IoMT innovation is straightforwardly advertised to shoppers for self-checking and gathering biometric data. E.g.: Smart thermometer which peruses temperature utilizing temperature sensors in cell phones or different gadgets. These days, there are gadgets in the showcase that can get an ECG (electrocardiogram) at home. Such gadgets help track and gather persistent data directly from their homes and aid telemedicine administrations
3.	Biometric sensors/ wearable sensors/ wearables	IoMT is often actualized during associated biometric sensors to be used in clinical or medical clinic scenery. Heart doses that screen heart connected readings; circulatory strain reading armlets so on will even be related to clinical perceptive gadgets organized at a far-flung space. Of late, advanced movable sceptered "auto refractor (this sensor is connected with patient eyeglass)" presentations for vision testing are manageable
4.	Fitness wearables	Wellness follower/wears or customer wearables which may gather information and direct the health system of every single are significantly required for our daily exercise/activities. These appliances, related to progressive cell sceptered applications, will track and provide a report on the health of someone
5.	Brain sensors/ neurosensor	Investigations are ongoing to grow leading-edge client centered on the bone wearable. IoT contraptions equipped for understanding brain

Table 1 Advance medical things in healthcare—cont'd

S.no	Advance medical things in health care	Description
		waves follow and transmit the state of mind erecting neuron signals and so forth are often used to assist screen the responsive well-being of patients. Nonclever neuron-tech (cerebrum wave scrutinizing/soundtrack) has also been inquired into, which may be used for dissecting remedy productivity
6.	Infant monitoring	IoMT aided wearables have been screening and transmit baby's actions, illness, body temperature, and the patterns of sleep to the defenders' through handheld gadgets like the cellular phone. The smartphone is one more innovative application technology in this smart environment for interchanging information. It helps guardians to be continually conscious of their child's healthiness and react in like fashion

restorative sensors/gadgets and keen things from one maker to have the option to impart with gadgets and servers from different makers all together such that the gathered information can be used completely for the analysis to provide a better solution for treatment in well-being applications [25].

Due to the IoT, there is a progressive change in the field of web correspondence; this has a ton of commitment to the development of many testing spaces, yet particularly in the field of therapeutic things. This is one of the significant motivations to close the hole between specialists, patients, and social insurance benefits by its simplicity, exactness, and adaptability. IoT empowers the specialists and medical clinic staff to do their work all the more unequivocally and effectively with less exertion and knowledge [22].

4. Security and privacy issues in IoMT

Remote health checking is personal comfort and medicinal information is gathered from a person in one zone and thereafter is transmitted to a supplier in an alternate area for use in the maintenance and related support. Thereby, the supplier can follow human services information for a patient once discharged to home or a consideration office, lessening readmission rates. This methodology can keep up people's well-being in their home and network, without making them physically go to the supplier's office. Social insurance data is very close to home and the choice of the necessary information needs to occur in light of the results. The information bodes well to improve the lives of patients and the association of social insurance over its different angles, for example, the capacity for specialists, experts, medical caretakers, and staff to settle on better choices quicker.

The different challenges recognized by the developers of the IoMT technology include: (a) a simple network for the smart applications and things is connected to accessing the services; (b) device management is improved by availability of things easily and reduces the cost of maintenance in health care, and (c) informative examination to gain understanding from enormous volumes of medicinal and human services information for better basic leadership [7]. Also, security and protection by configuration should be a piece of any IoT use case, venture, or organization. Utilizing the IoT and information means to improve and decrease mistakes and expenses explained in Fig. 2 [16, 27]. Ensuring that it does not get uncovered or utilized for inappropriate reasons is critical. As referenced in different articles, individual social insurance information should be dealt with uniquely in contrast to a security and consistency point of view. Various security and privacy issues are confronting IoMT-based devices that are used in medical care, both by developers and health-care service users. Producers of IoMT gadgets are looking at security worries about the protection of clients' information gathered with IoT gadgets, considering the possibility that the information is hacked. Cybercriminals gather personal information about patients and it causes virus attacks and malfunctions in smart health-care technology.

Gadgets furthermore causes a glitch along these lines, hurting patients' lives. The utilization of IoT in social insurance poses numerous security challenges which, if not appropriately overseen, could obstruct the full selection and use of IoT in the well-being part, and thus the urgent need to fundamentally recognize and assess unmistakable IoMT security and protection challenges including current and anticipated security issues, security prerequisites, vulnerabilities, danger models, and different ways to deal with give progressively powerful security [28]. Developers in the advancement of the IoMT security and protection framework will consider the effect of various factors, to show signs of improvement stability amongst them. In order to accomplish the superior refuge condition, a few tests requires exceptional consideration as shown in Table 2 [28, 29].

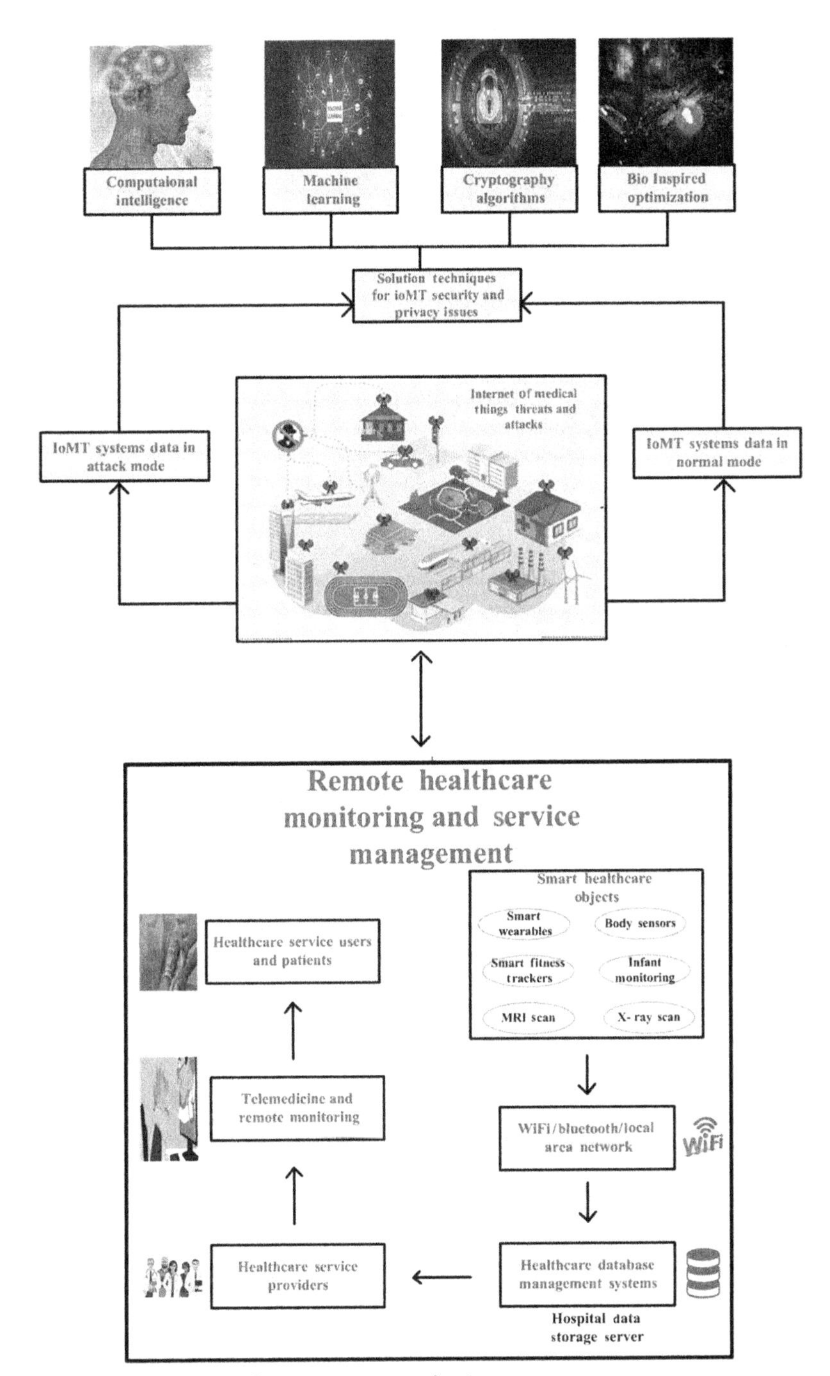

Fig. 2 The architecture of IoMT security and privacy.

Table 2 Security challenges in IoMT.

S.no	Security challenges in IoMT	Description
1.	Data modification	If a patient's restorative information is blocked by a malevolent either from the source hub of an IoMT-based gadget or during information trade between hubs, he/she could adjust the information, subsequently exhibiting a wrong data to guardians who react dependent on the wrong information; this could create calamity for the user whose well-being is observed utilizing this gadget
2.	Impersonation	Each hub on the system has a personality and IoMT-based organize gadgets are no special case as they all have their special personalities which potentially may contain a portion of the patient's data; when an interloper takes this personality, he could utilize it to keep an eye on the patient's well-being records
3.	Replay attack	An assailant can retransmit the information traded between hubs on the system and this may probably prompt treatment breakdown
4.	Eavesdropping	IoMT gadgets utilize remote channels to convey, which makes it simpler for an interloper to have the option to tune into the correspondences between hubs, in this way compromising the secrecy of the Patient's evidence, utilizing such information for additional perilous assaults than taking the patient's private data
5.	Insecure network	A huge number of software services and smart devices are heavily relying on wireless communication networks with more convenience and less money, for example, WiFi. These kinds of networks are defenseless with different interruptions, such as hacker-in-the-center attacks, traffic injections, unauthorized router accessing, denial of facility attacks, spoofing attacks, and brute-forcing attacks
6.	SQL injection	An SQL infusion assault is a place an assailant attempts to assault the backend database associated with the application by embedding a distorted SQL articulation. This assault represents a noteworthy hazard to the IoMT gadgets, particularly in the medicinal services part, as a fruitful SQL infusion assault can bargain touchy patient information or change basic information. In IoMT, SQL injection exposure can create more health-oriented problems and activate unapproved insulin injections to the patient

Table 2 Security challenges in IoMT—cont'd

S.no	Security challenges in IoMT	Description
7.	Brute force attack	The attack is attempting toward figure contributions like the secret word by attempting every single imaginable mix. The IoMT applications are helpless to animal power assaults since pretty much no assurance is set up to defeat such assaults in smart health-care gadgets. The threat is ascribed toward the sensors' counterfeit calculation control. Modernizers are helpless against this outbreak
8.	DOS attack	The current day's IoMT is used by most of the peoples, but services are denied by offenders. On the Internet of Medical Things, this kind of attack aims to interrupt the connection between readers and RFID tags. In 2017, security researcher Ruben Santamarta was clearly showing the problem on Norwegian Airlines flight's Wi-Fi service [26]

A Digital refuge firm announced a malicious code assault arranged by therapeutic gas blood analysis. These assaults were completed distantly by embedding indirect access. In this assault, enemies misused the feeble security instrument in the wired or remote organized arrangement of a human services supplier [5]. The indirect access was answered to convey touchy information to outside nations. As indicated by our scientific categorization, this assault is a break into the privacy of the delicate information. In the subsequent model, genuine vulnerabilities have likewise been distinguished in a significant tranquilize siphon framework that produces the best methods on health care. This model is, the assailant can greatly abuse the firmware, then give a deadly portion presenting a lifetime chance to a huge number of patients around the globe [30].

5. Major security and privacy requirements for IoMT

In this section, we discuss that the IoMT frameworks require certain safety efforts to ensure security, protection, information uprightness, and privacy of sick person health-care proceedings on all occasions. In a supporting BAN framework, all the features are implemented with proper security procedures and guarantee [31]. Safety and protection of persevering data remain the two significant highlights aimed at inside every wireless framework. Safekeeping suggests that information is shielded from unapproved clients when being moved, gathered, handled, and remains securely put away. Then again, protection proposes the position to control the social occasion and utilization of one's close to home data. For example, a patient might be requiring his health information to not be shared with health insurance organizations but hackers restrain this info from her or his attention [32]. All the more explicitly, strategic information inside a health care framework is very delicate, that whenever spilled to the unapproved workforce could prompt a few ramifications for the patient; for example, losing the activity, open degradation besides mental flimsiness.

Another model is when the gatecrashers get to the data through physically catching the hub and changing the data; along these lines, bogus data will be approved to the doctor that could bring about the sick people's demise. Someone can search the patient's health records with their having medicinal records for hacking as explained in Table 3 [41]. Therefore, the additional consideration should be given also reserved for the protection of serious data about the patients and challenging information from unapproved usage, changes, and access. An authorized person only can

Table 3 Security and privacy requirements in IoMT.

S.no	Security and privacy requirements in IoMT	Description
1.	Data confidentiality [33, 34]	Information confidentiality is suggesting that the safety of healthcare users secret data from unauthorized access in IoMT. Since wireless body area network (WBAN) pivots applied in therapeutic circumstances are needed and depended on to transmit sensitive and personal knowledge regarding the standing of a patient's prosperity, therefore their info should be secure so that the healthcare service user information should be secured from unapproved access that might be more dangerous to the patient's life. This vital moved info may be "fixed" throughout the broadcast which will either hurt the patient, the provider, or the framework itself. Secret writing will provide higher privacy to the current delicate info by giving a typical key on a verified correspondence frequency between verified BAN hubs and their organizers

Table 3 Security and privacy requirements in IoMT—cont'd

S.no	Security and privacy requirements in IoMT	Description
2.	Data integrity [35]	Information trait alludes to the procedures reserved to secure the substance of a memorandum, its truth, and steadiness. It applies mutually with every single message including streams of messages. In any case, info privacy does not defend information from outer changes, as knowledge is often illicitly modified once the info is communicated to an unsure WBAN as a foe the health service user's confidential data is changed by hackers before it reaches the system controller on the WBAN. All the more explicitly, changes can be made by coordinating a few pieces, controlling information inside a parcel, and afterward sending the bundle to the PS. This interference and alteration can prime to severe well-being apprehensions and even death in dangerous cases. Therefore, it is basic that information cannot be open, then altered through a possible foe by smearing validation conventions
3.	Data availability [36]	It recommends a restorative professional with operative admission to persistent data. Since such a framework conveys significant, exceptionally delicate, and possibly lifesaving data, the system must be accessible at all the occasions for patients' utilization if there should arise an occurrence of a crisis. For this, it is basic to change the tasks to another WBAN if there should be an occurrence of accessibility misfortune
4.	Data authentication [37, 38]	Restorative besides nontherapeutic submissions may involve information verification. In this way, hubs inside a WBAN necessity are skilled toward check; this data is shown from a realized trust focus and not a faker. In this manner, the system and organizer hubs for all information figure Message Authentication Code (MAC) always shared the undisclosed keys. Precise computation of an authentication cipher guarantees to the system organizer that the meaning is being directed through a reliable hub
5.	Accountability [39]	In the restorative field, it is important for human services suppliers to safely monitor the quiet well-being data. On the off chance that a supplier does

Continued

Table 3 Security and privacy requirements in IoMT—cont'd

S.no	Security and privacy requirements in IoMT	Description
		not verify this data, or more regrettably, manhandles their obligation regarding it, then the person ought to be made responsible for this to dishearten extra misuses. The creator talked about the responsibility issue and proposed a procedure to guard against it
6.	Privacy rules and compliance requirement [40]	The essential to verify nonpublic well-being information could be a worldwide apprehension. In privacy one of the most significant measure is to protect patient's sensitive information with proper privacy policies and rules for authenticated access. A few guidelines and acts are enrolled in human services arrangements. At present, there are numerous arrangements of guidelines/regulations for cover in every place all over the planet. The yank insurance movability and irresponsibleness Act enclosed a lot of bearings to aim at specialists, medicinal services suppliers, and emergency clinics and is intended to guarantee that a person's well-being and therapeutic records are secure. IoMT plots point-by-point insurances that have to be taken to harmless gatekeeper tolerant data once utilized for restrictive or correspondence desires
7.	Data anonymization [29]	Quiet delicate data may be divided into three categories: unequivocal appropriators, semiappropriators, and protection characteristics. The first category in data anonymization is an explicit identifier it can specify the patient's with unique information, they are mobile number, identification number, and name. A mix of semiappropriators will likewise remarkably demonstrate a patient, as an example, age, birth details, and address. Protection knowledge alludes to sensitive attributes of a patient, as well as malady and pay. Within the method of data distribution, while considering the distribution qualities of the primary data, it is necessary to ensure that the individual characteristics of the new-fangled dataset are fitly ready, to make sure of the patient's privacy

retrieve the final data about the patients by using protected mechanisms with their unique ID from the collected information and the different points of the network communications [39]. Giving security needs counteracting access to knowledge or different objects by unauthorized purchasers too as a means of protection against unapproved alterations or pulverization of clients' knowledge. The standard and more essential security model CIA Triad (confidentiality, integrity, and availability) is developed to monitor and protect data security in a single organization [12].

6. Solutions for IoMT problems

In this section, we discuss IoMT technology and it has been accepted that smart technology is part of daily life in the future. Restructuring and digitalizing techniques are used to exchange patient healthcare information with a better understanding of improved efficiency and significant money savings. There are numerous traditional safety actions that providers and producers of smart devices should take in encryption of algorithms for securing smart health care. The machine's control is protected and modified by the safety boot in medical smart applications [42]. The design of IoMT should incorporate security measures; it conducts the risk valuation of the gadget before it is released in the market for use and builds the authentication measures on medical gadgets. Monitor and verify the object-to-object communication when the access control is limited and there is lack of authentication in applications. Various security layers should be implemented in depth to secure the things from the risk of harm. Need to ensure the appropriate access control limits unauthorized access in the IoMT technology and systems as shown in Fig. 3. Before the production test, the security measures of health-care devices and the security of the device are monitored accurately. The culture of safety is established according to the reorganization of vulnerabilities in medical applications [43].

This work the instructs IoMT shoppers (e.g., patients, therapeutic specialists, and so forth) to frequently have a coffee degree of attentiveness regarding the IoMT security problems and the way to deal with users. The advantages of this work do not seem to be simply restricted to adopters. This technique will also be advantageous to IoMT arrangement suppliers in measuring their things and entrust them with different IoMT arrangements. This supports a lot of helpful and simple challenges among arrangement suppliers. Also, the lawmakers/regularization groups and investigators are

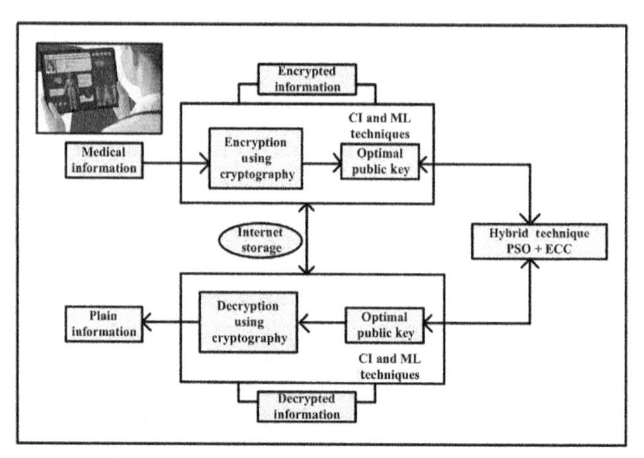

Fig. 3 Privacy and security solution techniques for IoMT.

designing better solutions and protocols for security risks through actual understanding [44]. Check the server name to place possible augmentation for their assortment and circulation to considering the great assortment and redundancy of data in health-care applications. This lessens the hazard bestowed by having a solitary purpose of disappointment which might injure your activities if the administration goes down. Within the event that you simply possess or work any IoMT gadgets, certify to alter their default passwords. Cripple with universal plug-and-play (UPnP) settings on your associated gadgets. This setting is sceptered after all IoMT gadgets and provides an open acknowledgment of malware to taint switches and near systems. Handicap any switch settings for remote management complete Telnet, as this element allows one computer to manage another from a far-off space, and has been utilized in past Mirai DoS assaults. As much as possible, the IoMT gadgets and network communication routers need to be updated with their security standards, firmware, and software for protection. Utilize a web device like Bull guard's IoMT Scanner, to verify the standing of your medical gadgets as so much of their unprotectedness to mirrai diseases. Within the event that shortcomings are found, connect with the maker, and boot scan for online fountains of safety patches.

6.1 Computational intelligence techniques for IoMT

Cyberattacks have increased in number as well as in resourcefulness too. For instance, one new assault framework utilizes apparatuses with computerized front closes that dig for data and vulnerabilities, joined with man-made consciousness AI-based examination to filter through big data. It is likewise now workable for cybercriminals to use AI and CI to change code on the fly of depending on what has been distinguished in dark cap labs to make these cybercrime and entrance devices increasingly sly and harder to recognize. A July 2016 overview directed by solutions found that human services are the business most now and again focused by malware, representing 88% from all findings in the globe. Also, this may simply be the start. It has been anticipated that we will before long observe custom malware composed totally by machines dependent on computerized powerlessness recognition, complex information examination, and mechanized advancement of the most ideal adventure got from the special qualities of a found shortcoming. For instance, the ongoing Reaper botnet malware utilized nine unique bundles that focused vulnerabilities in IoMT gadgets from seven distinct makers. A collector is exploiting a bigger assault surface, and these sorts of developing IoT botnets will undoubtedly hit IoMT gadgets [45].

Like many new fields, AI and CI bring a lot of fascinating antiques, related subtleties, and admonitions with them that represent a test from a security point of view. The two pictures underneath (from the Microsoft Azure documentation) show the common ancient rarities and partner bunches required at different phases of a start to finish the AI/CI work process. Computational Intelligence depends on organically propelled bioinspired computational algorithms. The key columns that make this field are neural systems, hereditary algorithms, and fluffy frameworks. Neural systems are calculations that can be utilized for function estimate or arrangement issues. They incorporate supervised, unsupervised, and support learning. Hereditary calculations are search algorithms enlivened by organic hereditary qualities. They depend on two principal administrators, hybrid and change [13]. Populations of people speaking to answers for the problem are made in more than a few pages. The calculation utilizes an irregular guided approach to enhance issues dependent on a wellness work. Fluffy rationale depends on the fuzzy set hypothesis to manage thinking that is liquid or surmised instead of fixed and definite. Downy rationale factors have truth esteems extending in gradation between zero and one which might affect the third truth. Machine intelligence strategies are effectively used in various global applications on an Associate in Nursing assortment of coming up with problems [14]. They will likewise be used to shield the protection and security of the shoppers of knowledge and correspondence technologies. Points of enthusiasm for this area include Location of primarily based services, applied math databases, applied math revealing management, denial of service attacks, Forensics, intrusion detection systems, country security, basic frameworks assurance, and access management. People's daily lives centralized with safety, business, privacy, and moral control, these regions are more important to develop the economy of the world and it supports smart technology user's trust [46].

6.2 Machine learning techniques for IoMT

ML in health care is a procedure that relies upon two phases: the preparation or knowledge stage and then the identification or challenging stage. The computational environment individualities are fully dependent on mathematical function and algorithms in the training phase, which is utilized for learning through regular data such as reference input data. At that point, in the identified organization, these qualities are utilized for discovery and characterization. Directed learning is one kind of AI procedure in which attributes of the preparation information set are utilized in the education stage toward making the characterization model, which is then used to arrange newfangled concealed cases. A kind of machine learning method is called an unsupervised learning approach, it is dependent on information features without grouped training information. An example of order strategy in AI relies upon design acknowledgment, while a solitary classifier technique relies upon a solitary AI calculation [47].

The learning prototype is the blend of operational smart security schemes and policy-based scheduling procedures with trust aware of data protection. This model guarantees the scale capacity, dependability, and security for the put away medical data and gets to administration [48]. In Reinforcement Learning (RL), no particular results are characterized, and the specialist gains from input in the wake of cooperating with nature. It performs a few activities and settles on choices based on the reward gotten. A specialist can be compensated for performing great activities or disciplined for terrible activities and input criteria can be used to boost the long-haul rewards. It is incredibly aroused by the learning practices of people and creatures.

Such practices make it an appealing methodology in profoundly unique applications of applied autonomy in which the framework figures out how to achieve certain assignments without unequivocal programming [49]. It is additionally very critical to pick reasonable reward work because the achievement and disappointment of the operator relies upon the gathered all-out remuneration.

6.3 Bioinspired optimization techniques for IoMT

IoMT demands a novel hypothesis to secure confidential information from the attackers; it includes a comprehensive approach to interactions and their actors. One of the more successful defense and interesting approaches in medical infrastructure is enhanced, bioinspired optimization models. This smart technology requires a strong and decentralized safekeeping mechanism. A security and privacy solution in the traditional approach is infeasible to provide exact results to the threats in IoMT. It has very less computing resources, and the devices can be damaged unpredictably and restrict memory in communication technology. To prevent and detect the intrusions in medical devices is a big challenge in remote health-care technology. These kinds of issues are resolved by using this approach. This review has proposed an enhanced bioinspired optimization framework for IoMT security and privacy in theory. Bioinspired optimization techniques are more efficient and effective than traditional approaches. It always finds optimal solutions to resolve complex problems in smart health-care technology. It is used to solve the security issues in IoMT applications also to provide decentralized safety. This health-care technology has huge applications and devices for providing service to the patients. It is an infeasible task to maintain billions of usernames and passwords in public communication technology securely. This approach is using artificial immunity-based safety mechanisms in medical applications remotely [50]. The IoMT technology is introducing more applications in healthcare to users, also it is facing more defects and risks in information safety and privacy. The major concentration of artificial immune system (AIS) is to protect the healthcare sensitive information, it offers not only parallelism but the reproduction of regulations, features, and changing aspects between biological systems and IoMT [51].

6.4 Cryptography techniques for IoMT

An elliptic curve is another technique to manage open key cryptography considering the arithmetic structure of elliptical corners over restricted fields. It is thought about as a creative technique with less size for medical image security but this technique is considered as challenging on healthcare video security in terms. This section additionally briefs the IoMT restorative footage within the medical half wherever computer code is enclosed [52]. These default steps are followed in the cryptographical safety process, like generating a key, data encryption, data decryption, and analysis of information. To upgrade the safety and privacy level of IoMT in tending frameworks, the associate improvement model is taken into account for key generation. Computer code elliptical regulation is crucial for undertaking rational and then capable implementation [53].

7. Analysis of IoMT usage and issues

Advancements coordinated to computerized health care are on the ascent. Novel gadgets and applications have entered the human services part, from wearables to genomics to customized medication, which guide in sickness conclusion, and their forecast, the executives and treatment. This model will be furnished clearly with huge high points where the specialist can look through his sick people from wherever he wants. Crises to send a crisis flag or message to the specialist with the patient's present status and full therapeutic data can likewise be developed. The implemented model of portable application turns the need for the patient all over the world as shown in Fig. 4. In this inexorably associated world and with the pervasiveness of information assortment, the social insurance division is utilizing progressed systematic procedures to advance patient consideration and doing as such in a savvy way. Between 2013 and 2019, the growth rate of IoMT technology is shown in the graph diagram.

Despite the very fact that mechanization in healthcare services monitoring would increment functioning productivity, it would gift real dangers

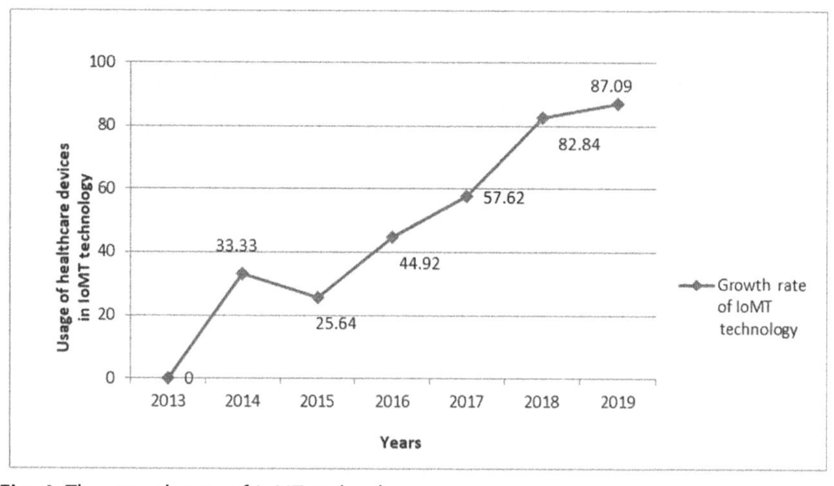

Fig. 4 The growth rate of IoMT technology.

throughout usage, for instance, info felony, shaky info moves, and unpredictable system associations. These difficulties, combined with body obstacles, are anticipated to drive development in IoT-based systems administration and knowledge solutions. Government activities, for instance, the Patient Protection and cheap Care Act that centers on valid electronic health records [EHRs], might improve the standard and proficiency of the mind and advance consistency in providing care. There is still an extension to boost convenience and international info pointers over the business, which might empower info, taking care of in an exceedingly steady vogue. Wondering about the benefits and connected difficulties, IoMT seems to be a promising account that improves healthful services checking and treatment results.

As the quantity of IoT gadgets develops, it is basic to consider the security vulnerabilities in connected gadgets explained in Fig. 5. Furthermore, the estimation of the data on these gadgets is likewise growing expressively. A few vulnerabilities incorporate access to a patient's medicinal data, releasing lethal portions of insulin without tolerant assent, and remote control of the gadget. We are at the precise minute in time when we have to consider introducing control instruments for IoMT devices. These systems fall into three classifications: developer, administrator, and device controls. Data administration arrangements should be refreshed to represent the

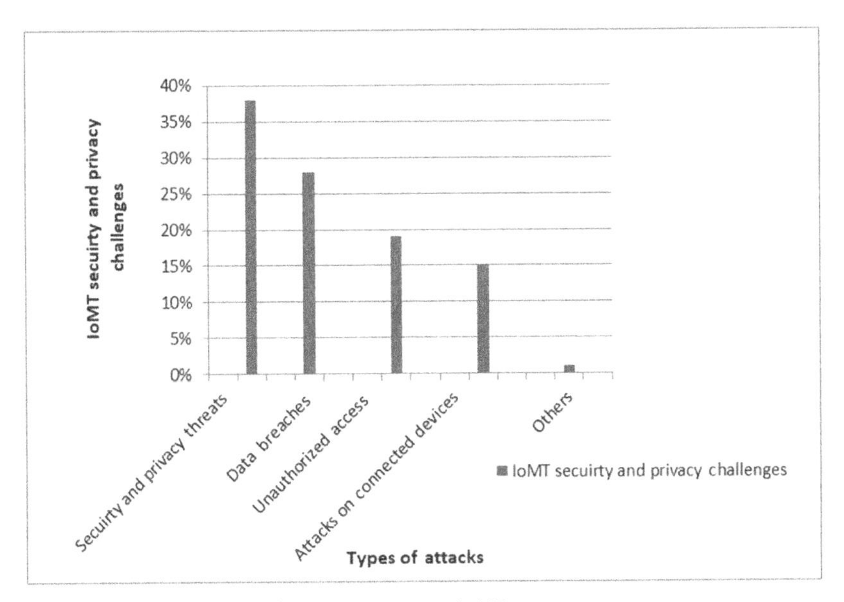

Fig. 5 Different security and privacy issues in IoMT.

expanding volume, natures, and wellsprings of information. These arrangements ought to organize data security and cautiously consider ramifications of new information streams. Remarkable changes would include: modifying access controls, improved anonymization of the information, the security of individual cell phones, and reevaluation of maintenance plans with respect for dangers.

8. Conclusion

In this review, we explained the IoMT technologies and services in remote health-care monitoring. It changes the way of service delivery in IoT health-care technology. IoMT has been given incredible consideration; be that as it may, the related norms and technical details are as yet improving; in particular, the superior application prerequisites of medicinal services and progressively fruitful investigation are required. This IoMT technology can improve the productivity of health-care technology. Patient information is more sensitive because it includes more personal details and we need to protect the privacy of users. But these concepts have numerous threats and susceptibilities about information safety and discretion. This broadsheet presents details about health-care challenges and solutions. This is essential to guarantee that the organizations run robotized work procedures and give snappy entrance to basic statistics while keeping the whole thing protected. These security control mechanisms are accomplished with enforceable safety rules and implement on malicious software barriers, exposures, assessments configuration, event, and activity monitoring for confidential information protection. These suggestions require the general people and social welfare to join hands and solve the difficulties that are part of WBAN use to ensure that the appliance in the conveyance of the title patient's healthful services is verified to the slightest degree levels. An assortment of therapeutic gadgets and programming applications are applied to enhance the character of healthful administrations and conjointly generate a great deal of data. At present, the importance of information is obvious. Directions to successfully guarantee information safety and protection in the slightest degree phases of the data stream will possess an important position in future related research. This hybrid theoretical framework discussed the numerous security and privacy solutions to the IoMT industry. Computational Intelligence, Machine Learning, and Cryptography Techniques provide an enhanced solution to the medical data and services to smart health-care users. The proposed approach is ensuring the security and privacy of confidential information in IoMT applications. The utilization of smart health-care gadgets in medicinal services is relied upon to keep expanding in the coming years. The fate of IoMT looks extremely splendid, and you should not pass up embracing this innovation.

References

[1] B. Al-Shargabi, O. Sabri, Internet of things: an exploration study of opportunities and challenges, in: 2017 International Conference on Engineering MIS (ICEMIS) (pp. 1–4). IEEE, 2017, May.

[2] K. Kaur, A survey on internet of things–architecture, applications, and future trends, in: 2018 First International Conference on Secure Cyber Computing and Communication (ICSCCC) (pp. 581–583). IEEE, 2018, December.

[3] M.B. Mollah, S. Zeadally, M.A.K. Azad, Emerging Wireless Technologies for Internet of Things Applications: Opportunities and Challenges, (2019).

[4] P. Asghari, A.M. Rahmani, H.H.S. Javadi, Internet of Things applications: A systematic review, Comput. Netw. 148 (2019) 241–261.

[5] Y. Sun, F.P.W. Lo, B. Lo, Security and privacy for the internet of medical things enabled healthcare systems: a survey, IEEE Access 7 (2019) 183339–183355.

[6] J. Rauscher, B. Bauer, Safety and security architecture analyses framework for the internet of things of medical devices, in: 2018 IEEE 20th International Conference on e-Health Networking, Applications and Services (Healthcom) (pp. 1–3). IEEE, 2018, September.

[7] K.L.M. Ang, J.K.P. Seng, Application-specific internet of things (ASIoTs): taxonomy, applications, use case and future directions, IEEE Access 7 (2019) 56577–56590.

[8] S.K. Polu, S.K. Polu, IoMT based smart health care monitoring system, Int. J. 5 (2019) 58–64.

[9] Y. Shaikh, V.K. Parvati, S.R. Biradar, Survey of smart healthcare systems using internet of things (IoT), in: 2018 International Conference on Communication, Computing and Internet of Things (IC3IoT) (pp. 508–513). IEEE, 2018, February.

[10] K. Sarwar, S. Yongchareon, J. Yu, A brief survey on IoT privacy: taxonomy, issues and future trends, in: International Conference on Service-Oriented Computing (pp. 208–219). Springer, Cham, 2018, November.

[11] Y. Yang, L. Wu, G. Yin, L. Li, H. Zhao, A survey on security and privacy issues in Internet-of-Things, IEEE Internet Things J. 4 (5) (2017) 1250–1258.

[12] M. Gulzar, G. Abbas, Internet of things security: a survey and taxonomy, in: 2019 International Conference on Engineering and Emerging Technologies (ICEET) (pp. 1–6). IEEE, 2019, February.

[13] CIS.IEEE, What is Computational Intelligence?, [Web log post]. Retrieved November from https://cis.ieee.org/about/what-is-ci 14.

[14] W. Duch, What is computational intelligence and where is it going? in: Challenges for computational intelligence, Springer, Berlin, Heidelberg, 2007, pp. 1–13.

[15] A.K. Rana, A. Salau, S. Gupta, S. Arora, A Survey of Machine Learning Methods for IoT and their Future Applications, (2018).

[16] M.A. Al-Garadi, A. Mohamed, A. Al-Ali, X. Du, M. Guizani, A survey of machine and deep learning methods for internet of things (IoT) security, (2018) arXiv preprint arXiv:1807.11023.

[17] R. Gour, Top 10 Applications of IoT, Retrieved 1 November 2019, from https://dzone.com/articles/top-10-uses-of-the-internet-of-things, 2018, October 30.

[18] J. Lin, W. Yu, N. Zhang, X. Yang, H. Zhang, W. Zhao, A survey on internet of things: architecture, enabling technologies, security and privacy, and applications, IEEE Internet Things J. 4 (5) (2017) 1125–1142.
[19] G. Lopez, J. Matanza, D. De La Vega, M. Castro, A. Arrinda, J.I. Moreno, A. Sendin, The role of power line communications in the smart grid revisited: applications, challenges, and research initiatives, IEEE Access 7 (2019) 117346–117368.
[20] i SCOOP, Internet of things (IoT) in healthcare: benefits, use cases and evolutions, https://www.i-scoop.eu/internet-of-things-guide/internet-things-healthcare/.
[21] A. Gatouillat, Y. Badr, B. Massot, Sejdi´c, E., Internet of medical things: a review of recent contributions dealing with cyber-physical systems in medicine, IEEE Internet Things J. 5 (5) (2018) 3810–3822.
[22] G.J. Joyia, R.M. Liaqat, A. Farooq, S. Rehman, Internet of medical things (IOMT): applications, benefits and future challenges in health-care domain, J. Commun. 12 (4) (2017) 240–247.
[23] A. Ahad, M. Tahir, K.L.A. Yau, 5G-based smart healthcare network: architecture, taxonomy, challenges and future research directions, IEEE Access 7 (2019) 100747–100762.
[24] Internet of Medical Things (IoMT), Connecting Healthcare for a Better Tomorrow, (2017) UST Global®.
[25] T. Saheb, L. Izadi, Paradigm of IoT big data analytics in health- care industry: a review of scientific literature and mapping of research trends, Telematics Informatics 41 (2019) 9.
[26] Finjan Team, IoT DoS Attacks—How Hacked IoT Devices Can Lead to Massive Denial of Service Attacks [Blog post], Retrieved from https://blog.finjan.com/iot-dos-attacks/, 2018, August 20.
[27] A. Limaye, T. Adegbija, HERMIT: A benchmark suite for the internet of medical things, IEEE Internet Things J. 5 (5) (2018) 4212–4222.
[28] N. Okafor, Uchenna, C. Opara, Chukwuma, Improving Security and Privacy of IOT in Healthcare, (2017).
[29] W. Sun, Z. Cai, Y. Li, F. Liu, S. Fang, G. Wang, Security and privacy in the medical internet of things: a review, Security Commun. Netw. 2018 (2018).
[30] F. Alsubaei, A. Abuhussein, S. Shiva, Security and privacy in the internet of medical things: taxonomy and risk assessment, in: 2017 IEEE 42nd Conference on Local Computer Networks Workshops (LCN Workshops) (pp. 112–120). IEEE, 2017, October.
[31] S. Al-Janabi, I. Al-Shourbaji, M. Shojafar, S. Shamshirband, Survey of main challenges (security and privacy) in wireless body area networks for healthcare applications, Egypt. Inform. J. 18 (2) (2017) 113–1225.
[32] M. Li, W. Lou, K. Ren, Data security and privacy in wireless body area networks, IEEE Wirel. Commun. 17 (1) (2010) 51–58.
[33] P.K. Dhillon, S. Kalra, Multi-factor user authentication scheme for IoT-based healthcare services, J. Reliable Intel. Environ. 4 (3) (2018) 141–1630.
[34] G. Wang, R. Lu, Y.L. Guan, Achieve privacy-preserving priority classification on patient health data in remote e-healthcare system, IEEE Access 7 (2019) 33565–33576.
[35] A. Zhang, L. Wang, X. Ye, X. Lin, Light-weight and robust security-aware D2D-assist data transmission protocol for mobile-health systems, IEEE Trans. Inf. Forensics Secur. 12 (3) (2016) 662–675.2.
[36] F. Firouzi, B. Farahani, M. Ibrahim, K. Chakrabarty, Keynote paper: from EDA to IoT e-health: promises, challenges, and solutions, IEEE Trans. Comput.-Aided Design Integr. Circuits Syst. 37 (12) (2018) 2965–2978.
[37] Y.K. Ever, Secure-anonymous user authentication scheme for e-healthcare application using wireless medical sensor networks, IEEE Syst. J. 13 (1) (2018) 456–467.

[38] K. Fan, S. Zhu, K. Zhang, H. Li, Y. Yang, A lightweight authentication scheme for cloud-based RFID healthcare systems, IEEE Netw. 33 (2) (2019) 44–49.
[39] M.S.A. Malik, M. Ahmed, T. Abdullah, N. Kousar, M.N. Shumaila, M. Awais, Wireless body area network security and privacy issue in E-healthcare, Int. J. Adv. Comput. Sci. Appl. 9 (4) (2018) 209–215.
[40] P. Knag, HIPAA: A Guide to Health Care Privacy and Security Law, (2002) Aspen Law Business.
[41] M. Wazid, A.K. Das, J.J. Rodrigues, S. Shetty, Y. Park, IoMT malware detection approaches: analysis and research challenges, IEEE Access 7 (2019).
[42] A. Chacko, T. Hayajneh, Security and privacy issues with IoT in healthcare, EAI Endorsed Trans. Pervasive Health Technol. 4 (14) (2018).
[43] F. Alsubaei, A. Abuhussein, V. Shandilya, S. Shiva, IoMT-SAF: internet of medical things security assessment framework, Internet Things 8 (2019) 100123.
[44] F. Alsubaei, A. Abuhussein, S. Shiva, A framework for ranking IoMT solutions based on measuring security and privacy, in: Proceedings of the Future Technologies Conference (pp. 205–224). Springer, Cham, 2018, November.
[45] A. Ala-Kitula, K. Talvitie-Lamberg, P. Tyrvainen, M. Silvennoinen, Developing solutions for healthcare—deploying artificial intelligence to an evolving target, in: 2017 International Conference on Computational Science and Computational Intelligence (CSCI) (pp. 1637–1642). IEEE, 2017, December.
[46] D.A. Elizondo, A. Solanas, A. Martınez-Balleste, Computational intelligence for privacy and security: introduction, in: Computational Intelligence for Privacy and Security, Springer, Berlin, Heidelberg, 2012, pp. 1–4.
[47] M.F. Elrawy, A.I. Awad, H.F. Hamed, Intrusion detection systems for IoT-based smart environments: a survey, J. Cloud Comput. 7 (1) (2018) 21.
[48] G. Manogaran, N. Chilamkurti, C.H. Hsu, Emerging trends, issues, and challenges in internet of medical things and wireless networks, Pers. Ubiquit. Comput. 22 (5–6) (2018) 879–882.1.
[49] F. Hussain, R. Hussain, S.A. Hassan, E. Hossain, Machine Learning in IoT Security: Current Solutions and Future Challenges, (2019) arXiv preprint arXiv:1904.05735.
[50] R. Banu, G.A. Ahammed, N. Fathima, A review on biologically inspired approaches to security for internet of things (IoT), in: 2016 International Conference on Electrical, Electronics, and Optimization Techniques (ICEEOT) (pp. 1062-1066). IEEE, 2016, March.
[51] S. Aldhaheri, D. Alghazzawi, L. Cheng, A. Barnawi, B.A. Alzahrani, Artificial immune systems approaches to secure the internet of things: a systematic review of the literature and recommendations for future research, J. Netw. Comput. Appl. (2020) 102537.
[52] M. Elhoseny, K. Shankar, S.K. Lakshmanaprabu, A. Maseleno, N. Arunkumar, Hybrid optimization with cryptography encryption for medical image security in Internet of Things, Neural Comput. Appl. (2018) 1-15.2.
[53] H.N. Almajed, A.S. Almogren, SE-Enc: a secure and efficient encoding scheme using elliptic curve cryptography, IEEE Access 7 (2019) 175865–175878.

CHAPTER SEVEN

Application of computational intelligence models in IoMT big data for heart disease diagnosis in personalized health care

Amos Orenyi Bajeh[a], Oluwakemi Christiana Abikoye[a], Hammed Adeleye Mojeed[a], Shakirat Aderonke Salihu[a], Idowu Dauda Oladipo[a], Muyideen Abdulraheem[a], Joseph Bamidele Awotunde[a], Arun Kumar Sangaiah[b], and Kayode S. Adewole[a]

[a]Department of Computer Science, University of Ilorin, Ilorin, Nigeria
[b]School of Computing Science and Engineering, Vellore Institute of Technology (VIT), Vellore, India

1. Introduction

In the diagnosis of diseases, investigations begin conventionally from the complaints presented by patients which are usually followed by clinical differential diagnoses [1]. Although this practice has become highly precise and contextual in health care through various innovative discoveries, yet it is not capable of addressing various multitudinous complexities of many diseases such as diabetes and heart diseases [2]. To address this challenge, recent researches focused in the direction of personalized health care. Personalized health-care systems comprise personalized, predictive, preventive, and participatory approaches where patients are diagnosed based on their clinical history and health profile. Personalized health care occupies an important position in modern diagnostics by efficiently addressing various dynamic complications [3] and plays a crucial role in transforming conventional disease-oriented treatments to personalized care [2]. Personalized health-care systems employ Computational Intelligence (CI) techniques to extract useful patterns and derive knowledge from patients' historical data.

CI is an evolving paradigm that is biologically and linguistically inspired. It involves processes and techniques that mimic biological processes and adopt human intelligent reasoning in developing intelligent systems. CI also includes the study of all nonalgorithmizable processes that humans (and

Intelligent IoT Systems in Personalized Health Care
https://doi.org/10.1016/B978-0-12-821187-8.00007-1

sometimes animals) can solve with a certain degree of competence [4]. Several techniques that make up CI include the neural network which is inspired by the interconnectedness in biological structure of the human brain; fuzzy systems that use approximate reasoning in processing vague or imprecise information; and evolutionary computation which is inspired by biological evolution with generation of population and selection of best performing sets from the successive population with the view to optimization [4a]. CI fundamentally entails the ability to learn patterns from observations in the form of data [5].

Machine learning (ML) techniques form a subset of CI and they entail learning from experience by way of analyzing the patterns inherent in data for prediction. The model so formed can be used to make a prediction from a new set of data not used in the learning process. Several studies have used ML techniques on medical datasets for the diagnosis and prediction of diseases [5a] and even other tasks in health-care data processing [5b]. Some of the ML techniques include logistic regression, linear regression, decision trees (DT), support vector machine (SVM), artificial neural network (ANN), and Naïve Bayes (NB) [5b]. The major tasks involved in the application of ML techniques are feature selection, classification, and clustering. Feature selection tends to optimize the dataset used by ML algorithms and their performance while classification and clustering are geared toward identifying and predicting disease classes. Some of the feature selection techniques used in disease diagnosis and prediction are relief, information gain, *t*-test, and chi-square.

Optimization is also a major computational intelligence in the process of disease diagnosis and prediction. Evolutionary algorithms such as genetic algorithm (GA) and particle swarm optimization (PSO) have been used in the task of feature selection and even selection of computational resources.

Big data analytics (BDA) is another subset of CI that is geared toward effectively processing huge and heterogeneous data collected from diverse sources using different devices ranging from stand-alone to mobile and wearable Internet of things (IoT) devices, which are usually very hard for conventional machine learning algorithms to process. It provides a robust architecture for storing huge amounts of data and powerful predictive algorithms for analyzing the data for real-time decision-making. These tools include Hadoop, Apache Spark, IBM Watson services, Apache Pig, HIVE, and STORM [6–8]. BDA has seen a lot of applications in medical diagnoses, disease monitoring, and health records management.

This study examined the application of the various CI techniques for heart disease diagnosis in the domain of personalized health-care systems, developed a taxonomy of the applied techniques, conducted a case study, and proposed a big data-based framework that can be used for personalized health care. The case study uses rule-based and tree-based approaches for heart disease prediction using a cardiovascular disease dataset of 70,000 instances. Three machine learning algorithms that have not yet been studied are investigated. These algorithms are JRIP, PART, and Random Forest.

The remaining part of this chapter is organized as follows. Sections 2 and 3 present a discussion of the concept of IoT and, by extension, IoMT, and heart diseases and medical record system, respectively. Section 4 discusses the application of computational intelligence in the diagnosis and management of heart diseases. Studies that have employed the use of intelligent systems in heart diseases diagnosis are reviewed in the section. Section 5 presents the case study conducted in this work. It includes the identification of the ML algorithms, description of the dataset, and performance evaluation used in the study. The results of the study are also presented and discussed in this section. Section 6 presents a proposed Big Data framework that can be used to improve on the application of IoMT in the diagnosis of heart disease. Section 7 concludes the work while and presents some open issues and future works.

2. The concepts of IoT and IoMT

IoT is a model for interconnected devices with tracking, sensing, processing, and analyzing capabilities. IoT is a paradigm that offers a new environment in which sensors and other actuators blend seamlessly with the habitat around us [9]. Jabbar et al. [10] defined IoT as a combination of distributed smart objects which have sensing capabilities and embedded identification through radio frequency identification (RFID) technology. It is a paradigm through which digital and physical objects can be interlinked and intercommunicated to provide some domain-specific services. IoT transforms real-world objects into smart objects which can sense environmental and physical quantity and communicate with them accordingly. With RFID technology, it is now possible to track people, objects, and animals by using RFID tags. In RFID technology, electronic product codes are encoded in RFID tags which can be used to track smart objects in IoT [11, 12].

Specifically, the integration of sensors, RFID tags, and communicating technologies underpins IoT. Many IoT devices combine embedded systems, mobile computing, cloud computing, low-price hardware, big data, and other technological advancements [13] in order to provide computing functionality, data storage, and network connectivity for equipment that hitherto lacked them. The ability of IoT to provide network connectivity for equipment has brought about new efficiencies and technological capabilities such as remote access for monitoring, configuration, and troubleshooting for the equipment. IoT also has the ability to analyze data about the physical world and use the results to better inform decision-making, alter the physical environment, and anticipate future events. IoT therefore addresses traceability, visibility, and controllability of smart objects. A wide range of IoT applications are now available which include environmental monitoring, health-care service [14], inventory and production management [15].

IoT architecture is basically made up of sensors, connectivity, and people/processes interconnected for effective communication. There are four main building blocks of an IoT system as depicted in Fig. 1 [16].

The implementation of IoT in an organization such as health-care service providers will require the use of the four-stage architecture [16]. Fig. 2 depicts the four-stage architecture. The first stage involves the configuration of the "things" to be networked and they include the wireless sensors and actuators. The next stage includes sensor data aggregation systems and analog-to-digital data conversion. The third stage involves data

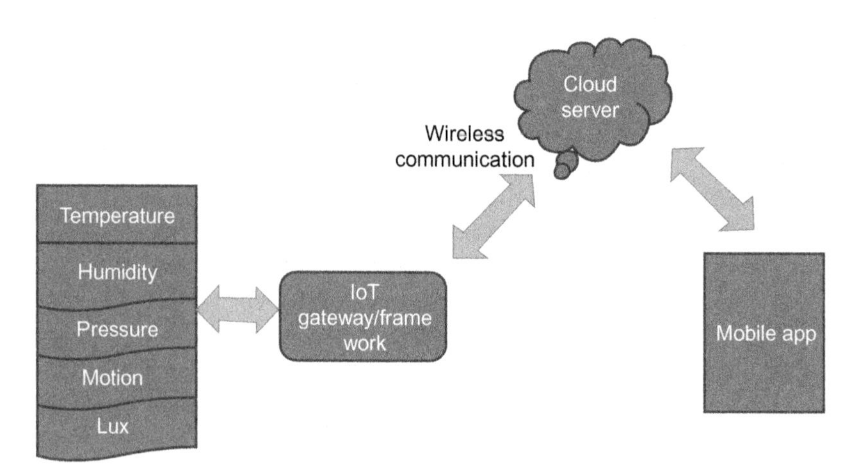

Fig. 1 IoT architecture. *(From A. Naval, How Do IoT Works?—IoT Architecture Explained With Examples, 2019, July 5.)*

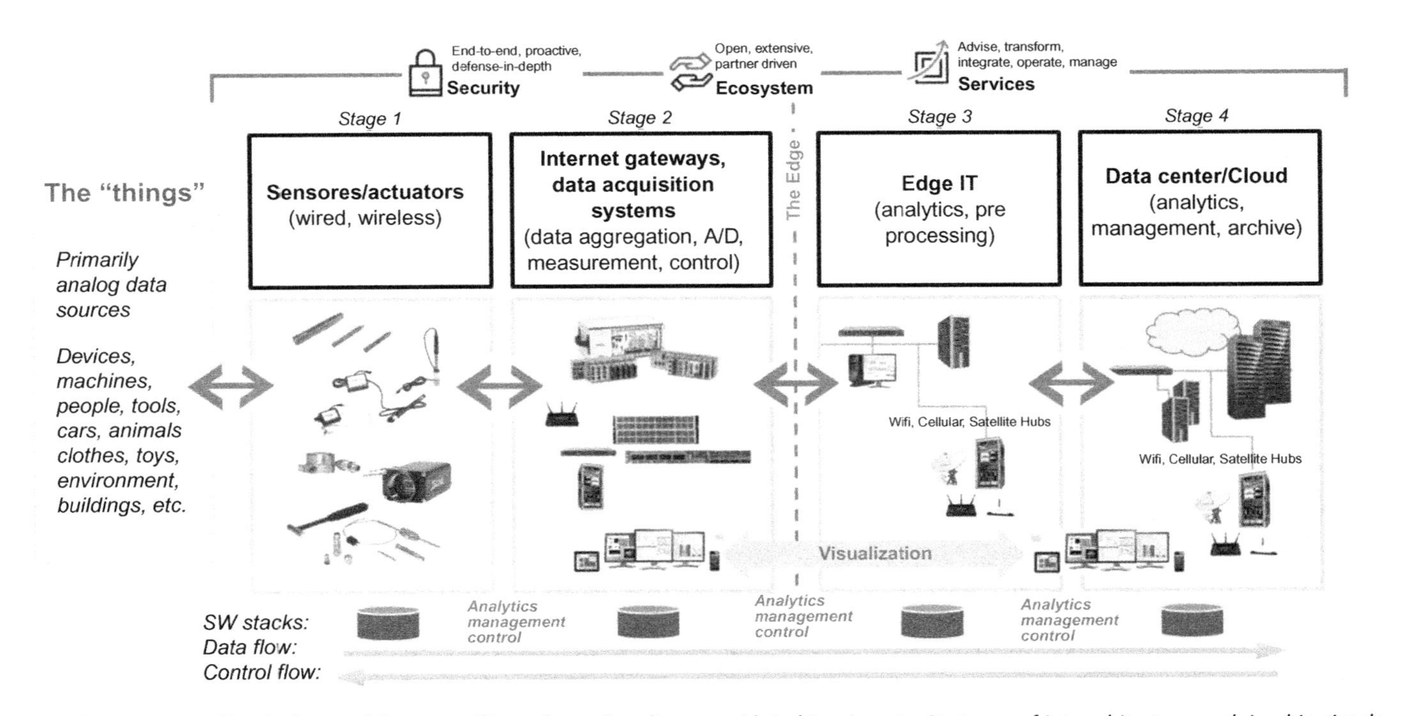

Fig. 2 Four-stage IoT solution architecture. *(From https://medium.com/datadriveninvestor/4-stages-of-iot-architecture-explained-in-simple-words-b2ea8b4f777f.)*

preprocessing for onward transmission to the cloud environment. The fourth stage involves the storage and analysis of data for carrying out tasks such as monitoring, detection, and prediction of diseases.

IoMT is thus defined as a connected system of medical devices and applications that collect data that is provided to health-care IT systems through online computer networks [16a]. IoMT consists of connected computing devices and objects for medical service management and delivery. The objects have identifications for unique referencing and communication. IoMT has the ability to transmit patient data over networks for storage and processing in cloud environment [17]. This system is necessitated by the need to provide health-care services to patients remotely and also reduce the astronomical growth of the cost of health care. Recent advances in the IoT have made it virtually possible to transform the health-care sector [18].

As rightly predicted by Ghanavati et al. [14], the adoption and diffusion of IoT could play a significant role in arresting the spiraling health-care costs without impacting the quality of care provided to patients. Qi et al. [19] showed that the rapid evolution of health care from a conventional hub-based system to a more personalized health-care system (PHS) is a result of the rapid proliferation of wearable devices, smartphone, and IoT enabled technology. Marr [20] is of the opinion that the IoMT will revolutionize safety and health since it is capable of monitoring, informing, and notifying not only caregivers but provide health-care providers with actual data to identify issues before they become critical or to allow for earlier intervention. Other additional factors responsible for the rapid growth of IoMT were given by Qi et al. [19]. The authors also pointed out that chronic diseases, especially diabetes, are on the increase and better treatment options at lower health-care cost make IoMT a welcome innovation. Also, high-speed internet access and favorable government policies have contributed to the growth of the adoption of IoMT. As this new revolution of the Internet called the IoT is rapidly gaining ground, new research topics in many academic and industrial disciplines, especially in health care, are emerging daily. It is obvious that IoT is improving health-care service delivery, most especially in the monitoring and diagnosis of diseases in personalized health-care systems [21].

Chronic illnesses, such as stroke, heart disease, diabetes, cancer, chronic respiratory diseases, are causing death in many parts of the world [22]. The number of people with a chronic disease is rapidly increasing, giving the health-care industry more challenging problems. There are several IoT-based health-care systems that can be used to intelligently supervise patients with chronic diseases for long-term care [23].

An IoT-SIM model for semantic Interoperability among heterogeneous IoT devices in the health-care domain was proposed by Jabbar et al. [10]. The model was proposed in order to provide interoperability among heterogeneous IoT devices. With the model, Physicians will be able to monitor their patients remotely anywhere, anytime, and without any constraint of specific vendors' devices. The model uses RDF to present patients' raw data into useful information. Physicians prescribe patients against diagnosed diseases using IoT devices; then this information is semantically annotated using RDF.

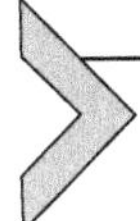

3. Heart diseases and electronic health record

3.1 Heart diseases

Heart disease is a debilitating disease in modern society. Cardiovascular disease can be described as a variety of conditions that affect the heart and blood vessels in the body, for instance, atherosclerosis caused by a buildup of plaque in the arteries. Plaque can become large and form clots that can block the smooth flow of blood in the artery and this can lead to either a stroke or a heart attack [24]. Other conditions include congenital heart disease, arrhythmia, coronary artery disease, dilated cardiomyopathy, myocardial infarction, hypertrophic cardiomyopathy, mitral regurgitation, mitral valve prolapse, and pulmonary stenosis.

As a result of aging, urbanization, and globalization, cardiovascular disease (CVD) is now a major global disease compared to other communicable diseases [25–27]. In 2005, an estimated 17 million people (30% of global deaths) suffered and died from this disease and among this number, 7.2 and 5.7 million deaths were as a result of heart attacks and stroke, respectively; the developing nations had a share in 80% of these deaths, and it was estimated by the WHO that if this trend is not arrested, about 24 million deaths could be recorded by 2030 [28, 29].

The common symptoms of heart disease include chest pain, breathlessness, and heart palpitations. Chest pain is commonly caused by a heart situation known as angina or angina pectoris which occurs when a part of the heart does not receive enough oxygen, which is triggered by stressful events or physical exertion. Other heart disease symptoms include a faster rate of heartbeat, dizziness, breathlessness, and clubbed fingernails.

Heart disease is caused by a number of factors which include: high cholesterol, smoking, overweight, diabetes, and even heredity. Other factors include age and a sedentary lifestyle. In spite of the efforts toward preventing, diagnosing, and treating CVD, it is one of the leading causes of disability and untimely death worldwide.

3.2 Electronic health records

Electronic health record (EHR) is an advanced electronic medical record (EMR) which is a computerized record of the patient's medical history in an organization and the record is used by specialists, pharmacists, and laboratory services of that specific organization. Unlike the EMR, EHR is made up of widespread patient ID linked to lifetime medical history that is valid and may be shared across numerous organizations [30]. This model provides the patient with the right to medical history which can be shared with other health-care providers [8].

Generally, the two ways by which EHRs are stored are the structured and unstructured formats. Structured data are an organized form of data which makes it easy for processing and analysis [31, 32]. For instance, structured CVD data will include information such as patient age, gender, lab results, drug doses, and ECG. Unstructured data are not organized in a schematic form like structured data, and thus it is harder to process and analyze [31, 32]. Commonly, EHR is stored in spreadsheets or databases. The increase in the level of variety and volume has moved data from a structured to an unstructured form. Big data can combine millions of patients' EHR collected from different sources and provide features for managing data in the three dimensions of volume, variety, and speed.

The significance of electronic health records cannot be overemphasized as they are a promising resource to improve the efficiency of clinical trials and to capitalize on novel research approaches. EHRs are useful data sources to facilitate effective and comparative research and new trial designs that may answer relevant clinical questions as well as improve efficiency and reduce the cost of clinical research. It is inspiring to say that early involvement through EHRs has been hopeful; besides, information gained through the application of EHRs will continue to transform clinical research [33]. The giant stride in EHRs skill has produced unique analytic capabilities.

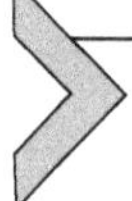

4. Computational intelligence technique for heart disease management

CI techniques can be broadly classified into single and hybrid methods. The single method involves the use of only one intelligence technique such as decision tree, logistic regression, and artificial neural network

(ANN) in a framework or system. The hybrid method involves the combination of two or more intelligent techniques such as neuro-fuzzy (NF) and fuzzy support vector machine (FSVM) [5]. This section discusses the various computational intelligence (CI) techniques that are applied for feature selection, detection, and prediction in the management of heart diseases in the context of IoMT enabled personalized health care.

4.1 Feature selection in heart disease diagnosis

IoMT for health management generates huge data for processing. This data is usually complex and contains both relevant and irrelevant data as well as redundant information. Thus, some selection of the data to remove the irrelevant and redundant ones is needed before further processing and analysis can be effectively carried out for the detection or prediction of diseases. This selection is done by using techniques referred to as feature selection techniques or algorithms. These techniques are classified into Filter, Wrapper, and Embedded approaches. Some survey has shown that the wrapper approach gives the best classification performance in health disease management such as predicting heart diseases. The following studies employed feature selection algorithms for selecting data for heart disease diagnosis.

Subanya and Rajalaxmi [34, 35] developed a Binary Artificial Bee Colony algorithm and K-nearest neighbor method (BABC-KNN) for selecting optimal relevant features and predicting heart disease, respectively. The study used data from the UCI machine learning repository. Compared to other feature selection and classifier combinations investigated in the study, BABC-KNN has a better accuracy result of 92.4% with only six features, which are the smallest number of features.

Song et al. [36] used linear discriminant analysis (LDA) for feature selection and support vector machine (SVM) classifier for arrhythmia classification. LDA extracted four features from 17 features. The classification accuracy of SVM using LDA for feature selection is better than that of using principal component analysis and also all the features without reduction. Furthermore, combining LDA with SVM performed better than combining it with a multilayer perceptron (MLP) and fuzzy inference system (FIS) classifiers.

The combination of simulated annealing (SA) and SVM for selecting optimal features was applied by Lin et al. [37] on datasets from the UCI repository. The results of the analysis showed that the combination of SA and SVM gave an optimal selection with better performance than without SA.

Gutlein et al. [38] compared the performances of the two wrapper approach to those of the feature selection, linear forward selection, and standard forward selection techniques. The experiments conducted showed that the linear forward techniques reduced the number of features in a shorter time without significantly affecting accuracy. The method also reduced the chances of overfitting in the training of a machine learning algorithm.

Esseghir [39] developed a wrapper-filter hybridized feature selection method using the Greedy randomized adaptive search procedure (GRASP). The study used five datasets in the UCI repository for validation of the method. The results of the study showed that the wrapper-filter method gives a robust performance. Also, Unler et al. [40] developed a filter-wrapper feature selection algorithm using particle swarm optimization and a SVM for classification. The performance of the proposed algorithm has also been compared with a hybrid filter-wrapper algorithm based on a genetic algorithm and a wrapper algorithm based on PSO. The obtained classification accuracy and utilization of computational resources showed that the PSO-based hybridization is equally good.

The study by Nahar et al. [41] investigated the performances of medical knowledge-driven feature selection (MFS) compared to Naive Bayes and sequential minimal optimization (SMO). The performances were based on the accuracy, true positive rate, F-measure, and even training time. The study used the UCI heart disease datasets for the analysis of the WEKA platform. The results showed that in the prediction of heart disease, a low performance most likely implies that the significant factors of heart disease were not captured in the process.

Shilaskar and Ghatol [42] developed a hybrid model combining the filter and wrapper approaches for SVM classification. The study used three feature selection models: Forward Feature Inclusion, Back-elimination Feature Selection, and Forward Feature Selection. Distance criterion is used for ranking and selecting features and a wrapper model is applied for classification and evaluation. These criteria generate all features' ranks as per their importance toward target class identification.

The Cleveland heart dataset in the UCI repository was used for the evaluation of the performance of the proposed swarm intelligence-based artificial bee colony (ABC) using SVM as a classifier by Subanya and Rajalaxmi [34, 35]. The performance of the proposed model supersedes that of the forward feature selection with reverse ranking. Also, good classification accuracy using only seven features was given by the proposed model.

Liu et al. [43] proposed a hybridized feature selection method that involved the use of ReliefF and Rough Set (RFRS) for selecting optimal features and passing the selected feature into an ensemble classifier. The proposed approach passes the output from RF into RS for further reduction. The study used the Starlog (Heart) dataset containing 270 observations of both healthy and ill-health heart records from the UCI repository. The result of the study compared the performance of the proposed ensemble method to those of individual classifiers of the C4.5 decision tree, Naïve Bayes, and Bayesian Neural Network. The proposed ensemble approach showed a better classification specificity, sensitivity, and accuracy compared to the individual classifiers.

Gokulnath and Shantharajah [44] proposed a wrapper approach to feature selection using GA and SVM. This model was compared to some other feature selection algorithms such as relief, info-gain, chi-square, gain ratio, and even ordinary genetic algorithm. The GA-SVM model gave a very high performance of 88.37% when compared to using SVM without selection.

Some other studies have used CIs for the selection of computational resources for processing Big Data in health care. The application of the cloud computing environment for the processing of health-care big data was studied by Abdelaziz et al. [45]. The proposed optimization of the selection of a virtual machine for the speedy processing of IoT Big Data was performed by using Particle Swarm Optimization (PSO) and Parallel PSO (PPSO). The performance of the proposed model was measured using waiting time, CPU utilization, and turnaround time. The performance of the PPSO model outperformed that of PSO.

Elhoseny et al. [46] studied the application of three optimization algorithms: genetic algorithm, PSO, and PPSO in the selection of cloud computing virtual machines for the processing of IoT big data for health services. The three models proposed in the study are the cloud-IoT model using GA, the cloud-IoT model using PSO, and the cloud-IoT model using PPSO. The three models were compared with the state-of-the-earth methods based on the speed of execution and efficiency in terms of a number of processors used. They outperformed the state-of-the-earth methods by 50% in terms of speed, and by 5.2% in terms of efficiency in real-time data retrieval.

4.2 CI techniques for classification of heart disease

Majumder et al. [47] used the J48 decision tree algorithm in the development of a wearable smart IoT system to predict cardiac arrest. The model uses heart rate, RR intervals, ST-segment, and body temperature in predicting the level of risk of cardiac arrest of users.

Manogaran et al. [48] in their proposed architecture of IoT and the big data system for smart health-care monitoring and alerting used MapReduce-based stochastic gradient descent with logistic regression for predicting heart disease. This computational intelligence part of the architecture was evaluated using patient data collected from the Cleveland heart disease database (CHDD), which contains 76 attributes, and for performance, accuracy, sensitivity, specificity, precision, recall, and f-measure are measured. The accuracy of the prediction model attained 72.82%.

ANN and fuzzy neural network (FNN) are hybridized for the classification of heart and diabetes diseases. The hybridization is based on the fact that medical datasets contain both crisp and fuzzy values. Two datasets used in the study are the Pima Indians diabetes and Cleveland heart disease datasets in the UCI repository for machine learning. The performance of the hybrid was evaluated based on accuracy, sensitivity, and specificity. K-fold cross-validation has used the training of the model. The accuracy of the model for the Pima Indians and Cleveland heart disease datasets is 84.24% and 86.8%, respectively.

K-Means clustering, genetic algorithm, and SVM were used in the study conducted by Santhanam and Padmavathi [49]. K-Means is used to remove noise from the data used in the study and GA selects an optimal set of data features from the cleaned data. SVM is used for classification. The performance of the integration of these models gave an average accuracy of 98.79% for the reduced dataset of Pima Indians Diabetes from the UCI repository.

Jain and Singh [50] reviewed some of the popular computational intelligence algorithms and models used in feature selection and classification of health-care data. The feature selection algorithms and models reviewed include ReliefF, Correlation-based feature selection, Chi-square, Information gain, *t*-test, genetic algorithm, randomized hill-climbing, Beam search method. Some others are decision trees, ridge regression, weighted Naïve Bayes, sequential forward selection, and feature selection using the weighted vector of SVM. Some of the classifiers reviewed include SVM, K-NN, Naïve Bayes Classifier, Neural Network, Bayesian Network, and C4.5. Also discussed is the parallel classification model in which several classifiers that could be of the same type or different types are used in parallel to perform classification and the resultant prediction is a combination of the output of the classifiers.

A federated IoT and cloud computing pervasive patient health monitoring system was proposed and evaluated by Abawajy and Hassan [51]. The proposed framework uses feature extraction and classifiers models in its analytic engine for predicting health status in the IoT context. For the classification process, the SMO-based algorithm was compared with the Bayesian network learning and classical Naïve Bayes algorithm. The results of the study showed that the SMO-based classifier outperformed the other two algorithms with accuracy ranging between 87% and 99% in different datasets.

4.3 Big data-based heart disease decision support system

Big Data has attributes that surpass the computational ability of commonly used hardware and software resources for processing within a reasonable time [48]. Basically, any dataset that is high in volume, velocity, and variety is regarded as big data [6]. These are regarded as the three V's models of big data as shown in Fig. 1.

Big data requires powerful tools for capturing, storing, and analyzing data. These technologies and techniques, referred to as Big Data Analytics (BDA), include Hadoop, MapReduce, [6], HDFS, Apache spark [7], IBM Watson services, Apache Pig and HBase [48], HIVE and STORM [8]. BDA is used to enhance decision-making, provide insight and discovery, and optimize processes.

The recent drastic increase in electronic health datasets and aggregation of health data collected from numerous IoMT sensors, internal and external health records sources such as EHR, EMR, and social media reports [8] that are difficult to process by the conventional techniques has called for the application of BDA in health-care decision support systems (DSS). The health-care informatics field is improved by the recent application of big data analytics tools to solve health-care problems including medical knowledge representation, clinical decision support systems, and diagnosis of diseases. This section focuses on the application of Big data in heart disease decision support systems.

A big data approach has been applied in condensing and aggregating the heterogeneous data generated from Electrocardiograph (ECG) sensors to a usable form for doctors in monitoring patients' heart rate, and researchers in developing heart disease diagnosis models. Page [52] proposes a system for automatically delineating key markers in ECG recordings, and displaying them in a novel way for identification of anomalies or patterns in the decision support system for diagnosing long QT syndrome (LQTS). The study developed an ECG clock for visualization of ECG biomarkers and built a hierarchical database that has the original data at its lowest layer, heart rate at the highest layer, and primitives such as "R peak locations" in between. Machine learning techniques based on supervised learning algorithms (such as SVM, decision trees, and KNN), Clustering techniques (such as GMM, K-means, and DBSCAN), and ANN implementation Scikit-learn python library were used and the methods show promising results in the tasks of genotyping, cardiac event prediction, and AMI detection.

IoT and big data analytics are increasingly gaining popularity for the next generation of eHealth and mHealth services [53]. Due to this, improved IoT-cloud architectures for big data have been proposed to effectively handle big data on the cloud for cardiovascular diseases. Elhoseny et al. [46] proposed a model to optimize virtual machines selection (VMs) in cloud-IoT health services applications to Big Data in integrated industry 4.0 with the aim of enhancing health-care system performance in terms of reducing the execution time of health-care requests, optimizing the storage of patients' big data, and providing a real-time data retrieval mechanism. To optimize the VMs selection, the Genetic Algorithm (GA), Particle swarm optimizer (PSO) and Parallel Particle swarm optimization (PPSO) were used to build the model. The proposed model performed better than other existing models by 50% in total execution time and by 5.2% in data retrieval efficiency. In the same direction, Abdelaziz et al. [45] also proposed a novel PPSO-dependent algorithm using the CloudSim package to solve task scheduling problems to support heart disease management in smart cities in order to minimize medical request execution time and maximize the usage of resources.

Manogaran et al. [48] proposed a big data hybrid fog-cloud computing model based on Meta Fog-Redirection (MF-R) with Grouping and choosing (GC) architecture for continuous monitoring of patients with heart diseases. The model employed big data Apache Pig and HBase for storing data collected from sensors and integrated fog computing from the network edge mobile devices with cloud computing using GC. The predictive model built with this architecture performed better compared to Random Forest, SVM, and multiple regression in terms of sensitivity, specificity, and F-measure. In the same category is the framework proposed by Tuli et al. [54] integrating ensemble-based deep learning in Edge computing devices for automatic Heart Disease analysis. The Fog-enabled cloud framework, FogBus, was employed to test the performance of the proposed model in terms of power consumption, network bandwidth, latency, jitter, accuracy, and execution time. The results show improved performance compared to other previous models that did not use deep learning.

Recently, Context-aware models based on big data analytics have also been proposed to detect and monitor a patient with heart diseases. One notable work identified was presented by Mohammed et al. [7]. The study applied Intelligent Hybrid Context-Aware Model for Patients Under-Supervision at Homes (IHCAF-PUSH) to classify the health status based on blood pressure disorders of patients into any one of the following—**normal class**, **warning class**, **alert class,** or **emergency class**. The distributed Weka and spark plug in tools on the WEKA platform were used for training and validating the model in parallel to prove the concept of handling big data using Spark in heart disease diagnosis. The study showed that Random Forest, decision trees, and Naïve Bayes have the highest accuracy while SVM has the worst.

To summarize the CI techniques that have been applied in the literature, Fig. 3 shows the extent of use of various CI techniques in the diagnosis of heart diseases as reviewed in this study. Both the single and hybrid

Fig. 3 The Vs of big data.

approaches to the use of machine learning and optimization techniques have been applied to heart disease diagnosis and prediction, and also an optimal selection of a cloud-based virtual machine for the processes heart disease dataset. Although Fig. 4 depicts the list of techniques and algorithms used, some of them are hybridized in the prediction and processing of the heart disease dataset.

Fuzzy neural networks and fuzzy inference systems have also been successfully applied. Various classification algorithms, such as function-based, instance-based, statistical-based, and tree-based models, have also been applied in the diagnosis of heart diseases as presented in the taxonomical structure in Fig. 3. Recent research focused on the application of distributed computing models such as big data analytics, fog computing, and edge

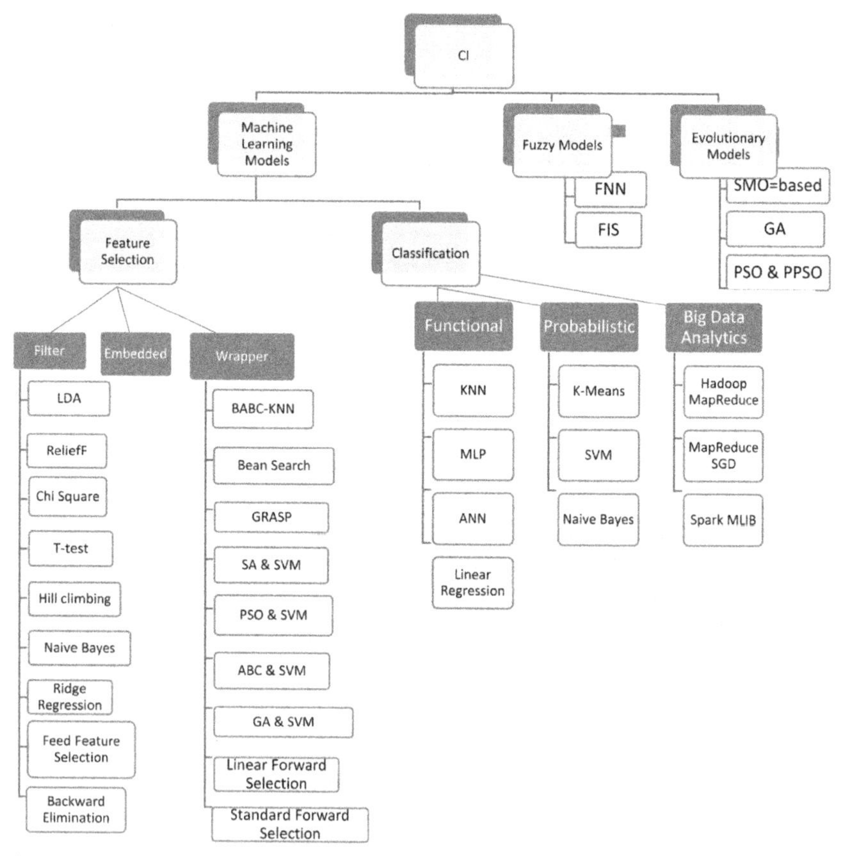

Fig. 4 A taxonomy of the application of CI techniques in heart disease diagnosis.

computing for providing robust and real-time processing capability for heart diseases diagnosis and management.

4.4 Rationale for CI application in IoMT

The computational intelligence technique requires historic data for computation and health-care systems generate such data that can be used in developing health-care intelligent systems. These systems can be in the form of personalized health-care models that use the IoT as a framework for their operations.

The diverse sources for the collection of medical data and the propensity of the exponential growth of the data require resources and platforms with high processing power and efficiency. Thus, the application of big data analytics and cloud-based architectures as computational intelligence techniques are suitable for processing and managing medical data. Another important rationale for adopting computational intelligence is that it is capable of replicating and mimicking expert knowledge, thereby complementing the effort of human experts and thus reducing the chances of error in medical diagnosis and disease management.

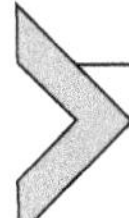

5. Case study on the application of ML for the diagnosis of heart diseases

In this section, we employ rule base learners: JRIP and PART, and tree-based ensemble learner Random Forest to classify and predict heart diseases using the new cardiovascular disease (CVD) dataset available publicly for use in the Kaggle repository. The rule-based and ensemble-based learners are selected for the experimental case study because they are rarely used models for heart disease diagnosis in the literature. The WEKA 3.8.3 tool for machine learning and data analysis was used to carry out the experiment.

5.1 Dataset description

The CVD dataset used is a new dataset that consists of 70,000 records of patients' data, containing 11 input features and a target output. The input features are of three types:

1. *Objective*: factual information about the patient.
2. *Examination*: results of medical examination.
3. *Subjective*: information was given by the patient.

A full description of the dataset is presented in Table 1.

Table 1 Description of the CVD dataset.

S/n	Label	Description	Feature type	Datatype
1	Age	Age of the patient	Objective	Integer
2	Height	The height of the patient measured in cm	Objective	Integer
3	Weight	The weight of the patient measures in kg	Objective	Float
4	Gender	Represents the gender of the patient	Objective	Categorical
5	ap_hi	The measured systolic blood pressure of the patient	Examination	Integer
6	ap_lo	The measured diastolic blood pressure of the patient	Examination	Integer
7	Cholesterol	The measured cholesterol level of the patient	Examination	Integer
8	Glucose	The measured glucose level of the patient	Examination	Integer
9	Smoke	Represents whether the patient is smoking or not	Subjective	Binary
10	Alco	The alcohol intake of the patient	Subjective	Binary
11	Active	The physical activity of the patient	Subjective	Binary
12[a]	Cardio	Presence or absence of heart diseases	Target	Binary

[a]Class attribute.

The dataset was collected from the Kaggle repository (https://www.kaggle.com/sulianova/cardiovascular-disease-dataset). The data were entered into the ML algorithms in the CVS form for processing.

5.2 The classifiers used

5.2.1 JRIP

JRIP is one of the most popular rule-based classification algorithms. JRIP is an implementation of a propositional rule model based on a Repeated Incremental Pruning to Produce Error Reduction (RIPPER) [55]. It examines classes by incrementing the size and an initial set of rules for the class. JRIP continues the process by treating all the instances of a particular judgment in the training data as a class and subsequently find a set of rules that covers all the members of that class, repeating this until all classes are covered [56]. The algorithm executes four main phases: growth, pruning, optimization, and selection.

5.2.2 PART

Projective adaptive resonance theory (PART) combines the partial decision tree approach and separate-and-conquer rule-learning technique to infer rules [57]. It builds a partial C4.5 decision tree at each iteration [55] and makes the best leaf into a rule. It depicts how good rule sets can be learned one rule at a time, without any need for global optimization. The algorithm is efficient because it avoids postprocessing and does not suffer from the extremely slow performance with which C4.5 is criticized [57].

5.2.3 Random Forest

Random Forest is a supervised learning algorithm that builds multiple decision trees systematically and merges them to produce a more accurate and stable prediction model. This merged model is referred to as a "Forest." The forest is an ensemble of decision trees, usually trained with the "bagging" method. The general idea of the bagging method is that a combination of learning models increases the overall result. Random Forest adds two levels of additional randomness to the model while growing the trees. Each tree uses a bootstrapped version of the training data by a randomly selected subset of the dataset to build the partial trees and searches for the best feature among a random subset of features [58]. This results in a wide diversity that generally results in a better model.

5.3 Performance evaluation

The performance of the selected algorithms is evaluated based on sensitivity, precision, recall, accuracy, and area under curve (AUC). The definitions of these performance measures are as follows:

(i) Accuracy: This is the fraction of predictions that the model gets right. Mathematically, accuracy can be calculated by

$$A = \frac{tp + tn}{tp + tn + fp + fn} \tag{1}$$

(ii) Sensitivity/recall: Sensitivity is a measure of the proportion of actual positive cases that got predicted as positive or (true positive). Sensitivity is also termed Recall. Therefore, in this case, it refers to the tests' ability to correctly detect patients who have heart diseases. Mathematically, sensitivity is calculated by

$$S_e = \frac{tp}{tp + fn} \tag{2}$$

(iii) Precision: Precision expresses the proportion of the data points the model predicts to be relevant, which is actually relevant. It is the number of true positives divided by the number of true positives plus the number of false positives. It is calculated mathematically by

$$P = \frac{tp}{tp + fp} \tag{3}$$

(iv) Area under ROC curve (AUC): AUC represents the degree or measure of separability. It is the measurement of the area under the ROC which is a probability curve plotted with TPR against the FPR. It tells how much a model is capable of distinguishing between classes. The higher the AUC, the better the model is at distinguishing between patients with the disease and without the disease. An excellent model has AUC that is close to 1, which implies that the model has a good measure of separability, and AUC value close to 0 implies a poor-performing model with a low measure of separability, i.e., the model is not differentiating the class properly.

5.4 Results and discussion

This section presents the preliminary performance evaluation results of the three classifier models on the preprocessed dataset using all 11 input attributes. The results of experiments with PART rule induction algorithm as a classifier are given in Fig. 5. It shows that the algorithm performance is

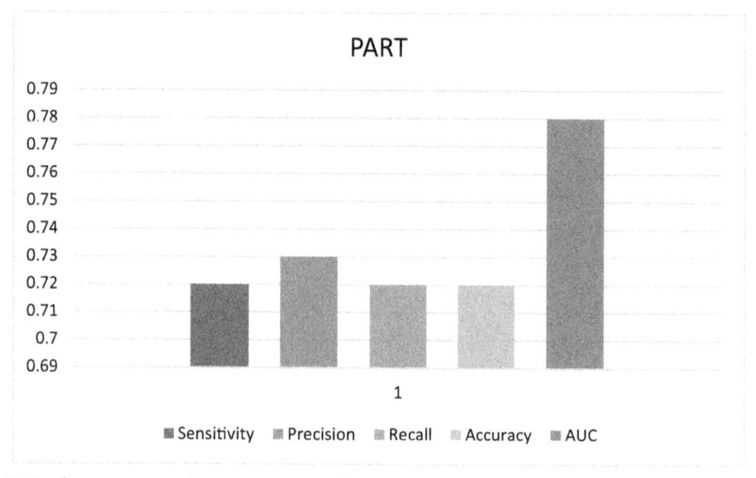

Fig. 5 Performance evaluation result of PART classifier.

good at recording a sensitivity, with a recall and accuracy value of 0.72. This result is competitive with previously applied algorithms in heart diseases. The classifier was able to correctly classify 72% of the test data and the sensitivity is balanced between the two categories of the target class. In terms of precision, PART performed effectively recording 0.73, showing its capability to give accurate results consistently. Furthermore, a 0.78 AUC value recorded by the classifier shows that it has a good measure of separability. This means it has a 78% chance of distinguishing between the presence and absence of heart disease from patient data accurately.

The JRIP rule induction classifier results as presented in Fig. 6 indicate 0.72 precision and recall values. This shows its good performance in determining the heart disease status of the patient accurately and consistently. The accuracy, sensitivity, and AUC values for the JRIP model are all 0.73, indicating good performance in terms of balance, effectiveness, and separation capability in predicting heart disease from patient data. It has a 73% power of distinguishing between patients with heart disease and without heart disease as revealed by the AUC value.

The ensemble-based Random Forest, on the other hand, produced a moderate result in terms of sensitivity, precision, recall, and accuracy recording 0.69 values in all the measures. This moderate performance may be due to the large size and the nature of the data. Its performance can be improved with attribute selection. However, the 0.74 AUC value recorded by the classifier indicates a very good performance capability in distinguishing

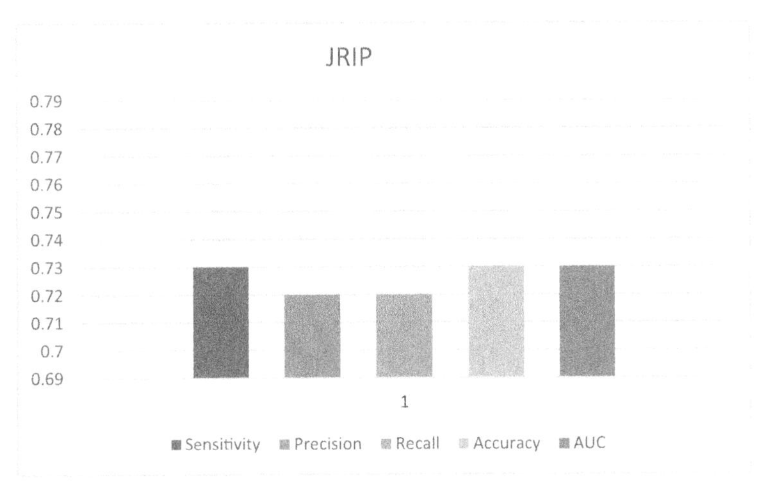

Fig. 6 Performance evaluation result of JRIP classifier.

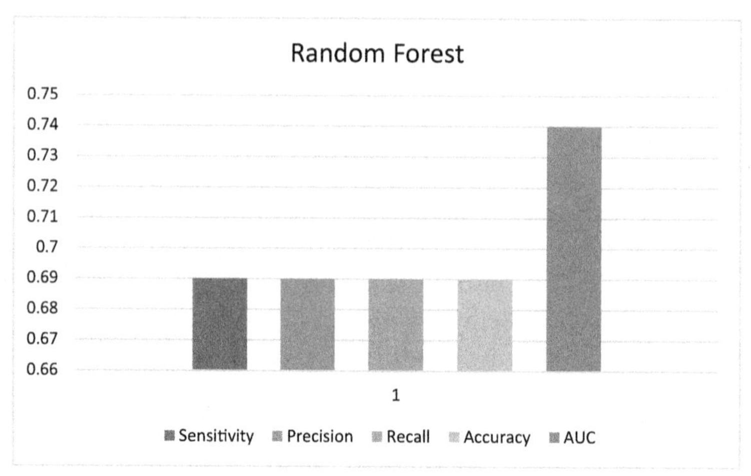

Fig. 7 Performance evaluation results of Random Forest classifiers.

between a patient with heart disease and without heart disease. These results are presented in Fig. 7.

To compare the performance of the three classifier models, Table 2 presents the results of all the classifiers considering all the evaluation metrics used.

As shown in Fig. 8, PART and JRIP recorded the best performance in terms of sensitivity and recall (0.72). In precision, PART produced the best result (0.73), followed closely by JRIP (0.72), and the least performance of 0.69 was produced by Random Forest. Contrary to the precision results, JRIP gave the best accuracy result (0.73), followed by PART (0.72). The least accuracy value was recorded by Random Forest. Considering the AUC performance measure, PART produced the significantly best result, followed by Random Forest. The least performing classifier in terms of AUC was JRIP.

Table 2 Performance results of the PART, JRIP, and Random Forest classifiers.

	PART	JRIP	Random Forest
Sensitivity	**0.72**	**0.72**	0.69
Precision	**0.73**	0.72	0.69
Recall	**0.72**	**0.72**	0.69
Accuracy	0.72	**0.73**	0.69
AUC	**0.78**	0.73	0.74

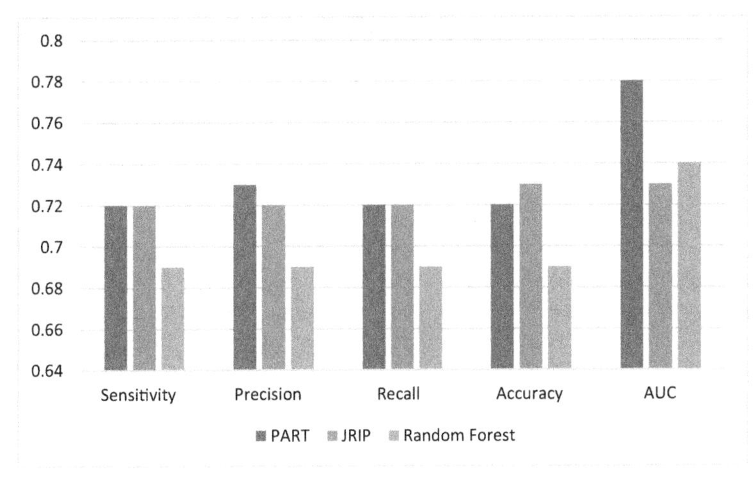

Fig. 8 Performance comparison of PART, JRIP, and Random Forest classifiers.

Consequently, it can be concluded that the PART algorithm performed better than the remaining classifiers (JRIP and Random Forest) in predicting the presence of heart diseases in patients. Also, the rule-based classifiers considered performed better than the tree-based ensemble models in predicting heart diseases. Although the general performance of the models is good, the performance can still be improved by employing feature selection techniques to reduce the dimension and complexity of the dataset.

With respect to real life applications, results from the experimental case study can be useful in developing simple rule-based expert systems for the diagnosis of heart related diseases in personalized health-care systems. Also, this proposed approach provides a data-driven approach toward heart disease diagnosis as rules are generated from medical data rather than from experts. This will therefore eliminate the need for expert knowledge in generating rules for expert systems in personalized health-care systems.

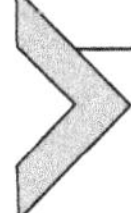

6. Proposed ensemble-based BDA framework for heart diseases diagnosis in personalized health care

Owing to the challenges of the data aggregation problem in the big data-based framework for heart disease diagnosis and unstable performance of the predictive models in personalized health-care systems, we propose the ensemble-based BDA framework that leverages on the fast and efficient processing of the SPARK big data tool, which employs hierarchical aggregation approaches by collecting heterogeneous data from different sources using IoMT and ensemble models.

The framework is composed of two components, the cloud component and the local component. The cloud component is used to collect big heterogeneous data from different IoMT devices and medical records to train the ensemble models. The local component is composed of users' data and mobile application that access the cloud component. It makes use of the built model to provide diagnosis services to heart disease patients using a mobile application. The hierarchical aggregation approach is composed of different storage levels and groups data of the same characteristics in the same storage level for easy processing. The framework employs heterogeneous ensemble models that combine predictive knowledge of different base learners (function-based, rule-based, and stochastic) to form improved and more robust metamodels for the diagnosis of heart diseases. The framework is presented in Fig. 9.

The proposed approach will provide real-time prediction and diagnosis support directly for patients from their own end without the cost of visiting the hospital; this is the goal of personalized health-care systems. It can also be

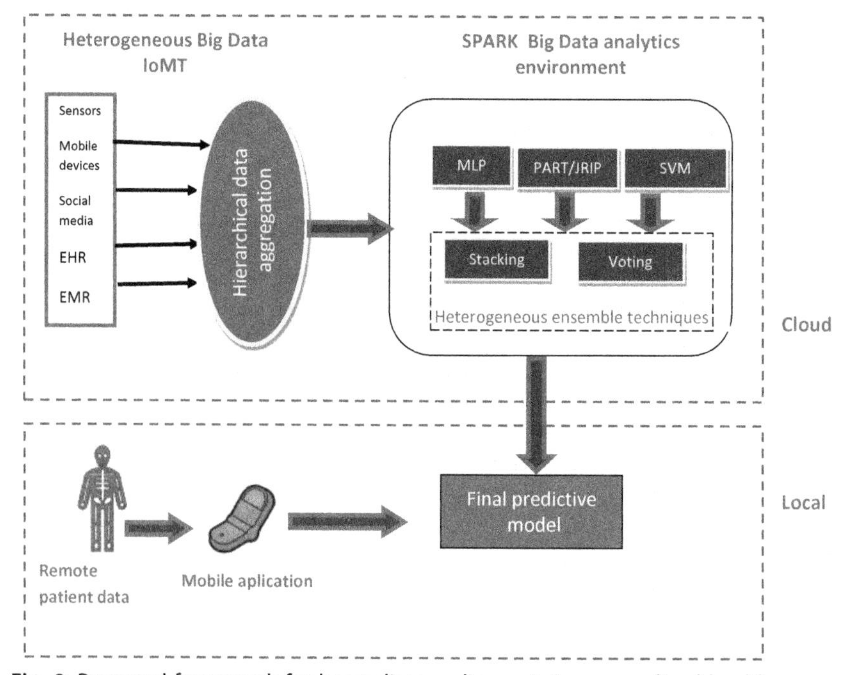

Fig. 9 Proposed framework for heart disease diagnosis in personalized health care.

integrated to health-care monitoring systems for doctors to remotely monitor patient heart status and provide real-time medical advices for managing the related diseases.

7. Conclusion and future works

Several computational techniques have been applied to medical data analysis for better health care. Machine learning techniques and algorithms are being trained for the diagnosis and prediction of cardiovascular diseases such as heart attack, arrhythmia, and coronary heart disease. Existing studies have demonstrated the extensive use of various computational techniques and algorithms majorly for feature selection and classification of medical data for prediction and diagnosis. Although not as extensive as the application of ML algorithms, fuzzy inference systems and big data analytics have also been used for heart disease prediction and diagnosis. Soft computing techniques such as fog computing and edge computing have also been recently adopted.

A taxonomy depicting the applied algorithms and techniques for the diagnosis of heart diseases was constructed. The taxonomy was used to guide in the selection of the algorithms employed in the experimental case study of this work. The algorithms selected are PART, JRIP, and Random Forest; they are the techniques not yet applied for the prediction of heart diseases. A framework was proposed for the study and used in running the experiments. Results from the experiments indicated that the rule-based classification technique PART performed the best in all the considered algorithms with an accuracy of 73% and AUC of 0.78. JRIP also gave a competitive result with an accuracy of 72% and AUC of 0.73. Random Forest also performed moderately good.

To improve the performance of the heart disease prediction models, a big data framework that uses heterogeneous ensemble methods has been proposed. It is believed that this framework will yield better results because of the diverse source of data that will be provided and the computational capability of big data analytics.

The aggregation of heterogeneous data collected from diverse sources such as medical health records, electronic health records, sensors data, stand-alone medical devices, and handheld devices is still a challenge in heart disease diagnosis systems. A robust framework for collating and organizing this multifaceted big data is required to enhance the processing of the underlying predictive models. Embedded feature selection for building predictive models is yet to be explored in the diagnosis of heart diseases. Embedded approaches such as LASSO should be considered as it will enhance the efficiency of the predictive models in the diagnosis of heart diseases. BDA approaches based on SPARK have not been really explored as this study only found one work that employed it. Findings have shown that the SPARK MLib is more efficient in processing huge big data than the mostly employed Hadoop MapReduce. More works are needed to explore this powerful tool for speedy diagnosis of heart diseases and quick response in terms of emergency. Another open area of research in IoMT-based heart disease management is the security and privacy issue. A robust security framework for IoMT data is needed to secure the patient's private medical information from unauthorized access and misuse.

In the future, we plan to implement the proposed framework with a hierarchical data aggregation approach and ensemble-based BDA approaches to build predictive models in IoMT heart disease diagnosis. Feature extraction and attribute selection would also be introduced to the BDA framework for heart disease diagnosis in the future.

References

[1] R. Snyderman, Personalized health care in 2013, N. C. Med. J. 74 (6) (2013) 478–484.

[2] K. Mahato, A. Srivastava, P. Chandra, Paper based diagnostics for personalized health care: emerging technologies and commercial aspects. Biosens. Bioelectron. 96 (2017) 246–259, https://doi.org/10.1016/j.bios.2017.05.001.

[3] C. Butts, S. Kamel-Reid, G. Batist, S. Chia, C. Blanke, M. Moore, ... K. Bonter, Benefits, issues, and recommendations for personalized medicine in oncology in Canada, Curr. Oncol. 20 (5) (2013) e475.

[4] W. Duch, What is Computational Intelligence and Where is it Going?. vol. 63, (2007) pp. 1–13, https://doi.org/10.1007/978-3-540-71984-7_1.

[4a] A.T. Azar, S. Vaidyanathan (Eds.), Computational Intelligence Applications in Modeling and Control, Springer International Publishing, Switzerland, Europe, 2015, https://link.springer.com/book/10.1007%2F978-3-319-11017-2.

[5] A. Kalantari, A. Kamsin, S. Shamshirband, A. Gani, H. Alinejad-Rokny, A.T. Chronopoulos, Computational intelligence approaches for classification of medical data: state-of-the-art, future challenges, and research directions, Neurocomputing 276 (2018) 2–22.

[5a] D.S. Watson, et al., Clinical applications of machine learning algorithms: beyond the black box, BMJ 364 (2019) 1–9.

[5b] T. Hothor, Machine Learning and Statistical Learning, http://cran.ms.unimelb.edu.au/web/views/MachineLearning.html.

[6] S.S. Kamble, A. Gunasekaran, M. Goswami, J. Manda, A systematic perspective on the applications of big data analytics in healthcare management, Int. J. Healthc. Manag. 12 (3) (2019) 226–240.

[7] K.H. Mohammed, I.E. Ali, M.E. Sally, M.S. Amany, Big data challenges and opportunities in healthcare informatics and smart hospitals. in: Security in Smart Cities: Models, Applications, and Challenges, 2019, pp. 27–45, https://doi.org/10.1007/978-3-030-01560-2.

[8] S. Shafqat, S. Kishwer, R.U. Rasool, J. Qadir, T. Amjad, H.F. Ahmad, Big data analytics enhanced healthcare systems: a review, J. Supercomput. 76 (2018) 1–46.
[9] J. Gubbi, R. Buyya, S. Marusic, M. Palaniswami, Internet of things (IoT): a vision, architectural elements, and future directions, Futur. Gener. Comput. Syst. 29 (7) (2013) 1645–1660.
[10] S. Jabbar, F. Ullah, S. Khalid, M. Khan, K. Han, Semantic interoperability in heterogeneous IoT infrastructure for healthcare, Wirel. Commun. Mob. Comput. 2017 (2017) 1–10.
[11] S. Hamilton, E. Gunther, R. Drummond, S.E. Widergren, Interoperability: a key element for the grid and DER of the future, in: Proceedings of IEEE PES Transmission and Distribution, 2006, pp. 927–931.
[12] G. Xiao, J. Guo, L. Da Xu, Z. Gong, User interoperability with heterogeneous IoT devices through transformation, IEEE Trans. Ind. Inf. 10 (2) (2014) 1486–1496.
[13] K. Boeckl, M. Fagan, W. Fisher, N. Lefkovitz, K. Megas, E. Nadeau, … K. Scarfone, Considerations for Managing the Internet of Things (IoT) Cybersecurity and Privacy Risks, (2018) National Institute of Standards and Technology, NISTIR 8228 (Draft.
[14] S. Ghanavati, J. Abawajy, D. Izadi, An alternative sensor cloud architecture for remote patient healthcare monitoring and analysis, in: Paper Presented at the 2016 IEEE Int'l. Joint Conf. Neural Networks, 2016.
[15] A. Agra, M. Christiansen, K.S. Ivarsøy, I.E. Solhaug, A. Tomasgard, Combined ship routing and inventory management in the salmon farming industry, Ann. Oper. Res. 253 (2) (2017) 799–823.
[16] A. Naval, How Do IoT Works?—IoT Architecture Explained With Examples, Available at:https://www.avsystem.com/blog/what-is-iot-architecture/, 2019, July 5.
[16a] B. Marr, Why the Internet of Medical Things (IoMT) Will Start to Transform Healthcare in 2018, [Online]. Available at: https://www.forbes.com/sites/bernardmarr/2018/01/25/whythe-internet-of-medical-things-iomt-will-start-to-transform-healthcare-in-2018/, 2018. Accessed 26 July 2018.
[17] A. Abdelaziz, A.S. Salama, A. Riad, A.N. Mahmoud, A machine learning model for predicting chronic kidney disease based internet of things and cloud computing in smart cities, in: Security in Smart Cities: Models, Applications, and Challenges, Springer, 2019, pp. 93–114.
[18] B.R. Ray, M.U. Chowdhury, J.H. Abawajy, Secure object tracking protocol for the internet of things, IEEE Internet Things J. 3 (4) (2016) 544–553.
[19] J. Qi, P. Yang, G. Min, O. Amft, F. Dong, L. Xu, Advanced internet of things for personalized healthcare systems: a survey, Pervasive Mob. Comput. 41 (2017) 132–149.
[20] B. Marr, Why the Internet of Medical Things (IoMT) Will Start to Transform Healthcare in 2018, [Online]. Available at: https://www.forbes.com/sites/bernardmarr/2018/01/25/whythe-internet-of-medical-things-iomt-will-start-to-transform-healthcare-in-2018/, 2018. Accessed 26 July 2018.
[21] P. Verma, S.K. Sood, Cloud-centric IoT based disease diagnosis healthcare framework, J. Parallel Distrib. Comput. 116 (2018) 27–38.
[22] R. Ani, S. Krishna, N. Anju, M.S. Aslam, O. Deepa, IoT based patient monitoring and diagnostic prediction tool using ensemble classifier, in: Paper presented at the 2017 International Conference on Advances in Computing, Communications, and Informatics (ICACCI), 2017.
[23] F. Ali, S.R. Islam, D. Kwak, P. Khan, N. Ullah, S.-j. Yoo, K.S. Kwak, Type-2 fuzzy ontology–aided recommendation systems for IoT–based healthcare, Comput. Commun. 119 (2018) 138–155.
[24] American Heart Association, What Is Cardiovascular Disease (Heart Disease)? American Heart Association, 2014, 2011.

[25] R. De Rosa, T. Palmerini, S. De Servi, M. Belmonte, G. Crimi, S. Cornara, ... A. Toso, High on-treatment platelet reactivity and outcome in elderly with non ST-segment elevation acute coronary syndrome-Insight from the GEPRESS study, Int. J. Cardiol. 259 (2018) 20–25.
[26] G. Galasso, S. De Servi, S. Savonitto, T. Strisciuglio, R. Piccolo, N. Morici, ... F. Piscione, Effect of an invasive strategy on outcome in patients ≥75 years of age with the non-ST-elevation acute coronary syndrome, Am. J. Cardiol. 115 (5) (2015) 576–580.
[27] S.C. Smith, A. Collins, R. Ferrari, D.R. Holmes, S. Logstrup, D.V. McGhie, ... K. Taubert, Our time: a call to save preventable death from cardiovascular disease (heart disease and stroke), J. Am. Coll. Cardiol. 60 (22) (2012) 2343–2348.
[28] A. Aggarwal, S. Srivastava, M. Velmurugan, Newer perspectives on coronary artery disease in young, World J. Cardiol. 8 (12) (2016) 728.
[29] P. Puska, B. Norrving, S. Mendis, Global Atlas on Cardiovascular Disease Prevention and Control, World Health Organization, Geneva, Switzerland, 2011.
[30] J. Zhao, P. Papapetrou, L. Asker, H. Boström, Learning from heterogeneous temporal data in electronic health records, J. Biomed. Inform. 65 (2017) 105–119.
[31] A. Silverio, P. Cavallo, R. De Rosa, G. Galasso, Big health data and cardiovascular diseases: a challenge for research, an opportunity for clinical care, Front. Med. 6 (2019) 36.
[32] G.M. Weber, K.D. Mandl, I.S. Kohane, Finding the missing link for big biomedical data, JAMA 311 (24) (2014) 2479–2480.
[33] M.R. Cowie, J.I. Blomster, L.H. Curtis, S. Duclaux, I. Ford, F. Fritz, ... M. Leenay, Electronic health records to facilitate clinical research, Clin. Res. Cardiol. 106 (1) (2017) 1–9.
[34] B. Subanya, R. Rajalaxmi, Feature selection using Artificial Bee Colony for cardiovascular disease classification, in: Paper Presented at the 2014 International Conference on Electronics and Communication Systems (ICECS), 2014.
[35] B. Subanya, R.R. Rajalaxmi, A novel feature selection algorithm for heart disease classification, Int. J. Comput. Intell. Informatics 4 (2) (2014) 117–124.
[36] M.H. Song, J. Lee, S.P. Cho, K.J. Lee, S.K. Yoo, Support vector machine-based arrhythmia classification using reduced features, Artif. Intell. Med. 44 (1) (2008) 51–64.
[37] S.-W. Lin, Z.-J. Lee, S.-C. Chen, T.-Y. Tseng, Parameter determination of support vector machine and feature selection using the simulated annealing approach, Appl. Soft Comput. 8 (4) (2008) 1505–1512.
[38] M. Gutlein, E. Frank, M. Hall, A. Karwath, Large-scale attribute selection using wrappers, in: Paper Presented at the 2009 IEEE Symposium on Computational Intelligence and Data Mining, 2009.
[39] M.A. Esseghir, Effective wrapper-filter hybridization through grasp schemata, in: Paper Presented at the Feature Selection in Data Mining, 2010.
[40] A. Unler, A. Murat, R.B. Chinnam, mr2PSO: a maximum relevance minimum redundancy feature selection method based on swarm intelligence for support vector machine classification, Inf. Sci. 181 (20) (2011) 4625–4641.
[41] J. Nahar, T. Imam, K.S. Tickle, Y.-P.P. Chen, Computational intelligence for heart disease diagnosis: a medical knowledge-driven approach, Expert Syst. Appl. 40 (1) (2013) 96–104.
[42] S. Shilaskar, A. Ghatol, Feature selection for medical diagnosis: evaluation for cardiovascular diseases, Expert Syst. Appl. 40 (10) (2013) 4146–4153.
[43] X. Liu, X. Wang, Q. Su, M. Zhang, Y. Zhu, Q. Wang, Q. Wang, A hybrid classification system for heart disease diagnosis based on the RFRS method. Comput. Math. Methods Med. 2017 (2017) https://doi.org/10.1155/2017/8272091.
[44] C.B. Gokulnath, S. Shantharajah, An optimized feature selection based on a genetic approach and support vector machine for heart disease, Clust. Comput. 22 (2019) 14777–14787.

[45] A. Abdelaziz, A.S. Salama, A. Riad, A swarm intelligence model for enhancing health care services in smart cities applications, in: Security in Smart Cities: Models, Applications, and Challenges, Springer, 2019, pp. 71–91.

[46] M. Elhoseny, A. Abdelaziz, A.S. Salama, A.M. Riad, K. Muhammad, A.K. Sangaiah, A hybrid model of internet of things and cloud computing to manage big data in health services applications. Futur. Gener. Comput. Syst. 86 (2018) 1383–1394, https://doi.org/10.1016/j.future.2018.03.005.

[47] A.K.M.J.A. Majumder, Y.A. ElSaadany, R. Young, D.R. Ucci, An energy-efficient wearable smart IoT system to predict cardiac arrest. Adv. Hum. Comput. Interact. 2019 (2019) 1–21, https://doi.org/10.1155/2019/1507465.

[48] G. Manogaran, R. Varatharajan, D. Lopez, P.M. Kumar, R. Sundarasekar, C. Thota, A new architecture of internet of things and big data ecosystem for secured smart healthcare monitoring and alerting system. Futur. Gener. Comput. Syst. 82 (2018) 375–387, https://doi.org/10.1016/j.future.2017.10.045.

[49] T. Santhanam, M. Padmavathi, Application of K-means and genetic algorithms for dimension reduction by integrating SVM for a diabetes diagnosis, Prog. Comput. Sci. 47 (2015) 76–83.

[50] D. Jain, V. Singh, Feature selection and classification systems for chronic disease prediction: a review, Egypt. Inform. J. 19 (3) (2018) 179–189.

[51] J.H. Abawajy, M.M. Hassan, Federated internet of things and cloud computing pervasive patient health monitoring system, IEEE Commun. Mag. 55 (1) (2017) 48–53.

[52] A. Page, Building Decision Support Systems From Large Electrocardiographic Data Sets, Retrieved from: https://urresearch.rochester.edu/institutionalPublicationPublicView.action?institutionalItemId=31019, 2016.

[53] F. Firouzi, A.M. Rahmani, K. Mankodiya, M. Badaroglu, G.V. Merrett, P. Wong, B. Farahani, Internet-of-things and big data for smarter healthcare: from device to architecture, applications, and analytics. Futur. Gener. Comput. Syst. 78 (2018) 583–586, https://doi.org/10.1016/j.future.2017.09.016.

[54] S. Tuli, N. Basumatary, S.S. Gill, M. Kahani, R.C. Arya, G.S. Wander, R. Buyya, HealthFog: an ensemble deep learning-based smart healthcare system for automatic diagnosis of heart diseases in integrated IoT and Fog computing environments, Futur. Gener. Comput. Syst. 104 (2019) 187–200.

[55] K.S. Adewole, A.G. Akintola, S.A. Salihu, N. Faruk, R.G. Jimoh, Hybrid Rule-Based Model for Phishing URLs Detection. vol. 1, (2019)https://doi.org/10.1007/978-3-030-23943-5_9.

[56] A. Rajput, R.P. Aharwal, M. Dubey, S.P. Saxena, J48 and JRIP rules for E-governance data, Int. J. Comput. Sci. Secur. 5 (2) (2011) 201–207.

[57] E. Frank, I.H. Witten, Generating Accurate Rule Sets Without Global Optimization, (Working paper 98/2) The University of Waikato, Department of Computer Science, Hamilton, New Zealand, 1998.https://hdl.handle.net/10289/1047.

[58] X. Zhu, X. Du, M. Kerich, F.W. Lohoff, R. Momenan, Random forest-based classification of alcohol dependence patients and healthy controls using resting-state MRI. Neurosci. Lett. 676 (2018) 27–33, https://doi.org/10.1016/j.neulet.2018.04.007.

Further reading

S. Alelyani, A. Ibrahim, Internet-of-things in telemedicine for diabetes management, in: Paper Presented at the 2018 15th Learning and Technology Conference (L&T), 2018.

P. Kaur, N. Sharma, A. Singh, B. Gill, CI-PDF: a cloud IoT based framework for diabetes prediction, in: Paper Presented at the 2018 IEEE 9th Annual Information Technology, Electronics and Mobile Communication Conference (IEMCON), 2018.

K. Ng, S.R. Steinhubl, C. Defilippi, S. Dey, W.F. Stewart, Early detection of heart failure using electronic health records: practical implications for time before diagnosis, data diversity, data quantity, and data density, Circ. Cardiovasc. Qual. Outcomes 9 (6) (2016) 649–658.

J. Sun, J. Hu, D. Luo, M. Markatou, F. Wang, S. Edabollahi, … W.F. Stewart, Combining knowledge and data-driven insights for identifying risk factors using electronic health records, in: Paper Presented at the AMIA Annual Symposium Proceedings, 2012.

CHAPTER EIGHT

An improved canny detection method for detecting human flexibility

Xu Lu and Yujing Zhang
School of Automation, Guangdong Polytechnic Normal University, Guangzhou, China

1. Introduction

Recent studies have shown that physical activity is one of the most important factors in maintaining human health [1]. Researchers in related fields generally believe that human flexibility is important for safe and effective exercise, as most physical exercise and sports programs include activities that promote flexibility [2]. Early researchers believed that human flexibility includes "static" flexibility and "dynamic" flexibility [3]. Specifically, static flexibility refers to the linear or angular measure of the actual motion limit of a single joint or complex joint, that is, the static flexibility measures a joint or a set of joint range of motion. On the other hand, dynamic flexibility indicates the ability to relax muscle tension or stretching passive impedance increase rate, the flexibility of the dynamic impedance changes to be construed as an extension over the entire range of motion. However, dynamic flexibility is related to the speed, strength, and coordination of motion, and has nothing to do with flexibility [1]. Therefore, the current flexibility mainly refers to static flexibility.

In physical education and sports medicine, the definition of flexibility may be the elasticity or stretch ability of soft tissues such as muscles and ligaments [4]. More specifically, it refers to the range of movement of human joints and its manifestation of exercise ability. It is an important physical quality of the human body and plays an irreplaceable role in sports training and daily life. Poor flexibility may lead to difficulties in performing and maintaining activities in daily life, causing chronic musculoskeletal injuries and back muscle problems, and even affecting the posture of the human body, walking gait, and falling [5]. At present, the most extensive method for measuring the static flexibility of the human body is the sit-and-reach

Intelligent IoT Systems in Personalized Health Care
https://doi.org/10.1016/B978-0-12-821187-8.00008-3

(SR) test [6]. This measurement method involves bending the hip, lumbar vertebrae, and thoracic vertebrae. The measurement index measures the distance of the finger or the head. Actually, the measurement of human flexibility is mainly achieved by measuring the anteflexion angle of the human body. However, the accuracy of the SR test results is usually affected by the length of the limb [7]. What is worse is the measurement of flexibility requires specific electronic equipment, which is inconvenient to measure, high in cost.

With the development of computer and artificial intelligence technology, the application of image-processing technology can be seen everywhere in life, such as license plate number recognition [8], face recognition [9], two-dimensional code (barcode) recognition [10]. Especially in the detection of human body images, it is also widely used, such as pedestrian detection [11], attitude estimation [12], gesture recognition [13], etc. These detection technologies are applied to security monitoring, driverless, intelligent transportation, and intelligent robots [14–17]. Inspired by this, edge detection and feature point extraction algorithm are applied to the human body flexibility detection. While the tester's sitting body flexion image is obtained by the imaging device, it makes the measurement of flexibility convenient and can play a certain role in promoting people's health and exercise.

In this chapter, a human flexibility fitness detection algorithm based on improved edge detection and feature point extraction is proposed. The proposed algorithm in this chapter is used to identify the human body's flexion angle and achieve flexibility detection. This method can solve the problem of inconvenient measurement of the conventional electronic measuring device and improve the reliability of the detection result. Our contributions are mainly summarized as follows:

(1) A human flexibility fitness detection technology based on image-processing technology is proposed in this chapter.

(2) An improved Canny operator is proposed, which uses improved filter instead of traditional Gaussian filter, and an improved gradient calculation method is presented, which can effectively suppress noise and retain edge information.

(3) For calculation of the human body anteflexion angle, a special detection point extraction algorithm is proposed.

(4) The proposed algorithm can be applied to the flexibility test in people's daily life, reducing the use of large electronic measuring equipment and reducing the cost of measurement.

The details of the design and experimentation of the algorithm are explained in Sections 2–5. Section 2 introduces related recent work. Section 3 describes the methods for recognizing the angle of body anteflexion and presents an image edge detection algorithm that initially eliminates noise and improves the contrast in body anteflexion images. In Section 4, the performance of the image analysis algorithm under different conditions is studied. Section 5 summarizes the chapter and looks forward to future work.

2. Related work

Flexibility fitness is recognized by the world as an important part of physical fitness. As an important factor in health, scholars around the world have conducted long-term research on its measurement methods. The most traditional way to measure the flexibility of the hamstrings and lower back of the human body is the SR test [18]. This method was originally proposed by Wells and Dillon [19]. The measurement procedure is simple, easy to manage, and does not require complicated training. However, this measurement method ignores the effect of different limb lengths on the measurement results. In order to eliminate this effect, a modified SR test [20] method was proposed. In the improved method, the distance from the finger to the box was combined to solve the influence of the length of the limb on the measurement result. However, this measurement had no evaluation system for the method at the time. In Ref. [5], Castro-Piñero et al. experimented with traditional SR test and modified SR test. In experiments with children and adolescents aged 6–17, it was shown that the modified SR test is no more effective than traditional methods. Although other measurement methods have been proposed, such as chair sit-back, V-SR test, back-saver SR test, and toe-touch test [21], these measurement methods are aimed at measuring different people and developing different evaluation criteria, but most of their evaluation criteria or measurement results need to be improved.

Studies have shown that the measurement of flexibility fitness is actually the angle at which joint motion is measured. Some researchers then studied the criteria for measuring through the protractor. The measurement result by this measurement method can improve the accuracy of the final detection result with respect to the previously proposed detection method. However, the use of the goniometer in the test has some difficulties, the operation is cumbersome, and to some extent, it brings inconvenience to the detection of flexibility, especially in public places such as schools.

The development of artificial intelligence technology, especially the wide application of image-processing technology, provides a new idea for solving the problem of human flexibility detection. In this chapter, the method of edge detection and feature point extraction techniques is used to achieve human flexibility detection. Classic edge detection operators include Robert, Laplace, Sobel, Prewitt, and Canny, and so on [22]. These detection operators are widely used in image processing, and researchers have improved them to achieve better detection results according to the actual situation, for example, Meng Yingchao et al. [23] proposed a locally adaptive Canny edge detection algorithm, which divides the image into many square sub-images and automatically finds the local high and low thresholds of each sub-image for further edge detection. It has more advantages than global thresholds. And also, in order to improve the adaptability of the Canny edge detection operator, Fu Fengzhi et al. [24] proposed a gray histogram of the image to automatically determine the threshold of image segmentation, and in the algorithm, anisotropic diffusion filter is used for filtering. The edge information is preserved while denoising. The algorithm has good noise resistance while detecting more edge details. In the improvement of the filter of the edge detection operator, there are also many related research results. Lin WeiChun et al. [25] proposed a quasi-high-pass filter for edge detection in medical images. The algorithm reexpress the mathematical formula of the detector into a quadratic form of the Toeplitz matrix, this form has rich spatial homomorphic symmetry and is robust to noise. On the other hand, the development of neural network technology also provides a new method for edge detection. This algorithm has strong robustness and can detect more edge features. Dorafshan et al. [26] use AlexNet network to detect cracks in concrete structures compared with traditional edge detection methods. The results show that the detection results using the neural network are much higher than the traditional detection method, and can detect very fine cracks; Nhat-Duc [27] also uses convolutional neural networks for road cracks. The detection results show that the accuracy of the method using a neural network is much higher than that of the traditional edge detection method. For feature point extraction, in addition to the classic corner detection algorithm, such as Harris, it is more necessary to design a specific feature point extraction algorithm according to the actual situation, such as in order to extract the key points of the weak texture region in the two-dimensional image matching. Tian Y et al. [28] proposed a linear multi-scale image enhancement matching algorithm based on image fusion technology. The algorithm mainly generates weighted images of the original image according to saturation, brightness, contrast, and other measurement factors enhances the distribution weight of the weak texture and fuses the Gaussian pyramid to fuse more details into the original image. In short, the algorithm can enhance the details of weakly textured areas, and then extract more and more accurate feature points. Zeng et al. [29] proposed a method of extracting feature points by contour detection and corner detection, which is based on the fact that the weld is regarded as a special line composed of feature points. In the past few years, due to the needs of people's material life and social development, the estimation of human body posture based on neural network and the study of clothing feature points have been proposed one after another, and also achieved good research results.

Canny operator is an optimal edge detection algorithm, and many modified algorithms are derived. These derived algorithms have also achieved better performance. In this chapter, we will improve the filter and gradient calculation algorithm of the Canny operator, and the Otsu algorithm is used to segment the threshold, and finally propose a feature point extraction algorithm based on human edge information. In this chapter, the method consists of four steps, as detailed in the further sections.

3. Recognizing the angle of body anteflexion

In order to recognize the human body angle, it is first necessary to obtain a human body image, and the acquisition device may be a smartphone or other photographic device. After the image is acquired, the image is processed using a related algorithm, as shown in Fig. 1. Firstly, the image is converted into a grayscale image, then the edge detection of the grayscale image is performed so that the edge information of the human body is extracted, and then the feature points are extracted according to the obtained edge information, specifically the feature points of the head, the buttocks, and the foot. Finally, through the coordinate information of the feature points, the magnitude of the body anteflexion angle is calculated, and the flexibility score is calculated according to the relevant flexibility level calculation formula.

3.1 Image preprocessing

Image preprocessing primarily aims to eliminate irrelevant information from pictures, restore real useful data, enhance the detectability of the relevant

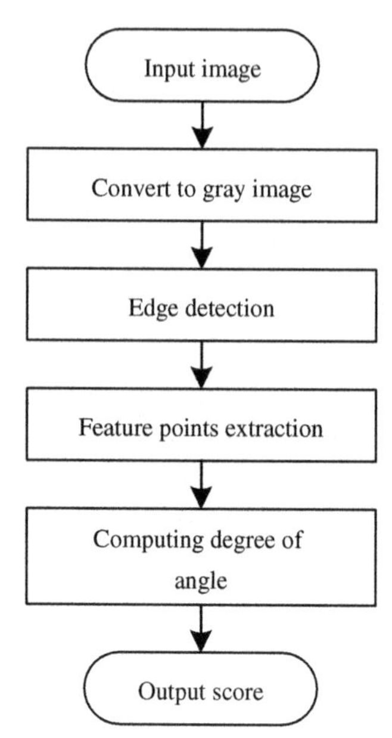

Fig. 1 Human body anteflexion angle recognition.

information, and simplify the data to the maximum extent to improve feature extraction and image segmentation.

A color image is made up of three different components, which are called a three-channel image. When processing color images, three channels are often needed to be processed, in turn, the time cost will be very large. Therefore, in order to improve the processing speed of the entire application system, it is necessary to reduce the amount of data that needs to be processed. In image processing, the graying of the image is the grayscale image of the data from the RGB three channels to the grayscale image of the single-channel data. In the RGB model above, if R, G, B have equal values, even three-channel data is a grayscale color in the representation, and the grayscale value is the value of R, G, and B. The grayscale range for each pixel position in a grayscale image is from 0 to 255. It can be found that when R, G, and B are equal, only one byte is needed to hold the grayscale value, and for R, G, and B different three-channel images, can also be grayed.

At present, there are four main methods of graying most commonly used, including single component method, maximum value method, average value method, and weighted average method. In this chapter, the weighted average method is used for grayscale processing, and its expression is shown in the following equation:

$$Y = 0.299R + 0.587G + 0.114B \tag{1}$$

where R, G, and B are the components of the color image.

3.2 Edge detection

Image preprocessing is necessary for the study to reduce desired image details and improve image quality. This process determines the accuracy of the final results. Then, a revised Canny edge detector is adopted for accurate detection.

The human body anteflexion angle detection algorithm in this paper is based on human limbic information. In the existing edge detector, the Canny edge detection operator is considered to be the optimal detection operator. The Canny edge detection operator was proposed by John F. Canny in 1986 [30]. Most importantly, he created the edge detection theory and explained how the technology works. The Canny edge detector satisfies the main evaluation criteria, namely, signal-to-noise ratio (SNR), localization accuracy, and single edge response criteria [31]. Among the edge detection methods proposed so far, the Canny edge detector is the most rigorously defined and is widely used. The popularity of the Canny edge detector can be attributed to its optimality according to the three criteria [32].

The steps of the traditional Canny edge detection algorithm are shown in Fig. 2, and the detail is expressed as follows [33].

(1) The image is smoothed with a suitable Gaussian filter to reduce desired image details.
(2) The gradient magnitude and direction of each pixel are calculated.

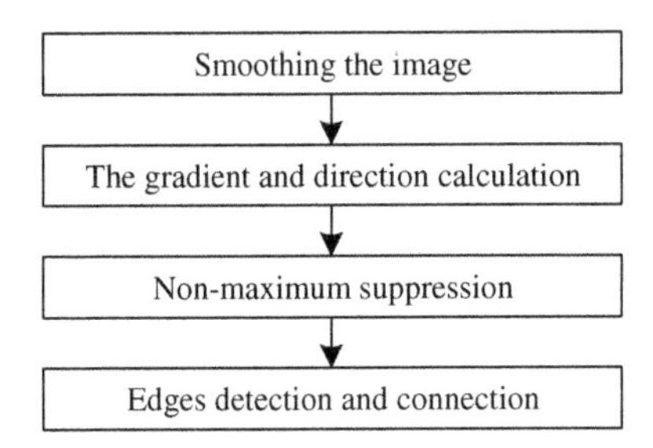

Fig. 2 The traditional Canny edge detection.

(3) If the gradient magnitude at a pixel is larger than those at its two neighbors in the gradient direction, then the pixel is marked as an edge. Otherwise, the pixel is deemed background.
(4) Edges are detected and connected with double threshold processing and connection analysis.

A new edge detector for aid in recognizing the human body anteflexion angle in this chapter.

3.2.1 Adaptive Gaussian filter

Gaussian filtering is a linear smoothing filter that eliminates Gaussian noise and is widely used in the noise reduction process of image processing. In general, Gaussian filtering is a process of weighted averaging of the entire image. The value of each pixel is obtained by weighted averaging of other pixel values in itself and in the neighborhood. However, as the edge and the noise both are high-frequency signal [34], the filter operator suppresses noise while the edges of the image are smoothed, which affects edge detection accuracy.

In this chapter, adaptive Gaussian filtering is used to perform filtering processing [35]. According to an image correlations rule, the pixels that cannot be polluted by noise must be correlated with statistical properties of its neighboring region, and its gray value should be closed to its neighboring region value in the image if it is a useful information point rather than a small probability event unexpectedly, unless the singularity represents the occurrence [36]. According to the difference between the pixel mean value in the neighborhood and the pixel value of the neighborhood center point, Gaussian filtering is performed. When the difference is relatively large, the neighborhood Gaussian filtering is performed. Otherwise, no filtering is performed.

Considering the two-dimensional input grayscale image as $f(x, y)$ and $f(x_0, y_0)$ is pixel of the center point in a neighborhood named A in the image. Set D as the difference between the gray value of the pixel at the point and the pixel of its neighborhood.

$$D = f(x_0, y_0) - \frac{1}{n}\sum_{A} f(x_i, y_i) \tag{2}$$

where n is the size of the neighborhood. If the whole pixel is regarded as a random variable the value of the expectation of D is 0. According to statistical theory [37], if A is a smooth region when the gray value of the pixel is not large, it can be approximated that the difference D between the gray value of the center point pixel and the mean value of the neighborhood pixel is approximately 0. And if A is the edge region, a part of the pixel gray value is smaller than the central boundary pixel gray value in the neighborhood, and the other part is larger than the boundary pixel gray value. It can be considered that the pixel positive and negative offset each other after the neighborhood center point pixel value and the neighborhood, the difference D of the pixel mean is also approximately 0.

Suppose the image interfers with the noise, it can be expressed as

$$f'(x, y) = f(x, y) + n(x, y) \tag{3}$$

where $n(x, y)$ is the noise signal. With noise, the difference between the gray value of the pixel affected by the noise and the neighboring pixel is defined as

$$D' = f(x_0, y_0) + n(x_0, y_0) - \frac{1}{n}\sum_A f(x_i, y_i) = D + n(x_0, y_0). \tag{4}$$

Since the interference-free D is approximately 0, the mean value difference between the noise-affected image pixel point gray value and its neighboring pixels, named D', is approximately equal to$n(x_0, y_0)$, which is the degree of noise hopping.

Based on the previous analysis, an adaptive Gaussian filter according to the absolute value of the neighborhood pixel point gray value and the local mean difference is proposed in this study.

$$h(x, y) = f(x, y) \odot \omega(x, y) \tag{5}$$

where $\odot$ is the convolution operator, $f(x, y)$ is the input image, and $h(x, y)$ is the output image. $\omega(x, y)$ is the adaptive filter operator.

$$\omega(x, y) = \begin{cases} \delta(x, y), & \left| f(x_0, y_0) - \frac{1}{n}\sum_A f(x_i, y_i) \right| \le T \\ \varphi(x, y) & \left| f(x_0, y_0) - \frac{1}{n}\sum_A f(x_i, y_i) \right| > T \end{cases} \tag{6}$$

where $\delta(x, y)$ is an impulse function, and T is the threshold of filtering, and $\varphi(x, y)$ is a binary Gaussian filtering function as follows:

$$\varphi(x, y) = \frac{1}{2\pi\sigma^2} \exp\left(\frac{x^2 + y^2}{2\sigma^2}\right) \tag{7}$$

where σ^2 is the variance.

The value of T is determined by the extent of the image affected by noise. Moreover, the influence extent of the image can be judged by the set of the absolute difference between the gray value of all pixel points and the mean value of the neighborhood, named DE. If the maximum, named $max(DE)$, in the DE is large, it can generally be considered that the point is more affected by noise, and conversely, the image can be considered to be less affected, so the threshold in this study is $T = \max(DE) \cdot \varepsilon$, where ε is approximately equal to 0.2 generally, the size of A is 3×3, the parameters of Gaussian function take 0.8, and max() is the maximum function.

To summarize, the revised Gaussian filter follows the steps.

S1. Calculate the difference between the grayscale value of each point in the image and the grayscale mean of the neighborhood in which it resides.

S2. Determine the absolute value of D and T. If $| \, |D| \leq T$, do nothing to the center point, the neighborhood is filtered by a defined filter operator.

S3. Cycle the first and second steps until all points have been processed.

3.2.2 Calculating the amplitude and direction of the gradient

The gradient of a certain pixel in the image represents the direction of the maximum change rate of the point. When Canny edge detection algorithm was just proposed, only the first partial derivatives of the x and y direction were calculated [38]. Some scholars improved the calculation method of partial derivatives, such as changing the size of difference operator from 2×2 to 3×3, or increasing the difference operator in 45° and 135° directions, so as to have the maximum response along the diagonal direction.

In this chapter, a 24-pixel neighborhood in the calculation of four directions is presented as shown in Fig. 3. For each pixel of the image point $I(x, y)$, the four 5×5 templates are used to calculate the size and direction of the

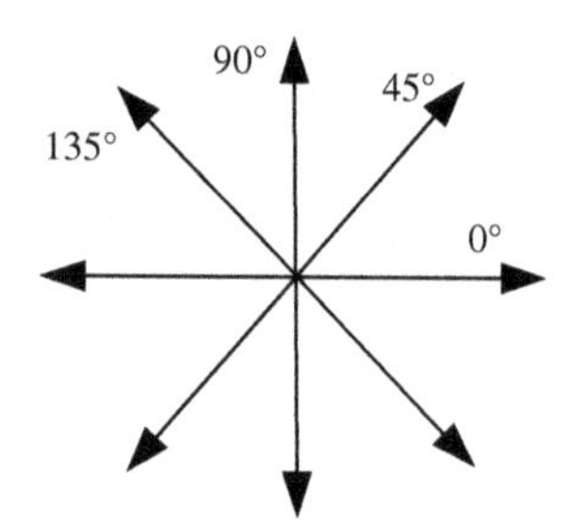

Fig. 3 Direction setting of Sobel operator.

gradient. The methods of calculation are the gradient amplitude, the edge location, and the requirements of noise suppression. In this experiment, this method achieves good results. The weight of each position in the templates is deduced based on Pascal triangle correlation theory as follows [39], here we take x direction as an example.

Set the smoothing filter function as the following equation.

$$S_m = \frac{(L-1)!}{(L-1-m)!m!} \tag{8}$$

where L is the size of the template, m is the point position of the x direction template, ! is the factorial operator, and S_m is the optimal smoothing operator coefficient, As shown in Fig. 4.

Difference function is

$$D_m = P_{\text{pascal}}(m, L-2) - P_{\text{pascal}}(m-1, L-2) \tag{9}$$

$$P_{\text{pascal}}(k, r) = \begin{cases} \dfrac{r!}{(r-k)!k!} & 0 \leq k \leq r \\ 0 & \text{otherwise} \end{cases} \tag{10}$$

where D_m is an optimal difference operator, as shown in Fig. 5.

The smoothing coefficient and difference coefficient of the traditional Sobel operator with a size of 2 × 2 templates are the factors with a size of two, respectively. The operator in this method corresponds to a coefficient

	Size of template
1 1	2
1 2 1	3
1 3 3 1	4
1 4 6 4 1	5

Fig. 4 Optimal smoothing operator.

	Size of template
1 −1	2
1 0 −1	3
1 1 −1 −1	4
1 2 0 −2 −1	5

Fig. 5 Optimal difference operator.

with a 5 × 5 template. To get the coefficient of the larger template, we can push it in the same way.

$$\text{sobel } X(\lambda) = \sum_{m=0}^{L-1} \sum_{n=0}^{L-1} S_n D_m \lambda \tag{11}$$

$$\text{sobel } Y(\lambda) = \sum_{m=0}^{L-1} \sum_{n=0}^{l-1} S_m D_n \lambda \tag{12}$$

where m and n are the point position of the X and Y direction template, respectively. S_m, D_m, S_n, and D_n are the optimal discrete smoothing operator coefficient and optimal difference system, respectively. And λ is adjustment factor, its value is 1 in this work.

The templates in the direction of 0° and 90° can be derived from Eqs. (11) and (12) as shown in Fig. 6A and C. The templates in the 45° and 135° directions are obtained after rotating the corresponding angles by 0° and 90°, respectively. The operation is not only based on the optimal smoothing on one axis and optimal difference principle on the other axis but also consider the distance between the window points and the center points in the template.

In the Sobel operator templates that have been improved above, each template consists of numerical symmetry that is opposite to each other and has an equal absolute value. The current pixel in the image has a specific template in the improved Sobel operator template. Next, the improved algorithm is illustrated by the template in the 0° direction as an example. In the following equation, H_0 is defined as the result of the calculation of pixels and template convolution in the image.

$$H_0 = \begin{bmatrix} f_{11} & 4f_{12} & 6f_{13} & 4f_{14} & f_{15} \\ 2f_{21} & 8f_{22} & 12f_{23} & 8f_{24} & 2f_{25} \\ 0 & 0 & 0 & 0 & 0 \\ -2f_{41} & -8f_{42} & -12f_{43} & -8f_{44} & -2f_{45} \\ -f_{51} & -4f_{52} & -6f_{53} & -4f_{54} & -f_{55} \end{bmatrix}. \tag{13}$$

$$\begin{bmatrix} 1 & 4 & 6 & 4 & 1 \\ 2 & 8 & 12 & 8 & 2 \\ 0 & 0 & 0 & 0 & 0 \\ -2 & -8 & -12 & -8 & -2 \\ -1 & -4 & -6 & -4 & -1 \end{bmatrix} \quad \begin{bmatrix} 6 & 4 & 1 & 2 & 0 \\ 4 & 12 & 8 & 0 & -2 \\ 1 & 8 & 0 & -8 & -1 \\ 2 & 0 & -8 & -12 & -4 \\ 0 & -2 & -1 & -4 & -6 \end{bmatrix} \quad \begin{bmatrix} 1 & 2 & 0 & -2 & -1 \\ 4 & 8 & 0 & -8 & -4 \\ 6 & 12 & 0 & -12 & -6 \\ 4 & 8 & 0 & -8 & -4 \\ 1 & 2 & 0 & -2 & -1 \end{bmatrix} \quad \begin{bmatrix} 0 & -2 & -1 & -4 & -6 \\ 2 & 0 & -8 & -12 & -4 \\ 1 & 8 & 0 & -8 & -1 \\ 4 & 12 & 8 & 0 & -2 \\ 6 & 4 & 1 & 2 & 0 \end{bmatrix}$$

(A) 0° (B) 45° (C) 90° (D) 135°

Fig. 6 Template of the revised Sobel operator.

The result of the 0° direction template convolution calculation of Sobel operator is shown as follows:

$$\begin{aligned}\Delta = |&f_{11} + 4f_{12} + 6f_{13} + 4f_{14} + f_{15} + 2f_{21} + 8f_{22} + 12f_{23} + 8f_{24} + 2f_{25} - \cdot \\ &(2f_{41} + 8f_{42} + 12f_{43} + 8f_{44} + 2f_{45} + f_{51} + 4f_{52} + 6f_{53} + 4f_{54} + f_{55}|\end{aligned} \tag{14}$$

Let α be the absolute value of the sum of all positive numbers and the corresponding gray value in the template,

$$\alpha = f_{11} + 4f_{12} + 6f_{13} + 4f_{14} + f_{15} + 2f_{21} + 8f_{22} + 12f_{23} + 8f_{24} + 2f_{25}. \tag{15}$$

Set β as the absolute value of the sum of negative numbers and corresponding gray levels in the template,

$$\beta = 2f_{41} + 8f_{42} + 12f_{43} + 8f_{44} + 2f_{45} + f_{51} + 4f_{52} + 6f_{53} + 4f_{54} + f_{55}. \tag{16}$$

The gradient magnitude can be calculated according to the typical Sobel operator:

$$\Delta = |\alpha - \beta|. \tag{17}$$

While the value of Δ' which calculated by the revised Sobel operator is

$$\Delta' = \frac{|\alpha - \beta|}{\alpha + \beta} \times 255. \tag{18}$$

This method is used to calculate the gradient magnitude for the following reasons.

On the one hand, the calculation of the revised operator will not overflow. Assume $\alpha \geq \beta$, then $0 \leq \frac{|\alpha-\beta|}{\alpha+\beta} = \frac{\alpha+\beta-2\beta}{\alpha+\beta} \leq 1$, and $\frac{|\alpha-\beta|}{\alpha+\beta} \times 255 \in [0, 255]$, the value of the calculation result is within the gray scale of the image.

On the other hand, the revised operator has stronger anti-noise ability. Suppose the value of the magnitude of noise is Φ, Canny's expression for magnitude with noise is $g_n = |\alpha + \Phi - \beta|$, and in the absence of noise is $g = |\alpha - \beta|$. Let θ be the ratio of $g$and g_n

$$\frac{g_n}{g} = \frac{|\alpha + \Phi - \beta|}{|\alpha - \beta|}. \tag{19}$$

For the revised operator, $g'_n = \frac{255 \times |\alpha + \Phi - \beta|}{\alpha + \Phi - \beta}$, $g' = \frac{|\alpha - \beta| \times 255}{\alpha + \beta}$, then

$$\frac{g'_n}{g'} = \frac{\dfrac{255 \times |\alpha + \Phi - \beta|}{\alpha + \Phi - \beta}}{\dfrac{|\alpha - \beta| \times 255}{\alpha + \beta}} = \frac{|\alpha - \beta| + \Phi}{\alpha - \beta} \cdot \frac{\alpha + \beta}{\alpha + \Phi + \beta} \leq \frac{|\alpha - \beta| + \Phi}{\alpha - \beta} = \frac{g_n}{g} \tag{20}$$

In brief, the specific steps of the improved Sobel operator are as follows:

S1. For each of the four-direction detector, the template center point pixel is corresponding to the current pixel position in the image, and calculate the value of equal$S' = \frac{|\alpha-\beta|}{\alpha+\beta} \times 255$.

S2. Get the maximum value of the previous four template calculation results, and take the gradient direction of the maximum value as the gradient direction of the current pixel, and use the maximum value as the new gray value of the current pixel.

3.2.3 Non-maximum suppression (NMS)

Non-maximum suppression (NMS) was designed to reduce multiple responses to a single edge [30], and it has played an essential role in the success of the Canny edge detector, thereby becoming a standard post-processing step for edge detectors in general.

The judgment of the measured point is the edge point or not can be determined if the point strength of the maximum value to the neighborhood is in the gradient direction. In this study, the method of the gradient direction of the central pixel is calculated by the improved Sobel template. As shown in Fig. 7, if the gradient magnitude at a pixel is larger than the two neighborhoods in the gradient direction, mark the pixel as a major edge. If the gradient magnitude at the pixel is larger than those pixels adjacent to it in any direction, mark the pixel as a minor edge. Otherwise, mark the pixel as the background. Its essence is to find the highest point in the edge data and verify whether all points in the edge data are the peak.

3.2.4 Adaptive threshold segmentation

Image segmentation is a process of grouping the pixels of an image on the basis of a set of criteria, such as intensity, color, and texture [40]. Threshold segmentation can be regarded as the most straightforward method for image

$(i-1,j-1)$	$(i-1,j)$	$(i-1,j+1)$
$(i,j-1)$	(i,j)	$(i,j+1)$
$(i+1,j-1)$	$(i+1,j)$	$(i+1,j+1)$

Fig. 7 Non-maximum suppression.

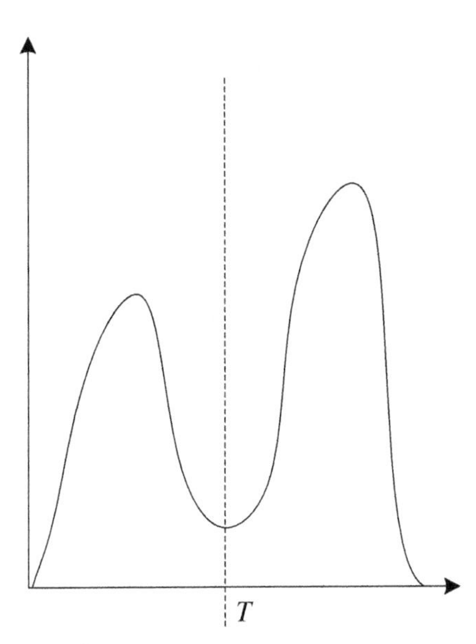

Fig. 8 Threshold segmentation.

segmentation and is one of the most critical and efficient techniques in image segmentation that is based on area image segmentation in arithmetic [41]. To extract the desired part from an image, the gray value of each pixel in the image is compared with the selected threshold, and a corresponding judgment is made.

As shown in Fig. 8, for a given image $I(x, y)$, a threshold T is selected; if the pixel $f(x, y)$ of the point in the image is larger than T, then the point is considered the object point. Otherwise, the point is regarded as a background point. Therefore, the segmented image $g(x, y)$ is presented by the following formulas:

$$g(x,y) = \begin{cases} 1, & f(x,y) > T \\ 0, & f(x,y) \le T \end{cases}. \tag{21}$$

The traditional Canny edge detection algorithm determines the threshold values by setting the proportion of the background area to the target region. The segmentation method based on the maximization of interclass variance, which was proposed by Otsu in 1979 [42], is the optimal method for segmentation and automatic selection. The image is divided into two categories, namely, background and target. The optimal threshold is obtained by the search and calculation of the maximum variance between classes [43]. The Otsu method is usually applied to calculate image gradient histogram so as to get high and low thresholds and those new values are applied to detect the edge [44, 45].

For the image $I(x, y)$, the segmentation threshold of the target and background is recorded as T, the proportion of pixel points belonging to the goal to the entire image is recorded as ω_0, and the average grayscale is μ_0. The ratio of the background pixels for the entire image is ω_1, the average gray level is μ_1, the total average gray level of the image is μ, and the interclass variance is g. The average value of the target in the image is calculated by the following formulas:

$$\mu_0(T) = \frac{1}{\omega_0(T)} \sum_{0 \le k < T} k \cdot p(k) \tag{22}$$

$$\mu_1(T) = \frac{1}{\omega_1(T)} \sum_{T \le k \le m-1} k \cdot p(k). \tag{23}$$

The total mean value of the image is given by

$$\mu = \omega_0(T)\mu_0(T) + \omega_1(T)\mu_1(T). \tag{24}$$

The interclass variance of the background and target pixels is defined as

$$G(T) = \omega_0(T)[\mu_0(T) - \mu] + \omega_1(T)[\mu_1(T) - \mu]. \tag{25}$$

The best threshold for the image is

$$g = \arg \max_{0 \le T \le m-1} [G(T)] \tag{26}$$

where g is used as the value of the high threshold for improved Canny edge detection. By analyzing the principle of threshold selection in the Canny algorithm and the histogram of the gradient image, the improved selection method is feasible.

3.3 Feature point extraction and computing angle

In the traditional method, the angle of joint movement is measured with the distance of the forward extension of the figure or the head during flexion [46, 47]. The sit-and-reach test is the most commonly used flexibility field test in epidemiological studies [48]. In order to measure the human body flexibility fitness, it is necessary to measure the human body anteflexion angle. However, the measuring methods and standards of the angle are different. In this recognize method, the feature points of human's head, buttock, and foot were extracted to compute the human body anteflexion angle.

Edge detection is a crucial step in the calculation of the human body anteflexion angle. Similarly, feature point extraction substantially influences the final result of the calculation. To compute the anteflexion angle of the human body, the feature points of the human's body head, buttocks, and foot are obtained. Concretely, the feature point of the head is the rightmost white pixel in the edge image, and the feature point of human's buttock is the left-most white pixel in the edge image. Finally, the feature point of the foot is extracted as the point at the bottom and the left-most point. The extracted feature points are shown in Fig. 9.

The obtained required feature points are then connected to form an angle with the buttocks' feature points as the vertices, and the magnitude of the angle is calculated by using the mathematical theorem.

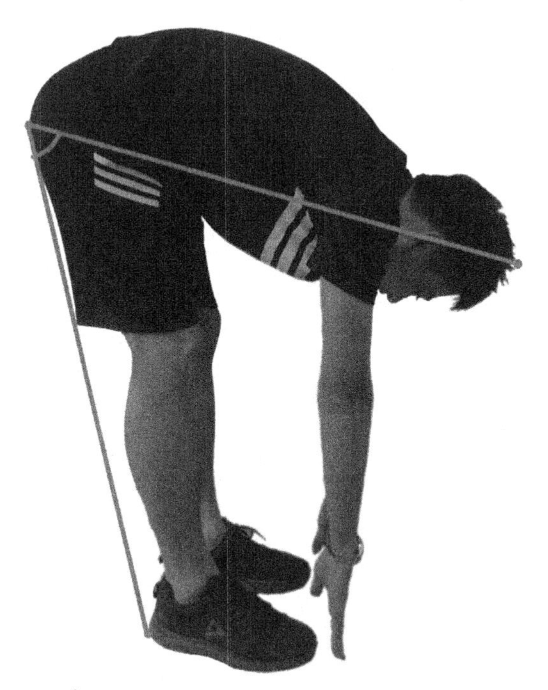

Fig. 9 Feature extraction.

Flexibility is calculated differently between genders. For males, assuming that the angle of the human body anteflexion is θ, the score of the flexibility of the human body can be calculated from the following formulas:

$$score=\begin{cases}0, & \theta\geq 68^{\circ}\\ 2.778(68-\theta), & 32<\theta<68^{\circ}\\ 100, & \theta\leq 32^{\circ}\end{cases} \tag{27}$$

And as for female, the method of calculation is as follows:

$$score=\begin{cases}0, & \theta\geq 61^{\circ}\\ 2.857(61-\theta), & 26<\theta<61^{\circ}\\ 100, & \theta\leq 26^{\circ}\end{cases} \tag{28}$$

4. Implementation and result

This experiment was carried out on the visual studio 2017 platform. Two human body anteflexion angle images are selected and then the improved Canny detection algorithm is used to extract the angle of the body. The improved Gaussian filter is compared with the traditional Gaussian filtering method. The improved Sobel operator is compared with the traditional Sobel operator, and the final result is compared with the traditional Canny edge detection algorithm.

4.1 Filter result analysis

Fig. 10A is an original image of the human body, Fig. 10B and C are images processed by Gaussian filter and improved adaptive Gaussian filtering

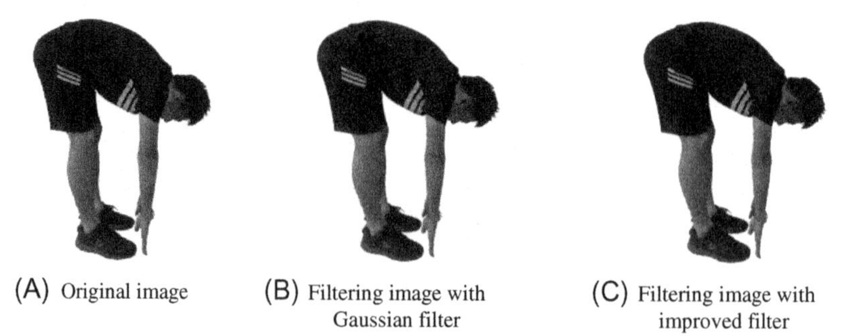

(A) Original image (B) Filtering image with Gaussian filter (C) Filtering image with improved filter

Fig. 10 Filter results.

algorithm after being converted into grayscale images and add Gaussian noise, respectively. It can be seen from Fig. 10 that the original Gaussian filtering algorithm treats the high-frequency edge signal as noise. The result of this filter shows that the edge is filtered as noise, which causes to the location of the edge is inaccurate. In the improved adaptive Gaussian filtering algorithm, we first determine whether the high-frequency signal is a noise signal, and corresponding processing is performed according to the judgment result. In the revised algorithm, not only the edge information of the image is well preserved, but also the noise filtering effect is better.

4.2 Gradient compute analysis

In order to make the test effective by the improved method, the images collected are also used to analyze the gradient calculation experiment without adding noise and with adding noise. The magnitude and direction of the gradient are calculated using a 5×5 Sobel template in four different directions. The results are compared with the 3×3 Sobel template results in two different directions.

It can be seen from the test results in Fig. 11 that the edge extracted by the traditional Sobel has noise interference, and the edge extraction is fuzzy, the edge extracted by the improved Sobel operator is clearer, and the continuity of the edge is better. As shown in Fig. 12 that for a noisy image, after the traditional Sobel operator is operated, a large amount of noise remains, and the extracted edge contour is unclear. However, the improved Sobel operator has a certain suppression effect and reduces the interference of noise in the image. Furthermore, not only the edge is continuous and clear, but

(A) Original image

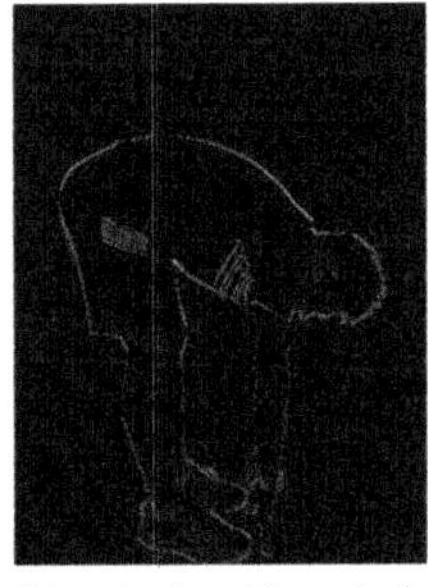

(B) Magnitude with typical sobel operator

(C) Magnitude with revised sobel operator

Fig. 11 Magnitude computation without noise.

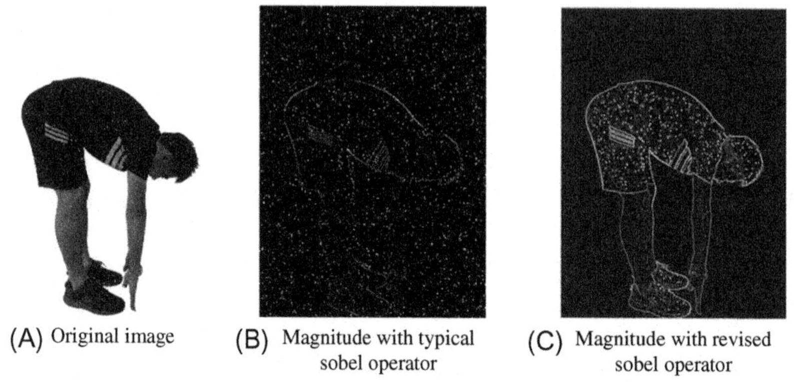

(A) Original image (B) Magnitude with typical sobel operator (C) Magnitude with revised sobel operator

Fig. 12 Magnitude computation with noise.

also the image has a smoothing effect, which verifies the anti-noise enhancement of the operator.

4.3 Edge detection and angle computation

Finally, in order to verify the effectiveness of the improved Canny edge detection algorithm mentioned in this chapter, three different human anteflexion angle images were collected for experiments. Experiments were performed in three cases: noiseless experiment, Gaussian noise experiment, and pepper-and-salt noise experiment. The performance is verified and compared with the results of the traditional Canny operator and the results of Budi's operator [49] and Kalra's operator [50].

It can be seen from the experimental results that in the absence of noise, in Fig. 13, compared with the same Otsu algorithm, more edge details can be detected by the algorithm proposed in this chapter, especially compared with the edge detection operator without the Otsu algorithm. For instance, the edge detected by the Kalra's operator is thicker, the detection result is inaccurate, and false edge information exists, while the traditional Canny operator and the continuous need to manually change the threshold of segmentation to achieve better detection result. In the case of adding Gaussian noise, it can be seen from Fig. 14 that the edge detection operator proposed in this chapter can filter most of the Gaussian noise. A median filtering is used to reduce noise in Kalra's operator. However, the detected edge information is still inaccurate. It can be seen from Fig. 15 that the algorithm proposed in this chapter also has a certain smoothing effect on pepper-and-salt noise, but it is not as good as the median filtering effect. However, the Gaussian filter in

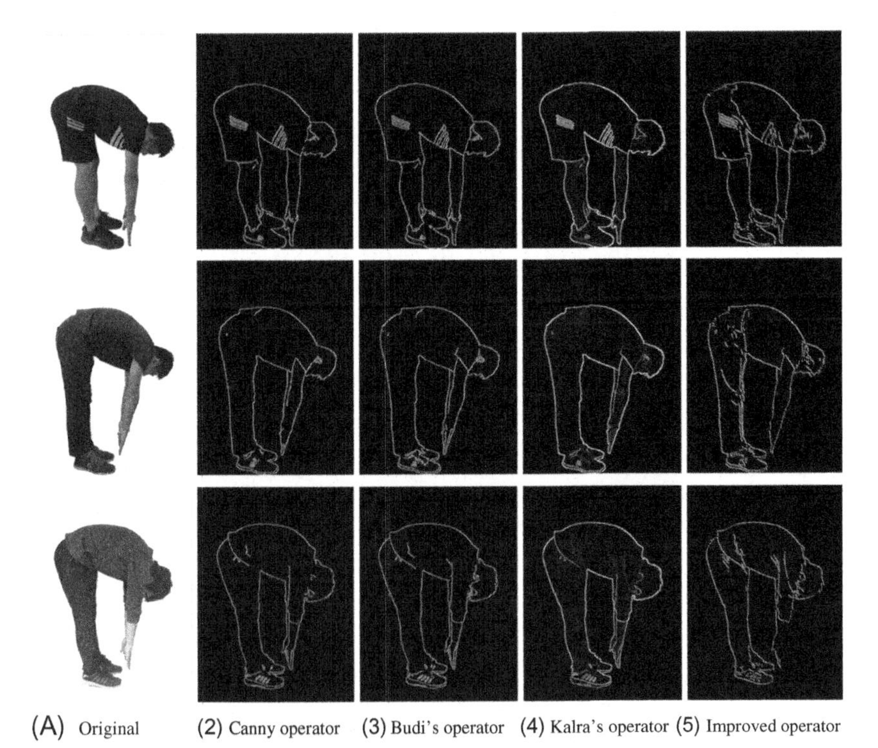

Fig. 13 Edge detection without noise.

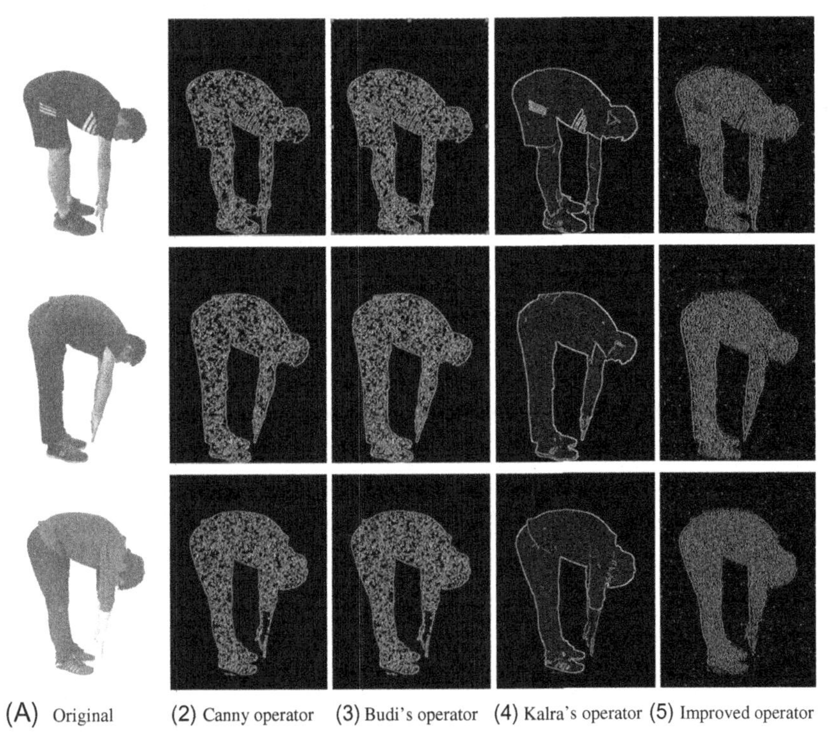

Fig. 14 Edge detection with Gaussian noise.

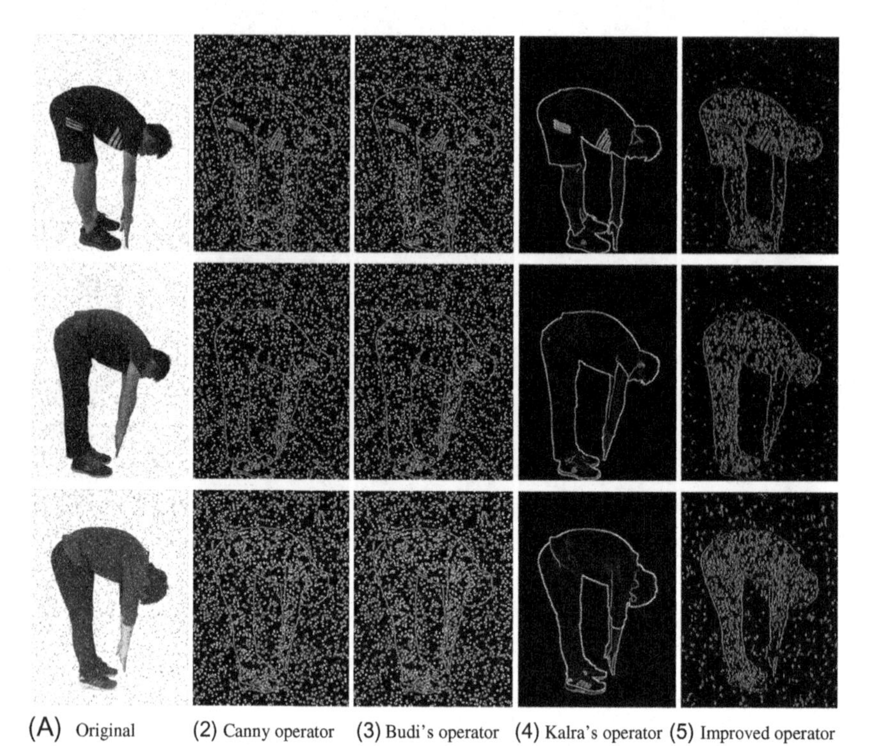

(A) Original (2) Canny operator (3) Budi's operator (4) Kalra's operator (5) Improved operator

Fig. 15 Edge detection with pepper-and-salt noise.

the traditional Canny operator and the operator proposed by Budi's operator is extremely sensitive to pepper-and-salt noise, which cannot effectively remove salt and pepper noise, and the noise detection result is not accurate.

The feature points of the hip, foot, and head were extracted after the edges of the human body were obtained. Fig. 16 shows that an angle was formed with the buttocks' feature points as the vertices for the three feature points. In the experiment, the improved edge detection operators are used to identifying the human body angles in Figs. 11A and 12A. The results of the two images are shown in Table 1.

As can be seen from Table 1, the detection results of the four edge detection algorithms are very approximate to the manual measurement values (the manual measurement error is not excluded). The human body flexion angle detection results of each detector are shown in Fig. 17. From the results of the graph and table, it can be seen that the improved Canny edge detector proposed in this chapter has better detection performances. Specifically,

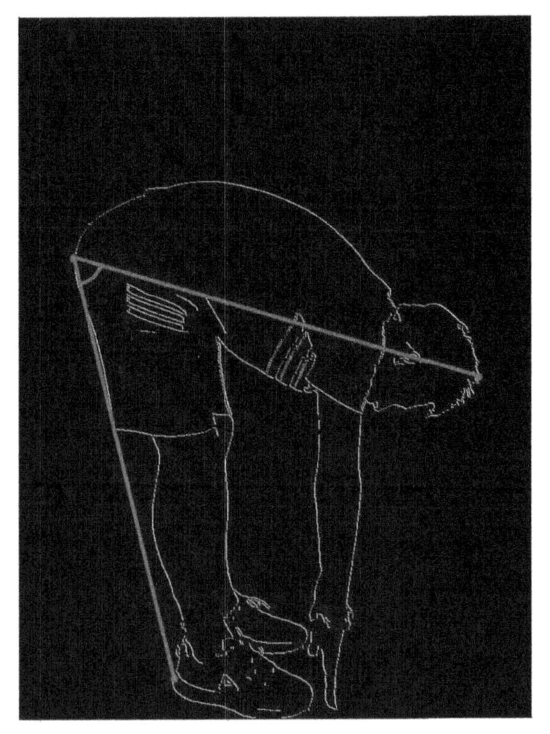

Fig. 16 Calculation of the angle.

Table 1 Results of the computing angle of the two algorithms.

	Canny operator	Budi's operator	Kalra's operator	Improved operator	Manual measurement(MM)
Fig. 13A1	59.69°	57.4428°	58.239°	59.0399°	60°
Fig. 13A2	71.3824°	68.7417°	77.0518°	69.399°	70°
Fig. 13A2	63.4033°	63.4033°	60.4642°	60.8542°	61.5°

statistics on the detection results of each detector are performed, and the average of the sum of the absolute values of the errors and the absolute values of the relative errors are calculated, respectively, as shown in Figs. 18 and 19. The experimental results show that the results of the improved algorithm are more consistent with the actual situation than the traditional Canny detector. Therefore, it can be considered that the improved edge detection algorithm has a certain improvement in the accuracy of the human body angle detection result.

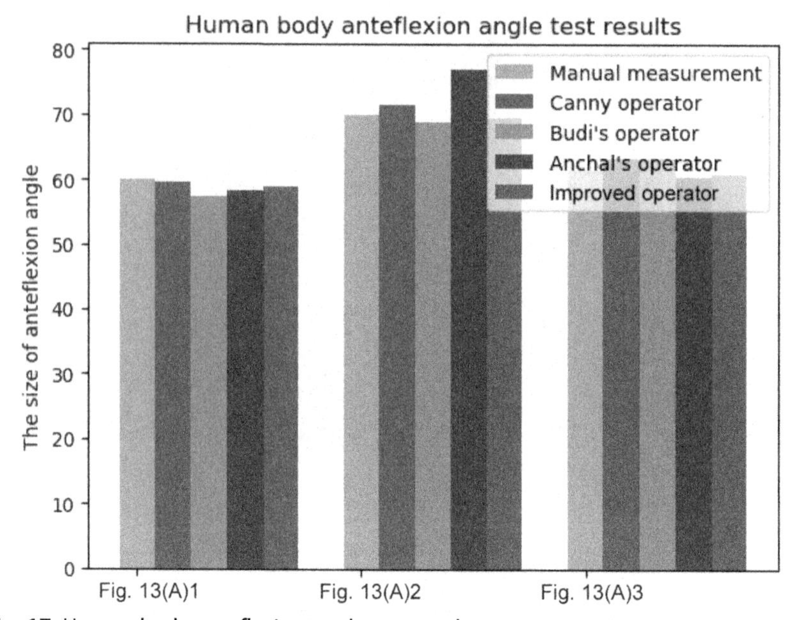

Fig. 17 Human body anteflexion angle test results.

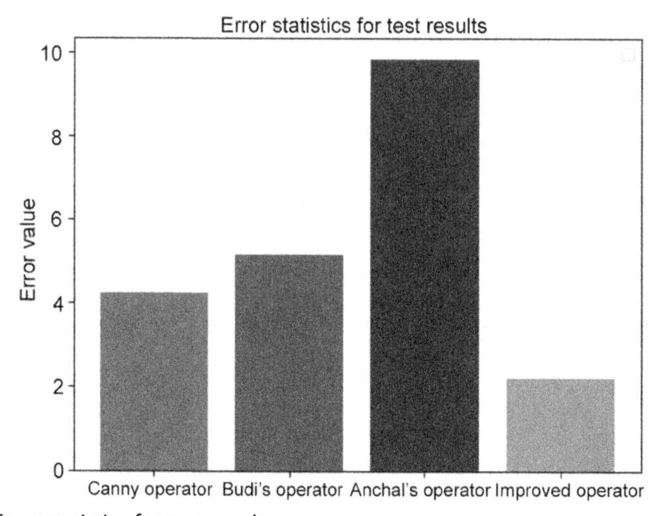

Fig. 18 Error statistics for test results.

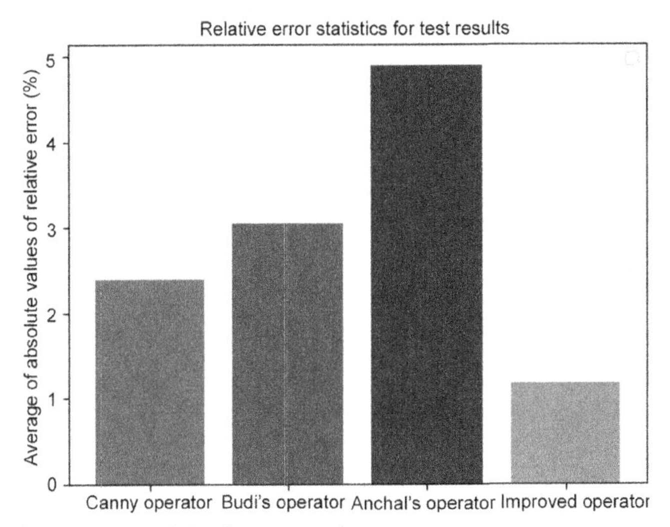

Fig. 19 Relative error statistics for test results.

5. Conclusion

In this chapter, a human anteflexion angle recognition algorithm based on edge detection and feature point extraction is proposed. An edge detection algorithm is used to obtain the edge of the human body. Since the traditional Canny operator is not processed in terms of filtering and gradient ideally, an adaptive Gaussian filter is proposed in this study, which is based on pixel point gray value and local mean difference, and an improved Sobel template is used to calculate the gradient amplitude and direction, which improves anti-noise performance of the algorithm. The non-maximal suppression is performed based on the obtained gradient direction to obtain a more refined edge. Eventually, the Otsu algorithm is adopted to calculate the threshold of the image, and double threshold processing of the image to obtain the final edge image. After the human body's edge is extracted, the three feature points are extracted to calculate the anteflexion angle of human body. Experiments show that the improved algorithm has a certain improvement in the final test results.

In this chapter, the algorithm is implemented on a simple background, so if the background of the image is more complex, the results of the detection will be greatly affected. As the feature point extraction algorithm is based on the human posture, it is susceptible to different postures of the human body, such as the position of the head. Therefore, in the future, we will focus on the background removal and feature point extraction algorithm.

Acknowledgment

This work is supported in part by the Key Grant Scientific and Technological Planning Project of Guangzhou, China (201704020113), the Grant Scientific and Technological Project of Guangdong Province, China (2017A040405059), the National Natural Science Foundation of China (61802073).

References

[1] D.V. Knudson, P. Magnusson, M. Mchugh, Current issues in flexibility fitness, Presid. Coun. Phys. Fitness Sports Res. Dig. 16 (49) (2000) S57–S73.
[2] A. Hedrick, Dynamic flexibility training, Strength Cond. J. 22 (5) (2000) 33.
[3] G.W. Gleim, M.P. Mchugh, Flexibility and its effects on sports injury and performance, Sports Med. 24 (5) (1997) 289–299.
[4] Y. Du, B. Zhang, Y. Shen, et al., The design of low-load human flexibility test system, in: International Conference on Man-Machine-Environment System Engineering, Springer, Singapore, 2017.
[5] J. Castro-Piñero, P. Chillón, F.B. Ortega, et al., Criterion-related validity of sit-and-reach and modified sit-and-reach test for estimating hamstring flexibility in children and adolescents aged 6–17 years, Int. J. Sports Med. 30 (09) (2009) 658–662.
[6] G. Baltaci, Comparison of three different sit and reach tests for measurement of hamstring flexibility in female university students, Brit. J. Sports Med. 37 (1) (2003) 59–61.
[7] J.E. Fernandez, N.B. Stubbs, Mathematical modeling and testing of the sit and reach test, Int. J. Ind. Ergon. 3 (3) (1989) 201–205.
[8] J. Wang, B. Boris, Q.Y. Wei, An effective method for plate number recognition, Multimed. Tools Appl. 77 (2) (2018) 1679–1692.
[9] H. Ran, W. Xiang, S. Zhenan, et al., Wasserstein cnn: learning invariant features for nir-vis face recognition, IEEE Trans. Pattern Anal. 41 (7) (2018) 1761–1773.
[10] L.L. Lu, W.Y. Yong, Z.W. Yong, Applications of two-dimensional bar codes and face recognition technologies in the management of college student archives, Int. J. Res. Edu. Studies 4 (5) (2018) 05–09 (ISSN: 2208-2115).
[11] J. Shen, X. Zuo, W. Yang, et al., Differential features for pedestrian detection: a Taylor series perspective, IEEE Trans. Intell. Transp. 20 (8) (2019) 2913–2922.
[12] L. Cao, Q. Dong, X. Chen, Laplace $\ell 1$ Huber based cubature Kalman filter for attitude estimation of small satellite, Acta Astronaut. 148 (2018) 48–56.
[13] L. Pigou, A. Van Den Oord, S. Dieleman, et al., Beyond temporal pooling: Recurrence and temporal convolutions for gesture recognition in video, Int. J. Comput. Vis. 126 (2–4) (2018) 430–439.
[14] L. Sun, S.Q. Wang, J.C. Xing, An improved harris corner detection algorithm for low contrast image, in: Proceedings of the 26th China Control and Decision Conference, IEEE, 2014.
[15] A. Broggi, P. Cerri, S. Debattisti, et al., Proud—public road urban driverless-car test, IEEE Trans. Intell. Transp. 16 (6) (2015) 3508–3519.
[16] J. Ren, C. Zhang, L. Zhang, et al., Automatic measurement of traffic state parameters based on computer vision for intelligent transportation surveillance, Int. J. Pattern Recognit. 32 (4) (2017) 1855003.
[17] W. Wan, F. Lu, Z. Wu, et al., Teaching robots to do object assembly using multi-modal 3D vision, Neurocomputing (2017) S0925231217302497.

[18] N. Miyamoto, K. Hirata, N. Kimura, et al., Contributions of hamstring stiffness to straight-leg-raise and sit-and-reach test scores, Int. J. Sports Med. 39 (02) (2018) 110–114.

[19] K.F. Wells, E.K. Dillon, The sit and reach—a test of back and leg flexibility, Res. Q. 23 (1) (1952) 115–118.

[20] W.W. Hoeger, D.R. Hopkins, A comparison of the sit and reach and the modified sit and reach in the measurement of flexibility in women, Res. Q. Exerc. Sport 63 (2) (1992) 191–195.

[21] P.A.L. Miñarro, P.S. de Baranda Andújar, P.L.R. García, et al., A comparison of the spine posture among several sit-and-reach test protocols, J. Sci. Med. Sport 10 (6) (2007) 456–462.

[22] H. Spontón, C. Juan, A review of classic edge detectors, Image Process. Line 5 (2015) 90–123.

[23] Y. Meng, Z. Zhang, H. Yin, et al., Automatic detection of particle size distribution by image analysis based on local adaptive canny edge detection and modified circular Hough transform, Micron 106 (2018) 34–41.

[24] F. Fu, C. Wang, Y. Li, et al., An improved adaptive edge detection algorithm based on Canny, in: Sixth International Conference on Optical and Photonic Engineering (icOPEN 2018). Int. Society Optics Photonics, Vol. 10827, 2018.

[25] W. Lin, J. Wang, Edge detection in medical images with quasi high-pass filter based on local statistics, Biomed. Signal Process. 39 (2018) 294–302.

[26] S. Dorafshan, R.J. Thomas, M. Maguire, Comparison of deep convolutional neural networks and edge detectors for image-based crack detection in concrete, Construct. Build Mater. 186 (2018) 1031–1045.

[27] H. Nhat-Duc, Q.L. Nguyen, V.D. Tran, Automatic recognition of asphalt pavement cracks using metaheuristic optimized edge detection algorithms and convolution neural network, Autom. Constr. 94 (2018) 203–213.

[28] Y. Tian, H. Yuan, G. Shaoyan, Extraction algorithm of image feature point based on multi-scale fusion information, in: Seventh International Conference on Optical and Photonic Engineering (icOPEN 2019). Int. Society for Optics and Photonics, Vol. 11205, 2019.

[29] J. Zeng, G. Cao, W. Li, et al., Feature point extraction based on contour detection and corner detection for weld seam, J. Phys. Conf. Ser. 1074 (2018).

[30] J. Canny, A computational approach to edge detection, IEEE Trans. Pattern Anal. (6) (1986) 679–698.

[31] P. Bao, L. Zhang, X. Wu, Canny edge detection enhancement by scale multiplication, IEEE Trans. Pattern Anal. Machine Intell. 27 (9) (2005) 1485–1490.

[32] R. Biswas, J. Sil, An improved canny edge detection algorithm based on type-2 fuzzy sets, Procedia Technol. 4 (11) (2012) 820–824.

[33] L. Ding, A. Goshtasby, On the Canny edge detector, Pattern Recogn. 34 (3) (2001) 721–725.

[34] B. Wang, S.S. Fan, An improved CANNY edge detection algorithm, in: 2009 Second International Workshop on Computer Science and Engineering. IEEE, 2010.

[35] R. Ke-Qiang, Z. Jing-Gran, Extraction of plant leaf vein edges based on fuzzy enhancement and improved Canny, J. Optoelectron. Laser 29 (11) (2018) 1251–1258.

[36] X.M. Zhao, W.X. Wang, L.P. Wang, Parameter optimal determination for canny edge detection, Imag. Sci. J. 59 (6) (2011) 332–341.

[37] J.S. Huang, D.H. Tseng, Statistical theory of edge detection, Comput. Vis. Graph. Image Proces. 42 (1) (1988) 133.

[38] G. Jie, L. Ning, An improved adaptive threshold canny edge detection algorithm, in: International Conference on Computer Science & Electronics Engineering, 2012.

[39] M. Nixon, A.S. Aguado, Feature Extraction and Image Processing. for Computer Vision, Academic Press, 2012.
[40] C. Mala, M. Sridevi, Multilevel threshold selection for image segmentation using soft computing techniques, Soft. Comput. 20 (5) (2016) 1–18.
[41] Y. Tan, Study on applied technology arithmetic of image threshold segmentation, Microcomput. Inf. 23 (24) (2007) 298–300.
[42] N. Otsu, A thresholding selection method from gray-level histogram, IEEE Trans. Syst. Man Cybern. 9 (1) (1979) 62–66.
[43] D.D. Han, T.C. Zhang, J. Zhang, Research and implementation of an improved canny edge detection algorithm, Key Eng. Mater. 572 (2013) 566–569.
[44] M.A. Ingle, G.R. Talmale, Respiratory mask selection and leakage detection system based on canny edge detection operator, Procedia Comput. Sci. 78 (2016) 323–329.
[45] Y. Zou, F. Dong, B. Lei, et al., Maximum similarity thresholding, Digital Signal Process. 28 (1) (2014) 120–135.
[46] F. Ayala, P.S.D. Baranda, M.D.S. Croix, et al., Reproducibility and criterion-related validity of the sit and reach test and toe touch test for estimating hamstring flexibility in recreationally active young adults, Phys.Therapy Sport 13 (4) (2012) 219–226.
[47] C.M. Mier, Accuracy and feasibility of video analysis for assessing hamstring flexibility and validity of the sit-and-reach test, Res. Q. Exerc. Sport 82 (4) (2011) 617–623.
[48] S.S. Hui, P.Y. Yuen, Validity of the modified back-saver sit-and-reach test: a comparison with other protocols, Med. Sci. Sports Exerc. 32 (9) (2000) 1655–1659.
[49] K. Anchal, R.L. Chhokar, A Hybrid approach using sobel and canny operator for digital image edge detection, in: 2016 International Conference on Micro-Electronics and Telecommunication Engineering (ICMETE). IEEE, 2016.
[50] S.B. Darma, A.N. Rusydi, K. Pradityo, Lake edge detection using Canny algorithm and Otsu thresholding, in: 2017 International Symposium on Geoinformatics (ISyG) IEEE, 2017.

CHAPTER NINE

Prediction and classification of diabetes mellitus using genomic data

Joseph Bamidele Awotunde[a], Femi Emmanuel Ayo[b], Rasheed Gbenga Jimoh[a], Roseline Oluwaseun Ogundokun[c], Opeyemi Emmanuel Matiluko[c], Idowu Dauda Oladipo[a], and Muyideen Abdulraheem[a]

[a]Department of Computer Science, University of Ilorin, Ilorin, Nigeria
[b]Department of Computer Science, McPherson University, Seriki Sotayo, Nigeria
[c]Department of Computer Science, Landmark University, Omu-Aran, Nigeria

1. Introduction

Diabetes mellitus (DM) is becoming a global epidemic, aggregating the occurrence of DM, affecting both genetic and environmental factors as well as a complex disease [1]. Recent research has shown evidence that many new loci and specific single-nucleotide polymorphisms (SNPs) are associated with increased risk of DM [1, 2]. Nonetheless, the predictive value is still controversial for a person's future production of DM using these SNPs [3–5]. Therefore, there is an urgent need to use both genetic and clinical data to develop an efficient predictive model for DM prediction. An efficient predictive model will assist in early diagnosis of DM to facilitate prompt blood glucose control and treatment [6–8]. The World Health Organization (WHO) has estimated that around 350 million people worldwide will suffer from diabetes by 2030 [9]. Prediabetes, gestational, type I, and type II DM are the four types of diabetes. Insulin-dependent diabetes is type 1 [10] if it is not produced by the hormone insulin. Non-insulin dependency diabetes is type II, where the brain is unable to use the body's insulin [10]. Diabetes during pregnancy is referred to as gestational diabetes [11]. It is prediabetes when blood glucose levels are higher than normal but not as high for diagnosis as diabetes [12]. DM causes various diseases such as kidney disease, nerve damage, damage to the vessels, damage to the blood vessels, blindness, and heart disease [13]. Hence, an effective analytical model is

Intelligent IoT Systems in Personalized Health Care
https://doi.org/10.1016/B978-0-12-821187-8.00009-5

needed to achieve better prediction, diagnosis, prevention, and self-management of DM [9, 14, 15].

The use of machine learning and statistical methods to forecast uncertain or potential effects is referred to as predictive analysis [9, 16]. It responds to the question as what is the next step? It also uses past and present data to predict actions, patterns, and activity in the future. The prediction is rendered using quantitative questions, automated machine learning, and statistical analysis [9]. Experts need to construct predictive models that are used for forecasting in predictive analytics [9].

DNA sequencing techniques have progressed rapidly following the completion of the Human Genome Project [17–19]. With these advances, the cost per base-pair sequenced has been dramatically reduced, making it exceptional in the last year. Because of this progress, there has been an abundance of individual genotype data and other forms of human biological genomic information [19, 20]. As a result of these technological developments, the idea of precision medicine or personalized medicine has gained worldwide support, transforming prognosis, diagnosis, prediction of illness, prognosis, and human contribution to preventive approaches [18, 19].

The field of information technology is constantly evolving and is now being applied to other disciplines that give way to multidisciplinary research, including medicine, agriculture, etc. Using fuzzy logic (FL) and artificial intelligence (AI) approaches [21], the advent of electronic health records has also made available information that can be analyzed and inferences in disease diagnosis and treatment. Genomic medicine is an area of medicine involving the use of genomic data to provide new approaches to the detection of disease research, prediction, and treatment resulting from an improved understanding of genetic causes of human disease [22]. The emergence of such an alliance has resulted in a decrease in the cost of genome sequencing, new forms of study and a growing interest in applying big data analytics in genomics, with achievements being reported in different medical fields, and the discovery of new disease therapies even in oncology [23–26]. Therefore, to address this peculiarity, a specialized approach such as a genomic predictive and classification model is desirable. The introduction of machine learning for prediction and classification of genomics data to detect diabetes is new and few authors have work in this direction, only the clinical approaches have been used to establish related biomarkers for diabetes using genomics data. The main contributions of this chapter are (1) to simulate genomics datasets using Monte Carlo experiment, (2) the use of AI techniques to improve sample selection bias common to past studies, (3) the use of GA to select the most informative gene biomarker for biological sample classification using DNN, and (4) the designed approach showed good performance for the prediction of type II diabetes.

The rest of this chapter is organized as follows: Section 2 discusses the prediction and classification of DM, AI, and the Internet of medical things (IoMT) and their research trends in DM and computational intelligence (CI) techniques. Section 3 presents an extensive discussion on genomic data and health-care systems, genetic risk factors, and commonly mutated genes in diabetes, and SNPs as biomarkers for DM. Section 4 highlights the rationale for choosing CI techniques for DM, an ensemble of AI and deep learning (DL) approaches for DM: challenges and opportunities, GA for selecting the most informative gene biomarker, DNN for the classification of biological samples, genomic data via Monte Carlo experiment and presents a practical case of DM prediction with CI techniques and results. Finally, Section 5 concludes the chapter and discusses future works for the realization of efficient uses of genomic data in health-care systems.

2. Prediction and classification of DM

Prediction is referred to as the procedure of recognizing the inaccessible arithmetical data for a novel inspection and this can also be a procedure of data investigation. This term can also be utilized to uncover a statistical outcome. Similar to classification, the teaching dataset comprises the contributions and equivalent statistical output values. As stated by the preparing dataset, the system derives the replica or predictor. Once the latest data are uncovered, the replica would discover an arithmetical outcome. Unlike in classification, this technique does not have set characteristics. The system forecasts an uninterrupted-valued result or systematic value [27–29]. Regression is mostly utilized for prediction. For instance, forecasting the value of a house depending on the shreds of evidence like the number of rooms, the total area, etc., is an example of a forecast. Another example is an enterprise discovering the total currency expended by the consumer during transactions [30, 31].

Prediction is the process of learning from historical data to make estimates about the future or unknown event. For health care, predictive analytics will enable the best decisions to be made, allowing for care to be personalized to each individual. When the prediction is done right in health care will help the medical scientist to see an increase in patient access, lower cost, increased revenue, increased asset utilization, and also improved patient experience. Predictive analytics tried to apply what doctors have been doing on a larger scale. The most changes brought to health care by predictive analytics are the ability to make sense of previously nonexistent behavioral, to better measure, quantitative, psychosocial, and biometric data, so the wheel is not reinvented [32–34]. It allows individual treatment to be personalized to each person and allows the best health-care decisions to be made. The medium goal of health-care predictive analytics is to reliably predict the unpredictable. Confidence in predictions affects the type of question posed with a high certainty of response. For example, it can be answered with a high degree of certainty to ask a historical question like "what did I eat today."

However, the certainty of prediction declines as we begin to ask questions about events that have yet to occur. Questions such as "will I get diabetes" and "how much weight I will gain" are becoming more difficult to predict with high self-confidence [32–34]. Applying predictive analytics in health care through data analytics helps improve the operational efficiency, planning, and execution of key care delivery systems, use of resources, staff schedules, and admission and discharge of patients. All of this leads to increased patient access to good health-care delivery and income, lower costs, increased use of assets, and improved patient experience. New data sources are used for predictive analytics, including those directly from patients. The simultaneous use of two or more forms of statistical analysis is called a predictive model and is a subset of concurrent analytics. Predictive modeling's main objective is to forecast action, outcome, or occurrence using a multivariate array of predictors [34]. The models were further used in health care by analyzing patient lifestyles, demographics, psychographics, and preferences that can then be used by medical scientists to produce results that are useful to both patient and physician, and theses can be delivered through specific channels.

Classification refers to the recognition of a group description of a novel inspection. Fundamentally, an established data are utilized as teaching data. The group of participation data and the equivalent outputs are specified to the algorithm. Hence, the teaching dataset comprises the input data and their related class descriptions. Employing the teaching dataset, the system obtains a replica or the classifier then this obtained replica can be a decision tree (DT), arithmetical method, or a neural network (NN). In the classification process, once an undescribed data are specified to the system, it would then discover the class which it fits into. The latest data delivered to the prototypical are the test dataset [27–29].

Classification can also be defined as the procedure of categorizing a report. An instance of classification is to scrutinize whether it is raining or not and the response can either be yes or no. Hence, there is a precise number of selections. Although occasionally there could be more than two groups to categorize and this is called multiclass classification [27–29]. An instance in real life is a bank requiring to consider whether granting a mortgage to a specific customer is uncertain or not. In this instance, a replica is created to discover the uncompromising label. The labels are uncertain [30, 31].

2.1 AI and IoMT

IoMT can be defined as the coupled structure of therapeutic gadgets and is also an application that accumulates data and gives to health-care IT schemes over virtual computer setups [35]. IoMT could deliver an advance means to take care of our aging and also possess incredible possibilities to assist deal with the increasing overheads of care. IoMT gadgets could assist trace crucial and heart performance, keep track of glucose, and other body systems, and activity and sleeping levels [35–37]. IoMT can also assist to remind personals in taking their medication and the documentation of the time the medications were taken would be reported down as well [36]. Furthermore, handy diagnostic gadgets can make repetitive blood and urine test stress free on our aging people and these aging are groups of individuals, where the ability to move is more demanding and who require to conclude these tests regularly than for younger patients [37, 38]. Transferrable diagnostic equipment can investigate and describe the outcomes of these tests without necessitating an appointment to the medical practitioner's office [36, 38]. A large number of prospects that exists in IoMT and this assist inaccessible guidance make certain the protection of their treasured persons with a wearable implement that study the consistent schedules of the individual who have on the gadgets and can give caution if something appears inappropriate as well as vigilant if elders have broken their borders, which is regularly of worry for memory-care patients [37–39].

The goal of IoMT is not to supersede medical contributors but to offer them with the data collected from gadgets aimed at offering enhanced diagnoses and management tactics as well as to decrease inadequacies and excess in the medical system. IoT can succor and support prevailing structures or construct a novel approach to guarantee the patient receiving the superlative attention needed by them. The responsibilities of the nurses, physicians, and staff turn out to be uncomplicated, well organized, cost effective, and harmless [35].

The IoMT fundamentals include: develop smart systems to assist supervisory, recommend personalized management, and guarantee acceptance with physiotherapy. Acquire amalgamated structures that can accumulate and investigate in health care increasing our intellect on illnesses and appraising doctor performance. Guarantee that entry to use the health care is obtainable from home not just in the hospital alone. Institute gadgets and radars that document medical data extensively and make them accessible. Compel examining people's medical data on a fundamental humanoid right. IoMT also safeguards medical data as well as concealment of patients' information to prevent misappropriation of them. Digitalize health care and guarantee everybody has entry to excellence, reasonable care while preventing the risk of pervasive entry to confidential medical data. Increasing the adoption of smart sensor technology in numerous systems improves the competency of IoT health-care market gadgets to accumulate additional data relating to patients' health. The accumulated datasets are utilized to improve scrutinizing procedures and evaluate them to discern novel treatments for several other ailments. These technological results hence assist connect the gap between the physical and the digital realm [37–39].

The IoMT refers to the assembly of medical gadgets and applications, which associate with health-care IT systems via virtual computer grids. Health-care gadgets equipped with wireless permit the machine-to-machine interaction, which is the foundation of IoMT. IoMT gadgets connected to cloud podiums that apprehend datasets can be warehoused and evaluated. Although the medical industry had been reluctant to embrace the Internet of things technologies than other businesses, the IoMT is self-assured to transmute ways we fix individuals' well-being and nutrition particularly as the request for results to lesser medical care overheads expansion in years to come.

The IoMT can render assistance in the aspect of scrutinizing, enlightening, and reporting not only to caregivers but also medical workers with definite data to recognize concerns earlier before they develop to being critical or to permit initial intervention. There was a prediction of dramatic growth in IoMT gadgets and this was given by Allied Market Research. The prediction is that IoT medical care market will reach $136.8 billion globally by 2021. Presently, there are 3.7 million health-care gadgets in the application that are associated with monitoring several human body parts and will help in medical decision-making [35–37].

Numerous experiences supported this intense development comprising the approachability of wearable gadgets and the reducing overheads of radars technology. Presently, the highest client mobile gadgets are furnished with near-field communication and radio frequency identification tags [40–42], they can interconnect with IT systems. Also, the rates of chronic ailments are on the increase and the request for improved medical attention options is also increasing and lower medical care overheads make it extra appealing to dabble with novel inventions that would render enhanced medical care results and competences. High-speed internet increase and access, as well as positive management governing guidelines, have also added to the advancement of IoMT implementation. Some instances of medical gadgets or wearables are discussed below according to Basatneh et al. [36], Corbin [37], and Parashar et al. [35].

i. *Smart health monitoring sensor devices*: These are interconnected to receive the virtual medical status as well as heart rate and blood pressure and enable them to provide remote medical care services.

ii. *Patient personal medication system*: This displays the images of all the medicines, which the patient is requested to take and also the comprehensive information and usage details.

iii. *iRobot*: This is a medical care robot, that is, located at the patient's home, which monitors movements of the patient within his home ad is connected with sensors to give information to the servers. This information can be shared with the medical practitioners with permission from the patient.

iv. *Medical kiosks and health spot*: Varieties of kiosks are provided in residential settlements that are fortified with diagnostics. The assistant is educated to deliver fundamental health necessities and also the medical practitioners can tenuously see the data and counsel appropriately.

v. *Infant monitors*: Supervise your utmost appreciated possession with coupled sensors transmitting information to your smart gadget in real time. The infant supervisor refers to parents' information in real time of the baby's breathing, skill temperature, heart rate, perspiration, and other important signs.

AI is an umbrella word that depicts various tools for knowledge and control. Without an AI instrument, the data from a wearable would be without any worth to the merchant as well as the consumer. For this purpose, wearable app inventors are progressively adding AI engine inside wearable medical apps. Furthermore, AI-aided data mining is also fundamental to the attainment of an intellectual medical care platform that connects many smartphones, website, IoT gadgets, and wearables together to congregate data and benefit fascinating medical comprehensions of an individual [37–39].

Contemporary technological and connectivity improvements have resulted in IoT and AI applications occurring in several companies. IoT is a collection of technologies that allow artifacts such as everyday consumer products or industrial machines to interact with each other and the Internet, to communicate information on their attributes (such as working condition, temperature, motion, location, etc.) and to provide instant data analysis and, ideally, smart action [43].

An environment in which IoT and AI make substantial impacts in the medical industry, individually or together, which is continuously under density to minimize overhead while addressing a rapidly growing detrimental population. These technologies can help medical care organizations gain the prospect of a world, that is, gradually interrelated and accessible [44].

IoMT refers to the collection of virtual computer networks of health devices and applications connected with IT systems for medical care. Wireless-enhanced health gadgets allow the machine-to-machine interaction, that is, the foundation of IoMT, connecting IoMT gadgets to cloud platforms such as Amazon Web Services, where apprehended data can be stored and evaluated. IoMT is also referred to as IoT health care. IoMT medical care resolutions include smart beds, smart rooms, scanners, remote medical care, asset management, and so on. All of these are related to the use of technology such as Bluetooth and wi-fi. Homeopathic private care requires smart pillboxes, digital pills (or small tablets), etc.

2.2 Computational intelligence (CI) techniques in DM

CI centers on complications that hypothetically only individuals and animals can unravel, difficulties needing intelligence. It is a section of computer science considering complications for which there are no operational computational algorithms. The term acts as patronage under which further approaches have been added overtime. CI methods have pulled an increased level of concentration in the investigation community. As described in several current studies, machine learning approaches have possibilities of providing an excellent precision in classification as likened to alternative algorithms for data classification [45–48]. Accomplishing conspicuous correctness in forecasting is significant because it could lead to an appropriate precaution programmer. Forecasting correctness may differ varying on diverse studying systems methodologies. Hence, it is fundamental to recognize devices proficient in offering extreme accuracy of projection in diabetes outbursts.

CI approaches are the best successful decision-making methods for the genuine world and systematic difficulties. The intention is also to examine the performance of diverse CI approaches for the categorization of diabetic and nondiabetic samples. Diverse categorization performance capacities have been employed for the assessment of the performance of CI approaches. Six commonly used CI approaches are as follows: artificial neural network (ANN), support vector machine (SVM), logistic regression, *k*-nearest neighbor (kNN), classification tree, and Naïve Bayes.

In the previous research works, machine learning was put to application in the biomedical field [46–51], for the involvement of heart disease and diabetes [52], in the investigation of diabetes proteins [53] among others. The best significant AI techniques that have been utilized by academics are ANN, SVM, FL systems, K-means classifier, and many others [45, 47, 48, 54, 55]. ANN is regarded as the most commonly used amid all. Survey work on utilizing ANN in homeopathic diagnoses was achieved by Amato et al. [54], and it has been demonstrated in Berglund and Sitte [53] that ANN can have numerous applications and can entertain diverse systems for teaching. Nearly, all research study utilized the multilayer ANN with feed-forwarded backpropagation (FFBP) algorithms to achieve DM classification. Zainuddin et al. [56] considered a research work that classified diabetic patients into insulin and non-insulin. The work was contingent on datasets gathered in India, called Pima Indian Diabetes Dataset [47]. Some other investigations employed the same FFBP algorithm to detect DM cases [57]. They gathered the dataset from the Sikkim Manipal Institute of Medical Sciences Hospital, Gangtok, Sikkim for diabetic patients.

An alternative type of AI device that had been applied by scholars to classify cases into type 1 or 2 diabetes is the FL classifier method [45]. The studies done by the scholars are dependent on a subordinate kind of data archive called Pima datasets. The accuracy of their studies was examined in contradiction to the rate of misclassified instances. Particular research applied DT which is also regarded as an AI device for diagnosing diabetes to accomplish classification and likened to ANN. These researchers concluded their findings that DT established enhanced correctness [49].

Numerous techniques and algorithms were also utilized to extract confidential information from biomedical datasets. Yegnanarayana [58] managed to categorize diabetes instances exploiting principal component analysis and neuro-fuzzy inference. The diabetes ailment dataset utilized in their research was from the Machine Learning Database (Department of Information and Computer Science, University of California) and the resulted classification accuracy was 89.47%. There was another classification conducted on diabetes cases and the researchers obtained 78.4% classification accuracy with 10-fold cross-validation utilizing evolving self-organizing map (SOM) [52]. A permutation method was surveyed to syndicate quantum particle swarm optimization (PSO), weighted least square, and SVM to diagnose type 2 diabetes. An additional study reported in this area is the one that applied the c4.5 algorithm for classification and attained a 71.1% accuracy rate [53].

AI is the capacity of devices that are normally in the computer hardware and software format to imitate or surpass individual intelligence in daily engineering and systematic undertakings connected with identifying, intellectual, and performing. AI is multifaceted since individual intelligence is multifaceted, hence including objectives that vary from knowledge representation and intellectual to acquiring, to ophthalmic observation, and language understanding.

SNPs with their association for a disease/phenotype at a population level can be used for the detection of critical diseases. Polygenic risk scoring and machine learning are typically used to construct genetic risk prediction. Hence, the integration of SNPs into a risk prediction model necessitates blending into models the score of an individual's genotype to enable the assessment of risk. Polygenic risk scoring leads to biased and less effective predictions due to its linear additive regression-based modeling, but machine learning algorithms robustly recognize patterns from strongly correlated data and nonnormally distributed by employ multivariate, nonparametric methods. Machine learning has to receive increasing levels of interest for complex disease prediction since the algorithms can model highly interactive complex data structures. Thus, the addition of high-dimensional genomic data within machine learning models can lead to enhancement in genetic risk prediction over other methods. Machine learning in a recent study has been used as a predictive model by integrating biochemical, cardiac physiological, and epigenomic biomarkers data to detect type II diabetic status. The model established the mitochondrial function, methylation status, and interconnectedness between diabetic classifications. Therefore, the chapter employed the use of machine learning for the prediction and classification of diabetes using genomics data and show that the methods can be used to provide new and more precise methods for identifying type II DM and the augment existing diagnostic standards.

3. Genomic data and health-care systems

In the years past, genomic sequencing was restricted to the research environment, it is now increasingly being used in the medical field, and genomic data from more than 60 million patients are expected to be produced in the health-care sector over the next 3 years [59, 60]. The health-care systems must, therefore, be prepared for the responsibility associated with genomic medicine and the need to share genomic, clinical, and epidemiological data on a global scale to optimize the individual's benefits and the associated complexity and volume of such data [59]. Because genomic sequencing is a revolutionary technology, the successful incorporation of genomic data into the health-care system needs system-wide change [60, 61]. The use of genomic technology in standardized clinical care has been facilitated by the number of improvements and innovations in human genomic science that are directly relevant to the detection, treatment, and prevention of diseases coupled with the decreasing cost of genome sequencing [62]. One of the major challenges concerning the widespread adoption of genomic medicine is the lack of evidence to show improved medical and economic results [63]. Beyond the technical requirements of sequencing and the ability to process samples in bioinformatics. The broad clinical application of genomic data in the health-care system has various obstacles, including policy participation and public acceptance, data interpretation and implementation, cost effectiveness, lack of evidence for medical use, ability and efficiency of the workforce, and ethical and legislative concerns [60, 61, 64, 65]. Implementation of the program framework for genomic medicine is both a single institution and a collaborative multi-institution [60, 61, 63], but translation of this information experience to transform the entire health-care system is rare and unusual [60]. Human datasets obtained by health-care providers and made available to scientists have enabled the rapid advancement of biological research while providing unparalleled opportunities [36]. Human beings have always been a priority of study in both studying basic life processes and hospitals, but attempts in this direction have been fairly reluctant and dangerous [59]. The advancement of human genomic sequencing on a population scale offers a groundbreaking and unprecedented prospect for both basic health-care systems and clinical practice research [59, 66]. There are a host of ongoing efforts around the world to develop national implementation strategies for genomic [67], but most of them were carried out in relative sequestration. The duplication of genomics-based medicine can be minimized by accelerated sharing of approaches, information, and standards, thus increasing progress in the detection and translation of genomics-based therapies into clinical applications [65, 67, 68]. Human genomic initiatives have propelled the development of disease biomarkers and drug targets, resulting in faster and cheaper techniques to question genomics, which also contributed to the creation of thousands of multinational biotechnology companies [62]. The pharmaceutical industry's acceptance of genome-enabled drug discovery and the molecular diagnostics market has grown rapidly [62]. In the recent years, genomics application has appeared in various stages of the disease. For example, care prognosis and risk stratification and diagnosis testing [62]. Genome-based technologies and methods have been developed and implemented for clinical practice and disease management in diseases such as chronic hepatitis C (HCV) infection, breast cancer, and HIV infection, but most medicine fields are sluggish in genomics implementation [62]. Genome-wide association studies (GWAS) and next-generation sequencing (NGS) were the force-enabling technologies that provided genomic science with transformative tools.

Human Genome Project laid the foundation for genomic medicine to define disease at the molecular level. Human Genome Project has guided significant biological developments at an unprecedented pace with the aid of advances in bioinformatics, systems biology, genotyping and sequencing techniques, and computational biology [62]. With the ultimate goal of personalized medicine [62], genomic medicine has incorporated these findings into clinical practice in the recent years. Clinical judgment can be driven by these genetic variants [64]. Use genetic data to diagnose, predict, and advise medical decisions is still slow and uncommon, yet increasing numbers of people are sequencing their DNA [64]. In comparison to extreme Mendelian conditions, genetic segmentation of complex and common diseases such as diabetes, cancer, and cardiovascular disease is more complicated. Such diseases are general by nature with a complex interplay of many gene variants as well as nongenetic factors. Clinical studies are used to evaluate whether genomic knowledge generates hands-on assistance. A fact now written in the US treatment guidelines is the variant called HLA-B*5701 which could help prevent toxic reactions to the AIDS drug, a study of almost 2000 HIV patients. Today, during a hospital stay or a doctor's appointment, genetic data are used to make it a part of routine care. The US National Human

Genome Research Institute has funded projects explicitly exploring the clinical implementation of genomic information through EMRs. The consortium of nine institutions formed the network of Electronic Medical Records and Genomics (eMERGE), using commercial and home-grown EMRs. Health systems can now carry out 'pragmatic clinical trials' as more genetic data are integrated into EMRs, randomizing clinician groups to study an experiment, while genetic knowledge compared to traditional practice is used to direct health-care treatments. Although more research remains to be done on how to assess and implement results, data collection from people with multiple disabilities and who are often less costly than traditional clinical trials except for trails [61, 64].

Protections against genetic discrimination promote genetic research and the medical use of genetics, as well as ensure that genetic data are used ethically [69]. After 10 years after passing the Genetic Information Nondiscrimination Act (GINA), the American Society for Human Genetics remains a credible supporter and proponent of the effective enforcement of GINA and other laws that strengthen community shields [69–71].

Around 5000–7000 rare genetic disorders, each with major medical changeability marinas [72], and each with individual characteristics. Most specific disorders are associated with unusual forms of genetic influences. Many genetic diseases have the foundation's enormous allelic diversity, so the consequences are amplified by situational genomic variations that have a greater impact on clinical presentation [72]. No single provider, medical center, laboratory, state, or even country will have sufficient knowledge to provide the best care for patients in need of care, taking into account the variability in the clinical presentation and molecular etiology of genetic disorders and their relative individual prevalence [72, 73].

The exchange of genomic phenotype and variant data will enhance the production of patient tests and treatments and provide comprehensive information needs for medical, product, and drug manufacturers [72]. This is very important as it will enhance treatment by providing useful key data for key medical characteristics that can be identified as the phenotype of those with genetic diseases. Data sharing should reinforce the link that needs to be formed between genetic diseases and the underlying causative genes. This will also strengthen the criteria used in the classification of variants, making it possible to make appropriate identification of variants of unknown importance and defining the classification of genomic variants as pathogenic across different parts. It helps overcome variations in laboratory variant perception. The exchange of genomics information will provide the best data for health-care providers, industry, and the scientific community that secondary research using these data is driven and web-based systems are based on integrated support for clinical decision-making.

3.1 Genetic risk factors and commonly mutated genes in diabetes

Genetic variants contribute to their risk in the development of common and complex diseases such as diabetes and cancer [74]. In the treatment of complex diseases, many variants contribute to disease risk [74, 75]. Each of these genetic risk factors contributes to any disease in various ways. Changes in many genes with a small effect may cause responsibility for many common diseases such as mental illness, heart disease, diabetes, cancer, and obesity [74]. The risk of disease may depend on an identified genetic change and multiple factors in people with a genetic predisposition. Genetic susceptibility is an increased likelihood that a certain illness may occur depending on the genetic makeup of an individual. This is often inherited from a parent and is the product of particular genetic variants. Such genetic changes are caused by the emergence of a disease but do not cause it directly. Even with the same parents, while others will, some will never get the disease with a predisposing genetic variation. Genetic variations with large or small effects on them can cause the likelihood of developing a particular disease. For example, the risk of a person developing breast cancer and ovarian cancer is largely caused by certain mutations in the BRCA1 or BRCA2 genes, as well as variations in other genes such as BARD1 and BRIP1, also increases the risk of breast cancer, but the overall risk appears to be a major contributor to a person's genetic changes [76, 77].

In addition to an established genetic change in people with a genetic predisposition, the risk of disease may sometimes depend on multiple factors. Certain genetic factors are referred to as modifiers as environmental factors as well as lifestyle. Multifactorial are diseases caused by a combination of the factor of genetics and alteration [75–77]. A person's genetic traits cannot be changed, although behavioral and environmental changes can reduce the risk of disease in people with a genetic predisposition, such as maintaining a healthy weight and more regular disease screening. Current research focuses on identifying genetic changes that have little effect on the risk of disease but are widespread in the general population [78]. While each of these variations increases the risk of infection only slightly, changes in several different genes can combine to significantly increase the risk of disease [76]. Changes in many genes, each with a small effect, may be vulnerable to many common diseases, including cancer, obesity, diabetes, heart disease, and mental illness [74, 75, 78].

There is no specific origin for any complicated disease, it is referred to as communication between genes, ecosystem, and lifestyle that ultimately leads to complicated diseases that can transpire [79, 80]. Complicated disease cases include cancer, heart disease, and diabetes. Although it is difficult to determine precisely how much impact any particular risk factor has, we can estimate the percentage of a disease or trait caused by genetic factors (heritability) and the percentage of a disease or trait caused by nongenetic factors such as lifestyle and climate [79, 80]. Genetic variants contribute to the risk that common and complex diseases like cancer and diabetes can develop. There are many variants in dealing with complex diseases that contribute to the risk of disease. Each of these genetic risk factors contributes significantly to your risk of disease. In some cases, the contribution to the risk of your disease will be low, while in other cases having a particular variant may have a significant impact on the risk of your disease. The overall gene-based risk depends on how many genetic risk factors you have and how they interact [79, 80]. Variations in additional 28 genes, for example, have been related to type II diabetes. To become diagnosed with type II diabetes, there is no known exact number of variations a person must possess. Nongenetic risk factors such as body mass index (BMI), partnership with genetic risk factors to produce ailment are not fully understood to be one of the genetic risk factors [79]. Genetic-wide association studies (GWAS) aid in assessing and evaluating human genome DNA variations and identifying risk factors for disease in some individuals. This is resolved by repeatedly transpiring variants of the solitary nucleotide polymorphism in the DNA sequences as separate base-pair modifications [81, 82]. The genotyping of the NGS and chip-based microarray [83] recognizes these disparities. Using this modern technology, epigenome-wide association studies (EWAS) showed that in addition to HLA-A, HLA-B, and HLA-DPβ1, HLA-DQα1, and HLA-DQβ1. HLA-DRβ1 haplotypes carrying HLA-DRβ3 alleles showed a greater risk than HLA-DRβ1 haplotypes carrying DRβ3 with two DRβ3 alleles in DRβ1/homozygotes [83–87].

For humans, MHC I molecules are HLA-A, HLA-B, and HLA-C, while MHC II molecules are HLA-DP, DQ, and DR, with the strongest association with T1D [88–92]. Two groups of genes from MHC II encode alpha-polypeptides and beta-polypeptides and together form the alpha-beta heterodimer functional group II. They form major DPα, DPβ, DQα, DQβ, DRα, and DRβ, plus minor DM and DO genes encoding the APCs with MHC II proteins are polymorphic genes of both alpha and beta polypeptide [93–96]. Due to their function in triggering autoimmune and inflammatory responses, these proteins are essential.

Diabetes is a chronic metabolic disorder that adversely affects the ability of the body to produce and use insulin, a hormone, that is, required to turn food into energy. For the roughly 16 million Americans affected by it, the disorder greatly increases the risk of blindness, heart disease, kidney failure, neurological disease, and other disorders. Type 1 or juvenile diabetes onset is the disease's most serious form.

Type 1 diabetes is what is referred to as a "complex trait," meaning mutations in several genes are likely to contribute to the disease. For example, it is now known that the locus on chromosome 6 of the insulin-dependent diabetes mellitus (IDDM1) may contain at least one susceptibility gene for type 1 diabetes. It is not clear exactly how a mutation contributes to patient risk at this locus, because gene maps to the chromosome 6 region that also has antigen genes (the molecules that usually warn the immune system not to attack themselves). The immune system of the body conducts an immunological attack on its insulin and the pancreatic cells that contain it in type 1 diabetes. Nevertheless, it is not yet known the process of how this occurs. Approximately 10 loci have now been found in the human genome that tends to be resistant to type 1 diabetes. Among these is (1) a chromosome 11 and 2 locus IDDM2 gene for glucokinase (GCK), an enzyme, that is, essential to glucose metabolism in chromosome 7 [97].

Conscious patient care and daily doses of insulin can be relatively healthy for patients. But to prevent the immune responses that often cause diabetes, we will need to further experiment with the disease's mouse models and advance our understanding of how genes on other chromosomes can contribute to the risk of diabetes for a person [93–96].

3.2 Single-nucleotide polymorphisms (SNPs) as biomarkers for DM

Normal biological processes are used as indicators for the detection and identification of symptoms, diseases, and/or another biological state of organisms and these processes are known as biomarkers. Biomarkers are of great medical significance because they can predict and diagnose diseases, to provide data on the pathophysiological features of such diseases, and to track pharmacological response to therapeutic remedies or to predict clinical outcomes [98, 99]. Factors such as cost effectiveness and productivity, easy access by noninvasive approaches, consistency in biofluids and body fluids, tolerance to changes in disease status, ability to detect disease before observable clinical symptoms, and distinction of disease pathologies are considered in ideal situations to classify appropriate biomarkers. Molecular biomarkers, easily measurable in biofluids such as a serum, can now be formed as whole blood and plasma thanks to the developments in molecular biology [98, 100, 101]. Commercial kits for SNPs are clinically readily available for a variety of diseases [102, 103], DNA methylation [104], and miRNAs [98, 103, 105].

SNPs refer to sequence variability in the genome at a specific location in the DNA sequence, which is one of the most common in the human genome with over 10 million SNPs [106]. Although in most cases SNPs are stealthy and do not affect gene expression, they do account for body reaction to treatment and susceptibility to various forms of disease and protein function alteration [107]. An important clinical research area is the quest and detection of such SNPs, which affects disease susceptibility and treatment response. According to the past studies [108], SNPs have been associated with various medical conditions that are metabolic such as cardiovascular disease, T2D, and obesity. Variants in more than 50 and 80 loci are associated with obesity [109] and T2D [110], respectively, which occur in genes that control homeostasis of glucose and insulin signals.

Genetic association studies have inherent limitations, particularly in polygenic and multifactorial disease studies such as gestational diabetes mellitus (GDM). These limitations, as noted above, include insufficient sample size to detect statistically significant associations, and differences in allele frequencies and disease etiology among ethnicities, which may explain why many genetic associations are not reproducible across populations. Also, the diagnosis of GDM is not globally standardized; thus, specific diagnostic criteria may have led to the discordant findings found between studies. Importantly, genetic variants do not only lead to complex disease formation, and it is widely believed that disease is caused by the relationship between genetic predisposition and environmental factors [103, 111, 112]. Biological and environmental factors such as maternal age and diet [103, 113] should, therefore, be considered in combination with genetic variants to evaluate the risk of GDM accurately [111]. Given the variable results obtained throughout the study, many of the variants found to be associated with GDM are also associated with T2D, which supports their biological plausibility [111]. Therefore, although GDM's etiology may vary from T2D, it is likely that the genetic mechanisms by which the symptoms manifest overlap. Only studies that profiled DNA-extracted SNPs from whole blood have been documented in this analysis. Nevertheless, it is known that less invasive genetic material sources, such as buccal swabs, are being used [111, 114]. Also, this analysis included only SNPs identified in two or more studies and may have omitted other important GDM-related SNPs [103, 111].

4. CI models for DM

Modern trends in the application of technology in medicine have computerized most of the complicated stages of diagnosis and treatments. The research discovery of digitalized medical machines for the automated treatment and awareness of different diseases in patients has been a vibrant research focus for modern medical scientists. Breakthrough in medical research for disease classification, prediction, and diagnosis has enabled the early detection of disease, with the consideration of doctor schedule, diagnosis and treatment cost, medical scientist consistency, and diagnosis competence. Research on disease classification, prediction, and diagnosis is practically important [115, 116].

The purpose of the computerization of medical processes in patients is to enable an easy explanation of the cause and effects of complex diseases and provide efficient countermeasures through patient history logs of signs and symptoms. Recently, many models have emerged for the grouping, detection, and analysis of DM. The two models commonly used in this regard can be summarized into medical and CI models [115]. The medical model uses the physical characteristics of diagnosed patients for precision diagnosis of an illness. The medical scientist using the FPG or the OGTT to check your blood glucose level to confirm the diagnosis of diabetes. The newest methods use hemoglobin A1c as a screening tool for prediabetes or diabetes. The test for several months is generally used to measure blood glucose control in diabetes patients. Countless medically inspired analytic techniques have been widely deployed in the prediction of diabetes in a patient. However, the medical inspired analytic techniques are complex to use and its operational processes cannot be precisely explained using mathematics notations. Hence, CI models are considered more efficient for addressing the complex and large-scale nonlinear problems without any statistical assumptions about the data [117, 118].

CI models have been developed to perform well with complex real-world data-driven problems using some naturally inspired models, where traditional statistical methods are unable to perform. CI-based techniques are systems based on the process inspired by a natural evolution [116, 119, 120] ranging from the ANN, swarm intelligence optimization, genetic algorithms (GAs), and genetic programming algorithms [121–123] that can be used for speedup computational and biological processes.

CI is extensively used in medical, engineering, and scientific research. This accounts for the popularity and simplicity of computational techniques and DL architectures [124]. CI-based techniques have been used in pattern recognition and learning of nonlinear functions, e.g., image processing [125–127], robotic vision [128, 129], voice recognition [130, 131], control [132], text categorization [133–135], data mining [136, 137], quality control [138, 139], robot control [140–143], medical and biological [54, 144–147], and environmental [148, 149] among others. Intelligent methods of optimization can also be called AI since it used in computer science and technology to mimic nature and human beings. CI can also be classified into three main types: fuzzy computing, evolutionary algorithms (EAs), and neural computing [150, 151]. Although important advances have been made, many new CI technologies have been introduced to solve several practical problems [152], such as a fuzzy neural network [150], DL network [153], and extreme learning machine (ELM).

CI-based techniques remain a hot science and technology space research subject, and it increases its dissemination into other jurisdictions such as industry, health care, and gaming. CI-based approaches have emerged as a capable tool for designing and implementing smart systems in the health-care system. Use CI in health care can enhance clinical disease management by incorporating smart approaches for prevention, diagnosis, treatment, and follow-up as well as administrative process review. By taking into account the complexity that characterizes health data and procedures, CI-based systems can learn from data and adapt according to changes in the environment.

The use of data mining and AI approaches was used to produce metrics that help the decision-making process through the use and identification of relevant information from collected hospital data using computer devices. CI-based paradigm, through the numerical representation of data, incorporated many methods and approaches to model information. Applications can be built using a CI-based approach that can classify high-risk patients with chronic diseases such as DM, thus becoming useful tools for implementing effective treatments and scheduling. Therefore, the number of hospitalizations of patients will be reduced. This can also result in a better diagnosis and rehabilitation system for an appropriate evaluation of readmissions, thereby enhancing the use of resources in the provision of health-care. Therefore, it can be used to determine the best treatment options by using the available evidence from CI applications that endorse medical decisions. The CI-based approach has been used to solve various health sector problems, particularly the most urgent problems, without reducing the quality of care it has reduced the patient's cost diagnosis and treatment. It used to reduce the length of stay in hospitals for patients, thus helping to prevent medical problems arising from hospital infections.

DL algorithms that provide effective biologically inspired computational modeling techniques and have attracted increasing attention to many tasks such as health care, robotic vision, object recognition from various sensory inputs, and signals [116, 154–158]. DL is the concept of using ANNs with multiple hidden layers, where learning is both unsupervised (bottom-up) to generate higher-level representations of sensory data that can then be used to train a classifier (top-down) based on the standard supervised training algorithms [159]. These techniques are based on the supervised methods such as recurrent NNs, convolutional NNs, DNN, and deep belief networks (DBN), which are nonsupervised methods that provide deep architecture incorporating structural elements of local receptive fields, mutual weights, and pooling aimed at mimicking. Recent review studies [115, 119, 120, 160, 161] have highlighted the potential of using DL techniques in identification, prediction, and diagnosis of diseases.

Using CI-based applications can also provide a valuable information to medical scientists and enhance the outcome of patients, which improves service quality and also reduces costs [162]. It is also helpful in recognizing and knowing, among others, patients with chronic signs and symptoms resulting in high health expenses [163]. CI techniques can add value to financial and health-related costs to health-care management [164]. Evolutionary CI approaches are effective in promoting and diagnosing chronic diseases such as Yang's heart [137, 165–167] and metabolic diseases [168–170] and diagnosis of cancer [171–174].

4.1 AI approaches for DM: Challenges and opportunities

Throughout the recent years, AI developments have continued to attract a lot of health-care workers. The digitization of health-related data and the use of computer hardware and software applications have brought about the growth and use of AI in medicine, thereby creating new possibilities and challenges and offering guidance for AI's future in health care. In healthcare, AI-based approaches have been used for various medical purposes such as diagnosis, identification, forecasting, and control of various diseases and diseases [175, 176].

The technologies that allow human intelligence to be adapted by computers and machines are called AI. AI data, including clinical, behavioral, environmental, drug, and biomedical data, were used to establish a decision support system, diagnosis, prediction, and patient classification from various diseases. Due to its ability to systematize several activities that currently require human intervention, AI has drawn a significant interest from various fields [176, 177]. Recently, AI techniques and methodologies are used to support, among others, in the processing of natural language, speech recognition, computer vision, and imagery.

AI provides new opportunities for human intelligence and mathematical models to evolve in order to extract information from health data. AI systems can, therefore, be used to simulate human intelligence in health care at various levels [178]. AI has been providing rapid development of computer hardware and software applications in the recent years to promote the digitization of health data to be used to build automated medicine systems for various uses. AI-based systems in medicine have always been considered both to reflect medical knowledge and to derive new knowledge from stored clinical data and to help clinical decision-making [179, 180]. AI-based has been described as a scientific discipline for applications and research studies aimed at supporting decision-making clinical tasks through computer-based solutions and extensive information data that ultimately supports and enhances the quality of medical scientists in human care providers [179, 180]. The health-care system has increasingly leveraged AI-based technologies to improve the delivery of health care at a reduced cost. AI's application to diagnosis, perdition, and identification is well established in the recent years, and AI is increasingly being used to inform decisions on health-care management [181].

The advantages of AI-based health care have been extensively discussed in the recent years to the point that it demonstrates that it is possible in the nearest future to replace human doctors with AI, thereby showing that AI can be used in health care in many ways: the ability of AI to learn features from a large volume of medical information and then to use the knowledge gained to assist clinical practice. The AI-based system is shown to be effective in collecting useful information from large patient populations and used to make real-time health risk warning inferences, diagnosis, and prediction of health outcomes. AI-based systems can also easily perform repetitive tasks such as X-rays, CT scans, or data entry. AI can be used to reduce unavoidable human medical practice clinical and diagnostic errors [182, 183]. AI-based can support doctors by presenting up-to-date medical information from clinical procedures, textbooks, and reviews to provide proper patient care. AI has shown significant improvements in the fields of record keeping and assessing the quality of an individual organization as well as the entire health-care system [184, 185]. AI leads to the development of new medicines and precision medicine based on the faster processing of mutations and disease linkages. Ultimately, health monitoring services and online consultations are provided by AI-based to the degree that they are medical bots or virtual nurses [182, 186].

To accept AI fully for clinical care, it will need to meet specific high standards to meet the needs and desires of physicians and patients. Though the AI approach has shown superiority over other approaches, it has shown some degree of error and it is not and never will be flawless, no matter how infrequent, it will drive major, negative perceptions [187]. AI-enabled system errors regardless of how small it would have a major worrying impact on any medical matter [187], so a suitable level of regulation and supervision is very important when incorporating AI into the clinical field. It is also important to assess the cost effectiveness of AI-based clinical efficacy [187, 188]. Huge investments in AI were made, similar to robotic surgery, with expected efficiencies and anticipated cost reductions in return. It is unclear, however, whether AI techniques can significantly reduce costs with their associated data storage requirements, data curation, design maintenance, upgrading, and data visualization. Such resources and associated needs can simply replace current costs with costs that are unique and theoretically lower [187, 188]. The following problems in the health-care system were still highlighted by AI-based techniques. Security is one of the health-care issues that have to apply aggressively when it comes to medical information as it is well established that security is very relevant in every sector [188–190]. For example, patient data are not allowed to leave Europe in European countries, most hospitals and research institutions are cautious about cloud platforms and prefer to use their own databases, making it difficult to access patient data to build AI-based systems [190]. In some cases, the use of standard application procedures to facilitate research based on the patient clinical data is much simpler for medical researchers. For example, AI algorithms intended to be used in health care (in Europe) must be licensed and controlled for proper use, CE labeling must be applied for. More precisely, according to the Medical Device Directive, they need to be marked. The Global Data Protection Regulation (GDPR) guidelines adopted in

May 2018 will also result in several new rules that need to be complied with and that are not clear-cut in some situations [191].

Throughout the recent years, DNN has performed poorly throughout health care, given their emerging capabilities. The technologies are still in the infant stage and resources needed to support them are still young, and few people have the requisite technical skills to handle the full range of information and software engineering issues. AI solutions are often faced with issues related to limited data and variable data quality, particularly in medicine. As new data comes into an AI algorithm, predictive models will need to be retrained, keeping a close eye on changes in data-generation processes and other real-world issues that can cause the data distributions to drift overtime. If multiple data sources are used to train models, additional "data dependencies" types are added, which are rarely reported or specifically treated [188, 190, 191].

For the AI-based medical system, the consistency of decision aid is very important. A doctor needs to understand the medical AI-based system and be able to explain why an algorithm has suggested a certain treatment. This includes the creation of methods for predictive analysis and more intuition. There is often a trade-off between predictive accuracy and accessibility of the model, particularly with the latest generation of AI techniques using NNs, which makes this problem even more urgent. Besides, the data contained in electronic health records are not appropriate for AI-based techniques in many cases [188, 192]. AI works best with high-quality data sources, while with large, restricted categories, electronic health records, and medical billing claims appear to be ill defined. Sociocultural is another health-care-based AI challenge because it can be challenging to get doctors to accept advice from an automated system, as doctors make decisions based on the previous experience and instinct, expertise gained, and problem-solving skills [188, 192]. It is, therefore, important to incorporate certain elements of AI literacy into medical curricula so that AI is not perceived as a threat to physicians, but as a medical knowledge aid and amplifier. Nevertheless, if AI is implemented in a way that empowers rather than displaces human workers, it could free up their time to perform more important tasks or provide more capital to hire more workers [188, 190, 192].

4.2 GA for selecting the most informative gene biomarker

The term biomarker is the shortest name for the biological marker, used as an indicator that there is or has been a biological process in the body. Work and interest in biomarkers linked to the disease have increased significantly in the recent years [193–196]. Diagnosis, detection, and identification of diseases are the possible therapeutic advantages of disease-specific biomarkers and the potential reduction in the size and length of clinical drug trials, which would improve the production of medicines. Many related medicinal biomarkers are used to show that the body has been subjected to a chemical, toxin, and other environmental impacts [195, 197]. A biomarker is something that can be correctly and reproducibly measured. According to the World Health Organization, a biomarker can be identified as any measurement that represents a biological system associated with a potential hazard that may be chemical, physical, or biological. The measured response can be functional and physiological, cellular-level biochemical, or molecular activity. Biomarkers are not a patient's medical signs or symptoms, such as a high temperature used by a patient to assess their well-being, but an increased heart rate is a biomarker as a result of physical activity. The physiological response to the exercise is the increased heart rate. To find out what will happen to you if you are using a particular biomarker for medication or if you are not using the drug and the risk of developing those medical conditions. A biomarker can be used to detect a heart attack in the blood, and testing the levels of blood hormones, proteins, and enzymes help a doctor to assess the extent of heart attack and how much heart damage has occurred. Levels of biomarkers can appear and disappear after an attack and change overtime at different stages. Biomarkers can be expressed differently between healthy patients and those with organ failure in such patient groups and between patients with organ failure As personalized health care is normal, biomarker use will become substantially more common [193-196, 198]. Therefore, people who are healthy express different types and/or proteins, metabolites, and gene amounts than those who are sick. Prostate-specific antigen associated with prostate cancer and the risk of coronary heart disease in blood cholesterol are well-known biomarker instances. Examples of biomarkers are gene expression profiling, blood pressure reading, and hemoglobin A1C, depending on the rate of glomerular filtration. Such biomarkers help to diagnose or track, predict, and provide insight into custom medicine [198].

Gene expression summarizing by microarray techniques appears to be capable of classifying and diagnosing various diseases, thus allowing the researchers to examine the expression of thousands of genes in a single experiment efficiently and quickly [199, 200]. Nevertheless, the limited number of specimens, dimensionality, and irrelevant and noise genes make the classification task difficult for a given sample endured by all microarray datasets [199]. Classification is an important task in machine learning, as it will be difficult to determine which gene is useful without prior knowledge. Relevant, obsolete, and redundant genes are, therefore, inserted in the database to provide a large number of genes. But in classification, not all irrelevant and redundant genes are required, so effective methods of selecting genes for DM are paramount. Gene selection is crucial to choosing the minimum number of relevant and insightful genes that are more reliable in the process of classification. For gene selection, an optimization algorithm is used to pick a subset of genes that have the most knowledge about classification from the original microarray data [201]. Heuristic approaches such as bioinspired EAs are the most widely used methods of gene selection since the optimal problem of gene selection is known as an NP-hard problem.

The structure and function of that cell or tissue are determined by a cell or tissue's gene expression pattern. The number of genes on a microarray chip is more than a thousand, as opposed to a small number of samples. Therefore, in microarray data analysis, the curse of noisiness, dimensionality, noisiness, and stochastic existence of this data are major problems, resulting in the implementation of numerous machine learning challenges and data mining. To improve classification accuracy, it is highly necessary to determine a small subset of specific genes in a given dataset for a high-dimensional problem [200, 202]. Besides, using other biological databases and bioinformatics resources (protein-protein interaction and pathway databases) can solve the stability problem.

Taguchi-GA and Taguchi-PSO [203–205], updated PSO and SVM [206], hybrid GA and SVM, hybrid GA and NN classification [204, 207], and K-NN implementations for microarray data [204, 205, 208, 209].

Selecting insightful genes is the most important task in computational biology or gene combinations with a high prognostic potency from the disease classification microarray data [210], so this can curse dimensionality and create the biggest problem in information on gene expression. It contains a huge number of lines for the number of samples of genes and columns. Selection methods are required to select significant genes for prediction and diagnosis. The problem with nondeterministic soft (NP-hard) polynomial time is the choice of features from the large dataset [199, 211]. The most are used to identify a subset of the original features that would perform better for the classifier built with the subset than a classifier built from the whole set of features. To increase the identification accuracy of the genes, delete obsolete and redundant features. The accuracy of the classification can be determined by the correct classification of diseases divided by the total classification number. This is as follows:

$$\text{Classification accuracy} = \frac{CC}{N} \times 100 \tag{1}$$

where N is the total number of instances in the initial microarray dataset. CC is the correct classified instances.

GAs are used to look for problems and construct optimization solutions and are very useful in solving related issues. It is a heuristic quest that imitates the natural selection process. GAs belong to the class of EAs used to produce solutions related to problems of optimization such as crossover problems, selection based on relative fitness, mutation, and inheritance inspired by a natural evolution [212]. To solve various problems in medical science, bioinformatics, engineering, business, technology, computational mathematics, etc., it has been applied in many fields. GAs could be used in areas too broad to be searched extensively [213].

The method is used to find solutions to problems and an iterative process involves a sample population as individuals, each represented by a finite string of symbols called the genome. Heuristically or randomly, an initial population of individuals is created. The individuals in the current population are decoded and evaluated in a generation according to a fitness function that describes the problem of optimization in the search space. The next generation is created by choosing individuals according to their fitness; a fitness method is used to calculate how close the person is to achieve the problem's set objectives [214].

Fitness-proportionate selection is a simple method of selecting individuals with a probability proportional to their relative fitness compared to other available methods of choice. Population quality depends on the number of times a person will be picked. Individual high fitness will give the community the ability to replicate and bring new people, but low fitness will not bring new participants to the population. Because GA is stochastic iterative processes, convergence is not guaranteed and the condition of stoppage may be defined as a total number of generations or a selected fitness level.

GA is an important optimization tool for selecting features based on Darwin's theory of natural selection and suitable survival. Through successive iterations (generations), it uses the market method and regulated variation to represent feasible solutions to particular problems. The competition process is used to choose the population chromosomes that fulfill the fitness associated to form new ones. Genetic operators commonly use fusion and mutation to create new chromosomes. There are three approaches involved in creating the next generation from the current population: recruitment, fusion, and mutation.

1. *Selection rules*: this produces single parents for the next generation to add to the community.
2. *Crossover rules*: it brings together two parents to create the next generation of children.
3. *Mutation rules*: children from changes to individual parents are randomly created.

The selection mechanism is used in the creation of a new population $P(t)$ with chromosomes in $P(t-1)$. Higher fitness chromosomes typically have a greater chance of contributing copies to $P(t)$ and depending on their fitness, the number of copies obtained for each chromosome. Instead, $P(t)$ is extended to both fusion and mutation. Fusion takes two different parents, by merging parts of the family, to produce two new individuals called offspring. After a randomly selected crossover point, the operator operates by exchanging substrings. The crossover operator is generally not applied to all pairs of chromosomes in the new population. Where the applied probability depends on a crossover rate specified likelihood. A sampling of new points in the quest space mutation is introduced to prevent premature loss of population diversity.

4.3 DNN for the classification of biological samples

Machine learning has shown a great potential in analyzing the broad and complexity of biological data. Biological information is required to successfully apply in practice the classification of biological samples using DL. In the recent years, DL has been used to derive valuable biological information from massive, complex datasets. Researchers have successfully used NGS techniques to track changes at different biological levels, such as gene expression and low abundance of RNA, epigenetic alterations, protein-binding motivations, genome-wide genetic variation, and high-throughput and cost-effective chromosome conformation. Data explosion problems have been long-standing methods for data analysis, particularly genomics data.

Most biological sample studies focus only on one particular aspect of the biological system, but biological systems are more complex. For example, using GWAS to concentrate on genetic variants associated with measured phenotypes. Complex biological portents, however, require both intrinsic and extrinsic, so a single data form cannot be fully explained. Further emphasis is, therefore, centered on incorporating the study of various types of data, thereby contributing to a comprehensive understanding of complex biological phenomena. It can, therefore, be difficult due to limitations of the inherently noisy existence of biological data, data dimensionality, and heterogeneous data, such as high-resolution genomics data. Because of costs and available sources, the number of specimens in biological experiments is often reduced and much smaller than the number of variables (e.g., replicates of plants and animals, cancer samples), resulting in the dimensionality of biological samples, resulting in multicollinearity, multiple trials, data sparsity, and surfeiting [215].

Driven by a massive increase in computational power and big data, DL can, therefore, be applied in many ways. DL-based has revolutionized many fields and can be controlled or unattended, and shows promise for genomics, medicine, and health-care applications. DL was used for identification, exploration, and prediction in biological studies. With the availability of more and different types of genomics data, the application of DL approaches has become more important and frequent. DL-based methods have been used to predict and identify genomic characteristics, particularly those that are difficult to predict using current or quantitative methods such as regulatory regions. Different machine learning methods, particularly DL-based methods, were used to predict the sequence specificities of DNA- and RNA-binding proteins, enhancers, and other regulatory regions [216–218] on data generated by one or more genomics methods; such as DNase I hypersensitive sites (DNase-seq), formaldehyde-assisted sequencing regulatory element isolation (FAIRE-seq), sequencing (ATAC-seq) transposase-accessible chromatin assay, and STARR-seq (self-transcribing effective regulatory area sequencing). Machine learning is very useful for building models in predicting regulatory elements and noncoding variant effects de novo from a DNA sequence [217, 218], thus testing/validating their contribution to the ultimate observable traits/pathologies and gene regulation. In addition to the prediction of regulatory regions, such as the classification of regions under purifying selection or selective sweeping, the DL-based approach has shown considerable potential in population and evolutionary genetics issues as well as more complex spatiotemporal issues [219]. The DL-based methods have recently been used to predict transcript abundance using datasets from known biological aspects such as genomes, transcriptomes, epigenomes, and metabolomics [220], failure to impute SNPs and DNA methylation [217, 221–223], calling variations [217, 224], and diagnosing and classifying diseases [217, 225].

Biological sample data are complicated and often overlooked, but data-rich fields are both biology and medicine. The DL-based approach will, therefore, be well suited to solve different problems from these fields and these algorithms have recently shown promising results in these capabilities. DL, therefore, has shown a significant transformation in the fields of biomedicine, medicine, and health care, providing a unique solution to some of its problems [217, 225].

4.4 Genomic data via Monte Carlo experiment

Experiments in Monte Carlo are carried out by random sampling to obtain quantitative results and are a broad category of computational algorithms. The approaches use randomness to solve problems that may be deterministic. Such methods are used when using other approaches in physical or mathematical problems that are impractical or difficult. The three problem groups that primarily used Monte Carlo methods were numerical integration, optimization, and generating draws from a probability distribution. The methods are useful for simulating systems in physics-related problems with many coupled degrees of freedom, such as cellular structures, disordered materials, liquids, and strongly coupled solids. Business risk estimation involves modeling events with substantial input uncertainty and testing multidimensional definite integrals with complicated mathematical boundary conditions, Monte Carlo-based forecasts of failure, cost overruns, and schedule overruns are consistently better than human intuition or alternative "weak" methods [226]. Monte Carlo methods can solve any problem that has a probabilistic interpretation. Similar approaches were used in various computational biology, biomedical, and health-care fields. Monte Carlo methods for genomes, membranes [227], and proteins [228] used in biological systems, also for Bayesian inference in phylogeny.

Monte Carlo simulation is a computerized computational tool that allows individuals to report in quantifiable analysis and decision-making for probability. This approach was used to overcome the challenges of analytical derivation in technical fields of negligible dispersal. The technique is also widely applied by specialists in numerous areas, including financing [229, 230], project management [231–235], energy [236–239], manufacturing [240–242], engineering [243–246], research and development [247–249], oil and gas [250–252], transportation [253–256], environment [257–261], and DM [262–267]. The approach provides decision-makers with a distinctive range of likely outcomes and possibilities for whichever option of action they will transpire. It shows the extreme possibilities of going for broke and the most conservative decision along with all possible implications for middle-of-the-road decisions. As the use of clinical data analysis in Monte Carlo advances, health-care participants should be able to hide boundaries and be able not only to prevent waste but also to drive enhanced outcomes through individualized treatments for each individual. Of example, cancer treatment, a medical practitioner may cork in the physiognomies of a patient such as past homeopathic history, heredity, and point of illness, undertaking thousands of different circumstances to assess the possible outcome with all treatment alternatives. Health-care administrators could go through quantitative analysis to explain all the possible results of medical treatment using numerical analysis and correct modeling.

Big data analysis in health care offers infinite incentives like analyze 250,000 clinical trials in 30 min. Evaluate nearly instantly the possible reaction of a person to a specific medication. Enter the symptoms and history of a patient's health and provide a prescribed course of action at the bedside. When correctly modeled and interpreted, health-care providers can use data to resolve personal biases that promote adherence to obsolete, expensive levels of testing, or discrimination toward certain methods of treatment. Using these Monte Carlo simulations can transform how companies along the supply chain of health care and continuum care testing, deliver and pay for care, and entail less emphasis on trial and error techniques.

4.5 Practical case of DM prediction with CI techniques

Machine learning was originally defined as a program that learns from data to perform a task or make a decision repeatedly instead of explicitly programming the behavior [268]. The concept, however, can cover almost any type of data-driven approach and is very specific. Consider, for example, awarding points to different factors and generating a number that predicts a cardiovascular risk score of 10 years by Framingham. If this is assumed to be an example of machine learning, it may be incorrect as the Framingham hazard score inspection shows that the response may not be as clear as it appears at first [268]. The rating was initially fashioned from more than 5300 patients by fitting a template of proportional threats to data, so the "law" was learned entirely from data. The labeling of a core risk as a machine learning algorithm may seem a strange idea, the uncertainty nature of machine learning techniques was revealed from the original definition [268]. ML-based techniques are regularly used in biomarker development examples such as locally weighted learning, LVQ, SOM, kNN, learning vector quantization, and SVM among others [216], resulting in the rapid growth of categorized data allowing for DNNs. DL-based architectures have outperformed traditional methods in solving a wide range of genomics, proteomics, image analysis, and transcriptomics [269].

Recently, the emphasis is on designing genetic risk models for reliable identification of individuals at risk [270–272] to achieve an accurate predictive model. SNPs require their association with a disease/phenotype at a population level, according to the GWA studies [3]. The predictive risk model, therefore, requires SNPs to be integrated into models that score the genotype of an individual to enable risk estimation. Models of genetic risk prediction typically involve (1) polygenic risk scoring or (2) machine learning [273, 274].

GWAS expose variations of genomic loci associated with multifaceted behaviors in the population in order to explore the genetic basis of DM [271]. GWAS findings show that SNPs act as candidate biomarkers for genes that could suggest the presence in individuals of complex diseases. This method is based on an analysis of a single locus in which each SNP is independently evaluated for association with a phenotype of interest, omitting interactions between loci [271]. Fig. 1 shows the workflow for creating a supervised machine learning model from a genomic (SNPs) dataset.

The heritability of DM is particularly strong for type II diabetes (T2DM), but it was difficult to use genetic models to clarify heritability of this kind [275]. To demonstrate CI in the prediction of DM, we tested deep neural network (DNN) to predict type II DM and GA was used to optimize datasets using Nurses' Health Study (3326 females, 45.6% T2DM) case–control study and

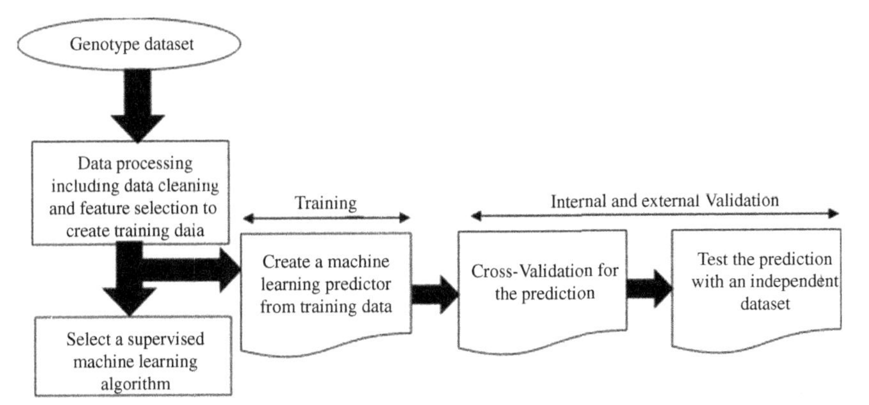

Fig. 1 Workflow for creating a supervised machine learning model from a genomic (SNPs) dataset.

follow-up study of health professionals (2502 men, 46.5% T2DM). SNPs are identified using GA 96, 214, 399, and 678. The two datasets were divided into 95% training and 5% test samples at 50–500 times and divided into 4:1 each arbitrarily to train prediction models and test their performance.

The significant genomic-wide associations between an SNP and the trait examined [271] are significant. Bonferroni correction can be used to calculate these associations as it is highly conservative in multiple test studies designed to minimize type 1 errors. Significant thresholds of 5 bis110-8 across the genome have currently been accepted as strong evidence of association [276]. Machine learning approaches such as SVM, FL, random forest, cellular automata, and NNs have been combined with positive inference algorithms such as multifactor dimensionality reduction (MDR) to detect epistasis interactions [277]. For genomic-wide association and massive candidate gene studies, this approach also suffers from computational complexity issues. Therefore, in genomic-wide studies, the combination of DL with feature selection algorithms will provide a successful data analysis.

4.5.1 Data normalization

Several thoughtful steps are required to preprocess genomics data due to diverse noises and errors. The high-dimensional nature of a typical genomics dataset has posed enormous challenges to computational biologists and statisticians. To eliminate the methodical discrepancy in genomics data, the preprocessing step is important and it deduces biologically relevant inferences [278]. A standard genomics dataset is composed of q records and response class information on n subjects, where $q \gg n$. Generally, the number of transcripts q measured on each biological subject is in the ranges between 1000 to more than 50,000 records, while the available experimental unit may fall below 100. Consequently, to analyze any genomics data, some preliminary steps must be taken before the feature selection stage and the classification of the data. Normalization and the preliminary feature selection are the two preprocessing employed in the chapter.

Normalization is used to remove the influence of nonbiological variations from the useful genomics data [279–281]. The z-score normalization where the values of a gene are normalized based on the average and standard deviation, thus the method was adopted in this chapter. The z-score measures the number of standard deviations of a particular data point from the mean.

Normalizing $n \times q$ matrix of genomic dataset with n arrays and vector $X = (X_1, \ldots, X_q)$ of q genes can be viewed as transforming all the expression patterns X_{ij} of jth gene across the n mSNP samples by

$$Z_{ij} = h\left(X_{ij}\right) - \frac{1}{n^* \epsilon\, n, q} \sum_{l \in i, j} h(X_l) \tag{2}$$

$h(.)$ represents the monotonically increasing Box–Cox family of transformations of X_{ij} given by

$$h\left(X_{ij}\right) = \frac{X_{ij^m - 1}}{m} \tag{3}$$

for some constant $m > 0$. If $h(X_{ij}) = (X_{ij})$ if $m = 1$, indicating no transformation except for a shift, and $h(X_{ij})$ becomes the square root transformation of (X_{ij}) if $m = {}^1/_2$, while it tends to the logarithm transformation as $m \rightarrow 0$.

From Eq. (2), the subtraction of the average expression levels of each gene from its expression levels for each data is the normalization of gene expression data across the collections, while normalization across the q genes is achieved by subtracting the mean expression levels of all genes for each selection from their respective individual expression levels. Normalization also involves data standardization. The genomic data are standardized to ensure that each selection has zero mean and unit variance.

4.5.2 Preliminary feature selection

From the previous researches, it has been established that only a few of the several genes of genomics data are relevant for any study [282–285]. Preliminary gene selection of the genomic data must be performed. The reasons for this preliminary feature selection are to minimizing the computational time that may occur during the analysis of this data, noise reduction, and prevention of noninformative genes from the Classifier [286–288]. Therefore, feature selection reduces system complexity, reduction of the processing time, and enhancement of the system performance. The introduction of irrelevant variables can reduce the effectiveness and add uncertainty to the predictions model. In this chapter, T-statistics was used for the preliminary feature selection.

By employing the student t-statistic method, the genes $X_j, j = 1, \ldots, q$, is divided into X_{0j} and X_{1j} considering the class labels $(0, 1)$ according to the sample sizes n_0 and n_1, respectively.

Equating the labels to be $\overline{X}_{0j}$ and $\overline{X}_{1j}$ are evaluated using the t-statistics

$$t_s = \frac{\overline{X}_{1j} - \overline{X}_{oj}}{\sqrt{\left(\frac{(n_0 - 1)s_0^2 + (n_1 - 1)s_1^2}{n_0 + n_1 - 2}\right) \times \left(\frac{n_0 + n_1}{n_0 n_1}\right)}} \tag{4}$$

Alternatively, the Welch t-test

$$t_w = \frac{\overline{X}_{1j} - \overline{X}_{0j}}{\sqrt{\frac{s_0^2}{n_0} + \frac{s_1^2}{n_1}}} \tag{5}$$

With a modified degree of freedom

$$v = \frac{\left(\frac{s_0^2}{n_0} + \frac{s_1^2}{n_1}\right)}{\frac{s_0^4}{n_0^2(n_0 - 1)} + \frac{s_1^4}{n_1^2(n_1 - 1)}} \tag{6}$$

j, $\overline{X}_{yj}, s_y^2$, and n_y means mean, variance, and the sample size for class label y, $y = 0, 1$, respectively.

The t-statistics are calculated using the Welch t-test by subtracting the average of one class and the average of all classes, where the within-class standard deviation standardizes the difference. t_w is calculated for all q genes and for each computation, a high positive t-score corresponds to high expression in favor of class 1, while the least negative t-score corresponds to the high expression for class 0.

The estimated t_w values are used to identify the top q^* genes sorted in descending order from the absolute values of the t-scores.

The cut point for the selection of the top q^* genes from the ordered list is determined either by a pre-specified p-value, p^* or its critical value equivalent for the upper tail of the student-t distribution. The higher the value of p^* chosen (i.e. as $p^* \rightarrow 1$ or as the cut-off $t_\alpha \rightarrow 0$), the higher the chance of retaining more genes and vice-versa. The p-value is the tuning parameter which is depends on data complexity. Therefore, a p-value that is suitable for a data structure may not be appropriate other data structure. If the data set is linearly separable, it may require minimal p level for excellent filtering, and if the data set is not linearly separable, it may take time to arrive at this level.

4.5.3 Feature selection using genetic algorithm

The feature selection was repeated because dealing with high-dimensional datasets, filtering is applied to reduce the dimensional space, and thereafter a wrapper method is applied. This repetition is required because, after the filtering process, the remaining features or genes might not be suitable for proper classification into the respected labels. Most relevant and informative genes from the pool of data must be extracted from the preliminary selection. The GA's feature selection process begins with the initialization of a population of size n. The GA randomly selects from the population chromosomes. The fitness function is used to check the fitness of each chromosome. Four of the most suitable chromosomes (parents) are randomly selected for reproduction (crossover and mutation) from these chromosomes. The crossover process is accomplished by randomly dividing and recombining the chromosomes into new offspring. A mutation is achieved by adjusting the binary value in the new offspring from its previous state. The mutation stage helps the algorithm to avoid the local minimum. Chromosomes are binary vectors in relation to the selection of features, they have the same length as the number of features in the datasets of the microarray.

The next steps are the GA pseudo-code, which is iterative.

Step 1: Selecting value m is the essential selection is made by selecting an initial population, $P(k)$, of a given size m, where $k = 1$ is made randomly from set Gn. If it is too big, the algorithm will not shift from an exhaustive search, but the algorithm may not find the optimal solution if it is too small.

Step 2: To determine the value of the fitness function $f(x)$ for each chromosome x in the population, measure the chromosome in $P(k)$ in terms of fitness.

Step 3: The natural selection procedure was used for the new population $Pn(k)$ from the given population $P(k)$. The only possible natural selection procedure is called deterministic sampling. For each x in $P(k)$, calculate the value $e(x) = mg(x)$ where $g(x)$ is a relative fitness by

$$g(k) = \frac{f(x)}{\sum_{n \in p^{(k)}} f(x)} \tag{7}$$

From the equation, the number of copies given by the integer part of $e(x)$ of each chromosome x in $P(k)$ is selected for $Pn(k)$. If the total number of chromosomes selected is less than m (the usual case), then the remaining $Pn(k)$ chromosomes with fractional sections of $e(x)$ are picked from the highest values down. This technique is aimed at duplicating those with high fitness and eliminating low fitness chromosomes.

Step 4: If you do not meet the criteria used for stop, go to step 5, otherwise stop.

Step 5: Repeat the $Pn(k)$ population chromosome surgery to create a population of new $P(k + 1)$ chromosomes. In biological systems, the phase attempts to mimic genetic activity observed. Perform the process again and again until the right answer is identified. The fitness measure relies on the model using the characteristics shown by the chromosome.

```
1. create an initial population size of C, and the probability of
   mutation p_m
2. randomly produce an initial set of binary vector m having each
   having a length of p
3. while the convergence criteria are not met
          repeat ( populationsize C/2) times
              Select two parents (chromosomes) based on the fitness
   criterion
              Crossover by randomly selecting a position and exchange
                              each chromosomes' genes.
              Mutate by randomly altering the binary values of each
                        gene in each new offspring chromosome with
                        probability, p_m
              Compute the fitness of the offspring
          end repeat
4. end while
```

Using the CARET kit in MATLAB programming, the GA for function selection is implemented (e.g., diseased or not) according to the rate of expression of selected sensitive genes, i.e., to model a variable pair classification rule (Y_i, X_{ij}), $i = 1, \ldots, n, j = 1, \ldots, q$ that will use subset x of the measured gene expressions X to correctly predict/classify any independent future subjects into either of the two biological groups $Y_i = y$, $y \in \{0, 1\}$ (binary classification) and more than two different groups (multiclass classification).

The binary classification is defined as classifying individuals in two groups of a given population. $Y_i \in \{0, 1\}, = 1, \ldots, n$, based on whether they have some property or not. $Y_i = 1$, if subject i has the outcome of interest, such as diabetes patient and $Y_i = 0$, if subject i does not possess the outcome of interest such as a normal patient.

This chapter illustrates how multilayer NNs are implemented.

Step 1: The informative genes are randomly divided into $n \times q$ from phase 2 into a training set $n_{t_R} \times q$ and test set $n_{t_E} \times q$ over s number of a s cross-validation runs $n = n_{t_R} + n_{t_E}$.

Step 2: Formulate r different NNs classification models with hidden layers, $h = 1, \ldots, r$ on the training set n_{t_R}, use these models to classify the class groups of the subjects in the test set n_{t_E}. This is done at different iterations.

Step 3: Repeat step 2 over s cross-validation runs and compute the average misclassification error rates (MERs) $\hat{\overline{\vartheta}}_{1,} \hat{\overline{\vartheta}}_{2,} \ldots, \hat{\overline{\vartheta}}_{r,}$ for the NN models.

Step 4: The model with m number of hidden layers is selected as the best if it has the least test sample MER with $\hat{\overline{\vartheta}}_{m,} = \left(\hat{\overline{\vartheta}}_{1,} \hat{\overline{\vartheta}}_{2}, \ldots, \hat{\overline{\vartheta}}_{r,}\right)$.

Algorithm 1 Restricted Boltzmann Machines (RBM) update (v_0, $\mathcal{E}$, W, b, c)

1. repeat for (all hidden units i)
2. calculate $Q(h_{1i} = 1/x_1)$ $(b_i + \sum_j W_{ij} x_{1j})$)
3. sample h_{1i} *from* $Q(h_{1i} = 1/x_1)$
4. end repeat
5. repeat for (all visible units j)
6. calculate $T(x_{2j} = 1/h_1)$ $(c_j + \sum_j W_{ij} h_{1i})$)
7. sample x_{2j} from $T(x_{2j} = 1/h_1)$
8. end repeat
9. repeat for (all hidden units i)
10. compute $Q(h_{2i} = 1/x_2$ $(b_i + \sum_j W_{ij} x_{2j})$
11. end repeat

$$W \leftarrow W + \epsilon\left(h_1 x_1' - Q(h_2 = 1/x_2) x_2'\right)$$
$$b \leftarrow b + \epsilon(h_1 - Q(h_2 = 1/x_2))$$
$$c \leftarrow c + \epsilon(x_1 - x_2)$$

where x_1 is an example of training distribution for restricted Boltzmann machines (RBM), $\mathcal{E}$ is the learning rate, W is the weight matrix, b is the hidden unit biases vector, and c is the input units biases vector.

Algorithm 2 is pretraining of the DBN; this is achieved by a layerwise greedy procedure that trains one RBM at a time until the last RBM.

Training the DNN with stacks of RBM modules using contrastive divergence.

Where $\hat{p}$ is the input training distribution for the network, $\mathcal{E}$ is the learning rate, L is the number of layers to train, $n = n^1, \ldots, n^F$ is the number of hidden units in each layer, W^i is the weight matrix for level i, for i from 1 to L, and b is the bias vector for level i, for i from 0 to L.

Algorithm 2 TrainUnspervisedDNN ($\hat{p}, \epsilon, L, n, W, b$)

```
1. Initialize b^0 = 0
2. for ℓ = 1 to F do
3.   initialize W^i = 0, b^i = 0
4. while (not stopping criterion) do
5.   sample g^0 = x  from p̂
6. for (i = 1 to ℓ − 1) do
7.   sample g^i from Q(g^i/g^(i−1))
8.   end for
9. RBMupdate (g^(ℓ−1), ℰ, W^ℓ, b^ℓ, b^(ℓ−1))
10.  end while
11. end for
```

Fine-tuning is the last stage of DBN's strictly supervised learning training process. All layer parameters are fine-tuned using a mean-field posteriors approximation $P(g^i|g^0)$ by changing the samples g_j^{i-1} at levels $i-1$ by their mean-field expected value μ_j^{i-1} with $\mu^i = sigm(b^i + w^i\mu^{i-1})$ proposed by Hinton et al. [157], Hinton [289], and Hinton et al. [290]. The propagation rules indicate that the whole network finally calculates internal representation as network input functions $g^0 = x$. The entire network can then be optimized by gradient descent in terms of a definite computable training criterion (supervised criterion C) that depends on these representations.

Fine-tuning enhances all the parameters concerning supervised criterion Z, employing stochastic descent.

Fine-tuning the DNN ($\hat{p}, C, \epsilon_C, L, n, W, b, V$)

1. Recursively express mean field $\mu^i(x) = E[g^i|g^{i-1} = \mu^{i-1}(x)]$, where $\mu^0(x) = x$, and $E[g^i|g^{i-1} = \mu^{i-1}]$ the anticipated value of g^i under the RBM conditional distribution $Q(g^i|g^{i-1})$, when the values of (g^{i-1}) are replaced by the mean-field values $\mu^{i-1}(x)$. In the case where g^i has binomial units, $E\left[g_j^i|\, g^{i-1} = \mu^{i-1}\right] = sigm\left(-b_j^i - \sum_k W_{jk}^i \mu_k^{i=1}(x)\right)$.
2. The network output function is expressed as $f(x) = V(\mu^L(x)', 1)'$.
3. Minimize the expected value of $Z(f(x), y)$ for pairs (x, y) samples from $\hat{p}$ by tuning parameters W, b, V.

where $\hat{p}$ is the supervise training distribution for the DBN (x, y), Z is the training criterion., $\mathcal{E}_{CD}$ is a learning rate for the gradient descent with contrastive divergence, $\mathcal{E}_Z$ is a learning rate for a gradient descent on supervised cost Z, $\mathcal{E}$ is a learning rate for the gradient descent in contrastive divergence,

L training layers, $n = n^1, \ldots, n^L$ number of hidden units in each layer, W^i weight matrix for level i, for i from 1 to L, b is the bias vector for level i, for i from 0 to L, and V is the supervised output weight matrix.

DNNs are implemented in MATLAB using the deepnet package [291].

The performance of the model was assessed using the following criteria.

Misclassification error rate (MER): The MER is the number of specimens misclassified, which estimates the error of misclassification in the test sample of any object, n_{TE}. It measures the sample size section in the test sets n_{TE} that were incorrectly classified by the classifier. For any r Monte Carlo cross-validation (MCCV) sample drawn to build a model, $\hat{\varphi}_i(x)$, the MER, $\hat{\theta}_r$ of the classifier is computed by

$$\hat{\theta}_r = \frac{1}{n_{TE}} \sum_{r=1}^{n_{TE}} \left[I\{\hat{\varphi}_i(x) \neq Y_i\} \right] \tag{8}$$

where the indicator function I (.) is 1 if it is true and 0 otherwise. Since the classifier is fitted over some repetitions of MCCV runs, the average MER for the classifier is computed by

$$\hat{\theta} = \frac{1}{R} \sum_{r=1}^{R} \hat{\theta}_r \tag{9}$$

which is the average estimate of several $\hat{\theta}_r$ obtained as R repetitions. In the MER given by Eq. (7), n_{TE} is the test sample drawn at each r repetitions, Y_i is the true class labels with $Y_i = 1$ if the subject is cancerous, and $Y_i = 0$ if the subject is normal. $\hat{\varphi}_i(x)$ is the classifier that uses the vector X of gene predictors for classification.

Similarly, Y_i , $\hat{\varphi}_i(x) = 1$, if the ith subject is categorized as cancerous and $\hat{\varphi}_i(x) = 0$, if the ith subject is classified normal by the classification rule.

The MER given was employed to assess the efficiency of the classifier [270].

4.6 Results

Nurses' Health Study (NHS) (3326 females, 45.6% T2DM) and Health Professionals Follow-up Study (2502 males, 46.5% T2DM). The datasets are distilled into four SNP function sets using P-value thresholds, thus specifying the number of SNP function sets to be: 96 (P-value: 1×10^{-5}), 214 (P-value: 1×10^{-4}), 399 (P-value: 1×10^{-3}), and 678 (P-value: 1×10^{-2}). The feature set with 678 SNPs (P-value: 1×10^{-2}) was used to construct a

DNN, DL model from 95% of the original genotypes as training, 20% as internal validation, and 5% as the testing sample.

The DNN DL system showed a significant predictive output in the research with type II DM having AUC $= 0.9537$ in male and AUC $= 0.9349$ in females and overall AUC $= 0.9705$. The Deepnet software package was used to perform the SNP genotypes feature of type II diabetes. With the use of the GA (using the CARET package) and also to control population structure, the selection of features has been improved. Correction for multiple tests was also considered using Bonferroni correction [276]. Bonferroni correction changes an alpha value, that is, widely used from $\alpha = 0.05$ to $\alpha = (0.05/n)$, where n is the number of statistical tests conducted. We, therefore, carried out an experiment to determine the associations of all SNP datasets with type II diabetes phenotype, in which we changed the total number of SNPs evaluated to establish the sense cut-off as P-value $< 5 \times 10^{-8}$ after Bonferroni's correction (P-value $= 0.05/5828$). As shown from the test, the number of SNPs included in the study had a significant impact on performance, hence the higher number of SNPs the better the predictive performance. Therefore, the model performance intensely enhanced with 399 and 678 SNPs, much larger than $\sim$90 variants identified in GWAS.

4.6.1 Misclassification error rate (MER)

The most informative SNPs genes were performed at 80%, 90%, and 95% of the whole sample size n as training samples. The remaining 20%, 10%, and 5% as the test samples. Having the sample size $n = 100$, $n_{TR} = 80\%$, 90%, and 95% were used to build the classifier; while each sample $n_{TE} = 20\%$, 10%, and 5% is used to assess the performance. MCCV is adopted to enhance generalization and to ensure the constancy of results and bias in the estimates. The random sample size $n_{TR} = 80\%$, 90%, and 95% are repeatedly drawn from the entire $n = 100$ samples without replacement at 1000 repetitions or iterations. To compute the MERs, each of the selected genes is used to predict the reaction class labels $y \in \{0, 1\}$ of the remaining left out $n_{TE} = 20\,\%$, $10\,\%$, and 5%samples. Table 1 shows the MER of the random splitting of the samples.

Table 1 shows the MER results of DNN at three different splitting ratios of training for both male and female datasets and the best prediction results of the simulated data were obtained at the random splitting of 95% training sample and 5% test sample. This shows that the higher the number of SNPs the better the predictive performance of the DNN. The best prediction value of the least mean MER occurred when DNN is trained with 95%

Table 1 The MER results of DNN at four different splitting ratios of training: test samples.

Splitting ratios	80:20	90:10	95:5
MERs (%)	8.2	12	1.05

of the whole sample and prediction performance is repossessed based on the remaining 5% of the sample. The results of sensitivity and specificity are both 100%. Therefore, the result of the DNN classification rules correctly classified 97% of the 100 subjects, whereas 3% were wrongly classified, thus translates to an average MER of about 3%.

4.6.2 Performance comparison of machine learning on SNPs

From Table 2, the proposed DNN system showed the significant predictive output of AUC = 0.954 in male and AUC = 0.935 in female, while Kim et al. [275] shown predictive out of AUC = 0.948 in male and AUC = 0.946 in female, their results also show that SNPs with clinical factors performed better with higher samples of 678 and DNN can be a handy instrument to predict type II diabetes using SNPs combined with clinical information, but the model fails to use a feature selection algorithm to select the most informative genes for the complex disease.

To validate the performance of our method, we compare our proposed system using DNN with some other recent methodologies in literature. As described in performance evaluation, it can be observed that the proposed method performs optimally than the other approaches with respect to sensitivity, specificity, and AUC expect [174] in Table 3, this is because the proposed system was used for the prediction of polygenic obesity, not types II

Table 2 performance comparison of the closest model.

Models	Male (AUC)	Female (AUC)
[275]	0.948	0.946
Our model	0.954	0.935

Table 3 Performance comparison of machine learning on SNPs genes for prediction of diabetes mellitus and polygenic obesity.

Models	Sensitivity	Specificity	AUC
[168]	0.93	0.87	0.94
[174]	0.96	0.97	0.99
Our model	0.95	0.93	0.97

diabetes. The superior performance is due to the use of GA for features selection and ANN for predictive and classification of SNPs gene for type II DM. The result showed that DNNs can be used to predict common and complex diseases using genomic-based data and is a very promising technique for machine learning to investigate genomic data, making it a healthy precise model for predicting type II diabetes.

5. Conclusion and future work

In this chapter, the fundamental concepts of genomic data and machine learning within the context of prediction and classification of DM management have been richly addressed. The use of genomics data and its applications in health-care practices is a new and fast-growing trend in health-related fields. From the evidence of research areas relating to the use of genomics data and CI algorithms in prediction and diagnosis of different diseases in medical case studies, it is established that both genomics-related data and CI have provided a wide range of new frontlines for the prediction, diagnosis, treatment, analysis, and management of different illness, especially for the prediction and diagnosis of these diseases in health-care environments. Hence, genomic data and the use of CI approach in the prediction, diagnosis, and management of a large volume of diseases in health care is a welcome development. The predictive capacity of some of the existing genetic markers is very weak since the approach omitting the existence of an interaction between loci and based on the single-locus analysis. Therefore, the use of machine learning to predict diseases using genomics-based data is required. Precision medicine has provided customized medical preventative, interventions, and treatments disease such as cancer and heart disease. The strength of machine learning data modeling in multifaceted disease prediction lies in its usage of collaborative high-dimensional data. Machine learning models are capable of classifying individual disease threats with high accuracy with huge datasets available in high-quality phenotyping at different stages in the life course. Machine learning could ultimately provide cost effective and proactive health care with great efficacy for instance, the predictors that include tissue-specific disease risks for individuals show greater promise of visions. A medicine learning approach for the prediction and classification of DM using genomic data has been proposed in this chapter to showcase some of the research opportunities of genomic data and CI. The framework comprises of both the

DNN architecture and genomic data for prediction and classification of DM has been implemented.

DNNs are multidimensional algorithms with parameters on the order of millions and computationally concentrated. Hence, such algorithms require data commensurate with the number of parameters for better convergence and optimal performance. Although no strict rules are governing the amount of data required to optimally train DNNs, empirical studies suggest that 10-fold more training data relative to the number of parameters are required to produce an effective model. The proposed system only used genomic data (SNPs) to predict type II DM, a GA was used for features selection and DNN for prediction and classification of the dataset. Therefore, future research should attempt to include more genomics data for better prediction and classification. Also, combining the strength of machine learning and association rule mining (ARM) will help to create a robust method for interpreting DL networks. Therefore, future research will combine machine learning techniques with ARM methods, thus increasing the performance accuracy of the system. Also, the attention on the preprocessing step remains consistently present, despite the increasing interest in genomic data analytics. As future work, hybrid feature selection methods like fuzzy PSO a feature selection algorithm to exhibit enhanced classification performance considering a high-dimensional genomic dataset will be considered and evaluate it within a genomic analytics process. Lastly, the introduction of clinical and environmental factors such as hypertension, hypercholesterolemia, physical activity, smoking, and alcohol drinking, age, BMI, family history of diabetes in first degree relatives with genomic dataset will greatly increase the prediction and classification of DM and phenotypic information like RNA levels or protein functions in ML approaches will improve prediction and classification efficiency and accuracy and also necessary to reach the goal of disease prediction.

References

[1] M. Murea, L. Ma, B.I. Freedman, Genetic and environmental factors associated with type II diabetes and diabetic vascular complications, Rev. Diabet. Stud. 9 (1) (2012) 6.

[2] S. Wild, G. Roglic, A. Green, R. Sicree, H. King, The global prevalence of diabetes: estimates for the year 2000 and projections for 2030, Diabetes Care 27 (5) (2004) 1047–1053.

[3] D.S.W. Ho, W. Schierding, M. Wake, R. Saffery, J. O'Sullivan, Machine learning SNP based prediction for precision medicine, Front. Genet. 10 (2019) 267.

[4] V. Lyssenko, A. Jonsson, P. Almgren, N. Pulizzi, B. Isomaa, T. Tuomi, … L. Groop, Clinical risk factors, DNA variants, and the development of type II diabetes, N. Engl. J. Med. 359 (21) (2008) 2220–2232.

[5] J.B. Meigs, P. Shrader, L.M. Sullivan, J.B. McAteer, C.S. Fox, J. Dupuis, ... R.B. D'Agostino Sr., Genotype score in addition to common risk factors for prediction of type II diabetes, N. Engl. J. Med. 359 (21) (2008) 2208–2219.
[6] S.A. El-Safty, M.A. Shenashen, Nanoscale dynamic chemical, biological sensor material designs for control monitoring and early detection of advanced diseases, Mater. Today Bio. 5 (2020) 100044.
[7] A. Nath, D. Deb, R. Dey, An augmented subcutaneous type 1 diabetic patient modeling and design of adaptive glucose control, J. Process Control 86 (2020) 94–105.
[8] S. Renner, A. Blutke, S. Clauss, C.A. Deeg, E. Kemter, D. Merkus, ... E. Wolf, Porcine models for studying complications and organ crosstalk in diabetes mellitus, Cell Tissue Res. (2020) 1–38.
[9] N. Jayanthi, B.V. Babu, N.S. Rao, Survey on clinical prediction models for diabetes prediction, J. Big Data 4 (1) (2017) 26.
[10] R. Sanakal, T. Jayakumari, Prognosis of diabetes using data mining approach-fuzzy C means clustering and support vector machine, Int. J. Comput. Trends Technol. 11 (2) (2014) 94–98.
[11] K. Lakshmi, S.P. Kumar, Utilization of data mining techniques for prediction of diabetes disease survivability, Int. J. Sci. Eng. Res. 4 (6) (2013) 933–940.
[12] P. Repalli, Prediction on Diabetes Using Data Mining Approach, Oklahoma State University, 2011.
[13] R. Motka, V. Parmarl, B. Kumar, A. Verma, Diabetes mellitus forecasts using different data mining techniques, in: Paper Presented at the 2013 4th International Conference on Computer and Communication Technology (ICCCT), 2013.
[14] R. Anichini, E. Brocco, C.M. Caravaggi, R. Da Ros, L. Giurato, V. Izzo, ... O. Ludovico, Physician experts in diabetes are natural team leaders for managing diabetic patients with foot complications. A position statement from the Italian diabetic foot study group, Nutr. Metab. Cardiovasc. Dis. 30 (2) (2020) 167–178.
[15] N.S. Elbarbary, E.A.R. Ismail, M.A. Zaki, Y.W. Darwish, M.Z. Ibrahim, M. El-Hamamsy, Vitamin B complex supplementation as a homocysteine-lowering therapy for early-stage diabetic nephropathy in pediatric patients with type 1 diabetes: a randomized controlled trial, Clin. Nutr. 39 (1) (2020) 49–56.
[16] D.E. Brown, A. Abbasi, R.Y. Lau, Predictive analytics: predictive modeling at the micro-level, IEEE Intell. Syst. 30 (3) (2015) 6–8.
[17] J. Jenkins, L.L. Rodriguez, Educational issues and strategies for genomic medicine, in: Genomic and Precision Medicine, Elsevier, 2017, pp. 45–58.
[18] S.G. Johnson, Genomic medicine in primary care, in: Genomic and Precision Medicine, Academic Press, 2017, pp. 1–18.
[19] Z. Laksman, A.S. Detsky, Personalized medicine: understanding probabilities and managing expectations, J. Gen. Intern. Med. 26 (2) (2011) 204–206.
[20] A.M. Spiegel, M. Hawkins, 'Personalized medicine' to identify genetic risks for type II diabetes and focus prevention: can it fulfill its promise? Health Aff. 31 (1) (2012) 43–49.
[21] J. Adler-Milstein, M.F. Furukawa, J. King, A.K. Jha, Early results from the hospital electronic health record incentive programs, Am. J. Manag. Care 19 (7) (2013) e273–e284.
[22] D. Kumar, Integrated genomic and molecular medicine, in: Clinical Molecular Medicine, Academic Press, 2020, pp. 535–543.
[23] International Human Genome Sequencing Consortium, Initial sequencing and analysis of the human genome, Nature 409 (6822) (2001) 860.
[24] J.S. Floyd, B.M. Psaty, The application of genomics in diabetes: barriers to discovery and implementation, Diabetes Care 39 (11) (2016) 1858–1869.
[25] H.L. McLeod, Cancer pharmacogenomics: early promise, but the concerted effort needed, Science 339 (6127) (2013) 1563–1566.

[26] J. Szustakowki, Initial sequencing and analysis of the human genome, Nature 409 (6822) (2001) 409.
[27] P. Radha, B. Srinivasan, Predicting diabetes by cosequencing various data mining classification techniques, Int. J. Innov. Sci. Eng. Technol. 1 (6) (2014) 334–339.
[28] R.J. Roiger, Data Mining: A Tutorial-Based Primer, Chapman and Hall/CRC, 2017.
[29] N. Ye, Data Mining: Theories, Algorithms, and Examples, CRC Press, 2013.
[30] T. Denœux, Handling imprecise and uncertain class labels in classification and clustering, COST Action IC 0702 Working group C, Mallorca, March 16, 2009.
[31] S. Yan, H. Wang, T.S. Huang, Q. Yang, X. Tang, Ranking with uncertain labels, in: Paper Presented at the 2007 IEEE International Conference on Multimedia and Expo, 2007.
[32] M. Alehegn, R. Joshi, P. Mulay, Analysis and prediction of diabetes mellitus using machine learning algorithm, Int. J. Pure Appl. Math. 118 (9) (2018) 871–878.
[33] H. Lingaraj, R. Devadass, V. Gopi, K. Palanisamy, Prediction of diabetes mellitus using data mining techniques: a review, J. Bioinform. Cheminform. 1 (1) (2015) 1–3.
[34] S.C. Newman, Prediction and Privacy in Healthcare Analytics (Doctoral dissertation), (2016).
[35] R. Parashar, A. Khan, A. Neha, A survey: the Internet of Things, Int. J. Tech. Res. Appl. 4 (3) (2016) 251–257.
[36] R. Basatneh, B. Najafi, D.G. Armstrong, Health sensors, smart home devices, and the internet of medical things: an opportunity for dramatic improvement in care for the lower extremity complications of diabetes, J. Diabetes Sci. Technol. 12 (3) (2018) 577–586.
[37] B. Corbin, When 'Things' go wrong: redefining liability for the internet of medical things, South Carolina Law Rev. 71 (1) (2019).
[38] C. Kotronis, I. Routis, E. Politi, M. Nikolaidou, G. Dimitrakopoulos, D. Anagnostopoulos, … H. Djelouat, Evaluating internet of medical things (IoMT)-based systems from a human-centric perspective, IoT 8 (2019) 100125.
[39] M. Singh, Evaluating Persuasive User Interfaces for Online Help-Seeking for Domestic Violence (No. M. Eng (Mechanical)), Deakin University, 2018.
[40] G. Orecchini, F. Alimenti, V. Palazzari, A. Rida, M. Tentzeris, L. Roselli, Design and fabrication of ultra-low-cost radio frequency identification antennas and tags exploiting paper substrates and inkjet printing technology, IET Microwaves Antennas Propag. 5 (8) (2011) 993–1001.
[41] T. Sanpechuda, L. Kovavisaruch, A review of RFID localization: APPLICATIONS and techniques, in: Paper Presented at the 2008 5th International Conference on Electrical Engineering/Electronics, Computer, Telecommunications, and Information Technology, 2008.
[42] L. Yang, A. Rida, M.M. Tentzeris, Design and development of radio frequency identification (RFID) and RFID-enabled sensors on flexible low-cost substrates, Syn. Lect. RF/Microwaves 1 (1) (2009) 1–89.
[43] D. Schatsky, J. Camhi, S. Bumb, Five vectors of progress in the Internet of Things, Channels 2 (B2B) (2018) B2C.
[44] R. Shah, A. Chircu, IoT and AI in healthcare: a systematic literature review, Issues Inf. Syst. 19 (3) (2018) 33–41.
[45] J.F. Baldwin, D.W. Xie, Simple fuzzy logic rules based on fuzzy decision tree for classification and prediction problems, in: Paper Presented at the International Conference on Intelligent Information Processing, 2004.
[46] R. Dey, V. Bajpai, G. Gandhi, B. Dey, Application of artificial neural network (ANN) technique for diagnosing diabetes mellitus, in: Paper Presented at the 2008 IEEE Region 10 and the Third International Conference on Industrial and Information Systems, 2008.

[47] A.G. Karegowda, V. Punya, M. Jayaram, A. Manjunath, Rule-based classification for diabetic patients using cascaded k-means and decision tree C4. 5, Int. J. Comput. Appl. 45 (12) (2012) 45–50.
[48] L. Liberti, C. Lavor, N. Maculan, A. Mucherino, Euclidean distance geometry and applications, Siam Rev. 56 (1) (2014) 3–69.
[49] E. Caballero-Ruiz, G. García-Sáez, M. Rigla, M. Balsells, B. Pons, M. Morillo, ... M. Hernando, Automatic blood glucose classification for gestational diabetes with feature selection: decision trees vs. neural networks, in: Paper Presented at the XIII Mediterranean Conference on Medical and Biological Engineering and Computing 2013, 2014.
[50] Z. Kurd, T. Kelly, J. Austin, Developing artificial neural networks for safety-critical systems, Neural Comput. Applic. 16 (1) (2007) 11–19.
[51] S. Samarasinghe, Neural Networks for Applied Sciences and Engineering: From Fundamentals to Complex Pattern Recognition, Auerbach Publications, 2016.
[52] A. Feizollah, N.B. Anuar, R. Salleh, A.W.A. Wahab, A review on feature selection in mobile malware detection, Digit. Investig. 13 (2015) 22–37.
[53] E. Berglund, J. Sitte, The parameterless self-organizing map algorithm, IEEE Trans. Neural Netw. 17 (2) (2006) 305–316.
[54] F. Amato, A. López, E.M. Peña-Méndez, P. Vaňhara, A. Hampl, J. Havel, Artificial neural networks in medical diagnosis, Appl. Biomed. 11 (2013) 47–58 Elsevier.
[55] V.A. Kumari, R. Chitra, Classification of diabetes disease using a support vector machine, Int. J. Eng. Res. Appl. 3 (2) (2013) 1797–1801.
[56] Z. Zainuddin, O. Pauline, C. Ardil, A neural network approach in predicting the blood glucose level for diabetic patients, Int. J. Comput. Intell. 5 (1) (2009) 72–79.
[57] A.B. Adeyemo, A.E. Akinwonmi, On the diagnosis of diabetes mellitus using artificial neural network model artificial neural network models, Afr. J. Comput. Ict. 4 (2011) 1–8.
[58] B. Yegnanarayana, Artificial neural networks for pattern recognition, Sadhana 19 (2) (1994) 189–238.
[59] E. Birney, J. Vamathevan, P. Goodhand, Genomics in healthcare: GA4GH looks to 2022, BioRxiv (2017) 203554.
[60] Z. Stark, L. Dolman, T.A. Manolio, B. Ozenberger, S.L. Hill, M.J. Caulfied, ... M. Lawler, Integrating genomics into healthcare: a global responsibility, Am. J. Hum. Genet. 104 (1) (2019) 13–20.
[61] C.L. Gaff, I.M. Winship, S.M. Forrest, D.P. Hansen, J. Clark, P.M. Waring, ... A.H. Sinclair, Preparing for genomic medicine: a real-world demonstration of health system change, NPJ Genom. Med. 2 (1) (2017) 1–9.
[62] J.J. McCarthy, H.L. McLeod, G.S. Ginsburg, Genomic medicine: a decade of successes, challenges, and opportunities, Sci. Transl. Med. 5 (189) (2013) 189sr184.
[63] T.A. Manolio, R.L. Chisholm, B. Ozenberger, D.M. Roden, M.S. Williams, R. Wilson, ... C. Eng, Implementing genomic medicine in the clinic: the future is here, Genet. Med. 15 (4) (2013) 258.
[64] G. Ginsburg, Medical genomics: Gather and use genetic data in health care, Nat. News 508 (7497) (2014) 451.
[65] T.A. Manolio, M. Abramowicz, F. Al-Mulla, W. Anderson, R. Balling, A.C. Berger, ... R.L. Chisholm, Global implementation of genomic medicine: we are not alone, Sci. Transl. Med. 7 (290) (2015) 290ps213.
[66] M. Gilmour, M. Graham, A. Reimer, G. Van Domselaar, Public health genomics and the new molecular epidemiology of bacterial pathogens, Public Health Genomics 16 (1–2) (2013) 25–30.

[67] B. Simone, W. Mazzucco, M.R. Gualano, A. Agodi, D. Coviello, F.D. Bricarelli, … M. Genuardi, The policy of public health genomics in Italy, Health Policy 110 (2–3) (2013) 214–219.
[68] S.F. Terry, G. Ginsburg, S. Shekar, Genomics-Enabled Learning Health Care Systems: Gathering and Using Genomic Information to Improve Patient Care and Research, in: A Workshop Summary, Institute of Medicine, National Academies Press, 2014.
[69] L. Slaughter, ASHG perspective, Am. J. Hum. Genet. 104 (2019) 6–7.
[70] D. Hellman, What makes genetic discrimination exceptional, Am. J. Law Med. 29 (2003) 77.
[71] Y. Joly, M. Braker, M. Le Huynh, Genetic discrimination in private insurance: global perspectives, New Genet. Soc. 29 (4) (2010) 351–368.
[72] ACMG Board of Directors, Laboratory and clinical genomic data sharing is crucial to improving genetic health care: a position statement of the American College of Medical Genetics and Genomics, Genet. Med. 19 (7) (2017) 721.
[73] D.N. Paltoo, L.L. Rodriguez, M. Feolo, E. Gillanders, E.M. Ramos, J.L. Rutter, … R. Baker, Data use under the NIH GWAS data sharing policy and future directions, Nat. Genet. 46 (9) (2014) 934.
[74] T. Pang, The impact of genomics on global health, Am. J. Public Health 92 (7) (2002) 1077–1079.
[75] D. Chasioti, J. Yan, K. Nho, A.J. Saykin, Progress in polygenic composite scores in Alzheimer's and other complex diseases, Trends Genet. 35 (5) (2019) 371–382.
[76] D.E. Beaudoin, N. Longo, R.A. Logan, J.P. Jones, J.A. Mitchell, Using information prescriptions to refer patients with metabolic conditions to the Genetics Home Reference website, J. Med. Libr. Assoc. 99 (1) (2011) 70.
[77] Genetic Alliance; The New York-Mid-Atlantic Consortium for Genetic and Newborn Screening Services, Understanding Genetics: A New York, Mid-Atlantic Guide for Patients and Health Professionals, Lulu.com, 2009.
[78] Genomes Project Consortium, A global reference for human genetic variation, Nature 526 (7571) (2015) 68.
[79] L. Scheinfeldt, T. Schmidlen, N. Gerry, M. Christman, Challenges in translating GWAS results to clinical care, Int. J. Mol. Sci. 17 (8) (2016) 1267.
[80] L.B. Scheinfeldt, N. Gharani, R.S. Kasper, T.J. Schmidlen, E.S. Gordon, J.P. Jarvis, … M.F. Christman, Using the Coriell personalized medicine collaborative data to conduct a genome-wide association study of sleep duration, Am. J. Med. Genet. B Neuropsychiatr. Genet. 168 (8) (2015) 697–705.
[81] Genomes Project Consortium, A map of human genome variation from population-scale sequencing, Nature 467 (7319) (2010) 1061–1073.
[82] D.J. Smyth, J.D. Cooper, R. Bailey, S. Field, O. Burren, L.J. Smink, … D.B. Dunger, A genome-wide association study of nonsynonymous SNPs identifies a type 1 diabetes locus in the interferon-induced helicase (IFIH1) region, Nat. Genet. 38 (6) (2006) 617.
[83] I. Kockum, C. Sanjeevi, S. Eastman, M. Landin-Olsson, G. Dahlquist, Å. Lernmark, Complex interaction between HLA DR and DQ in conferring risk for childhood type 1 diabetes, Eur. J. Immunogenet. 26 (5) (1999) 361–372.
[84] J.K. DiStefano, D.M. Taverna, Technological issues and experimental design of gene association studies, in: Disease Gene Identification, Springer, 2011, pp. 3–16.
[85] H.A. Erlich, A.M. Valdes, S.L. McDevitt, B.B. Simen, L.A. Blake, K.R. McGowan, … Type 1 Diabetes Genetics Consortium, Next generation sequencing reveals the association of DRB3* 02: 02 with type 1 diabetes, Diabetes 62 (7) (2013) 2618–2622.

[86] O.L. Griffith, S.B. Montgomery, B. Bernier, B. Chu, K. Kasaian, S. Aerts, … M. Haeussler, ORegAnno: an open-access community-driven resource for regulatory annotation, Nucleic Acids Res 36 (Suppl. 1) (2007) D107–D113.
[87] B.P. Koeleman, B.A. Lie, D.E. Undlien, F. Dudbridge, E. Thorsby, R.R. De Vries, … J.A. Todd, Genotype effects and epistasis in type 1 diabetes and HLA-DQ trans dimer associations with disease, Genes Immun. 5 (5) (2004) 381.
[88] J. Howson, N. Walker, D. Smyth, J. Todd, Analysis of 19 genes for association with type I diabetes in the Type I diabetes genetics consortium families, Genes Immun. 10 (S1) (2009) S74.
[89] J.A. Noble, A.M. Valdes, Genetics of the HLA region in the prediction of type 1 diabetes, Curr. Diab. Rep. 11 (6) (2011) 533.
[90] M. Polydefkis, J.W. Griffin, J. McArthur, New insights into diabetic polyneuropathy, JAMA 290 (10) (2003) 1371–1376.
[91] J. Precechtelova, M. Borsanyiova, S. Sarmirova, S. Bopegamage, Type I diabetes mellitus: genetic factors and presumptive enteroviral etiology or protection, J. Pathogens 2014 (2014) 1–22.
[92] A. Sinclair, T. Dunning, L. Rodriguez-Manas, Diabetes in older people: new insights and remaining challenges, Lancet Diabetes Endocrinol. 3 (4) (2015) 275–285.
[93] J.R. Gruen, S.M. Weissman, Human MHC class III and IV genes and disease associations, Front. Biosci. 6 (2001) D960–D972.
[94] J.A. Noble, H.A. Erlich, Genetics of type 1 diabetes, Cold Spring Harb. Perspect. Med. 2 (1) (2012) a007732.
[95] R. Redon, S. Ishikawa, K.R. Fitch, L. Feuk, G.H. Perry, T.D. Andrews, … W. Chen, Global variation in copy number in the human genome, Nature 444 (7118) (2006) 444.
[96] I. Santin, D.L. Eizirik, Candidate genes for type 1 diabetes modulate pancreatic islet inflammation and β-cell apoptosis, Diabetes Obes. Metab. 15 (s3) (2013) 71–81.
[97] National Center for Biotechnology Information (US), Genes and Disease [Internet]. Bethesda (MD): National Center for Biotechnology Information (US), Available from:https://www.ncbi.nlm.nih.gov/books/NBK22183/, 1998.
[98] A. Berezin, The single nucleotide polymorphisms in the C-reactive protein gene: are they biomarkers of cardiovascular risk? Int. Biol. Biomed. J. 4 (2) (2018) 122–125.
[99] K. Strimbu, J.A. Tavel, What are biomarkers? Curr. Opin. HIV AIDS 5 (6) (2010) 463.
[100] A. Etheridge, I. Lee, L. Hood, D. Galas, K. Wang, Extracellular microRNA: a new source of biomarkers, Mutat. Res. Fundam. Mol. Mech. Mutagen. 717 (1–2) (2011) 85–90.
[101] P. Sahu, N. Pinkalwar, R.D. Dubey, S. Paroha, S. Chatterjee, T. Chatterjee, Biomarkers: an emerging tool for diagnosis of a disease and drug development, Asian J. Res. Pharm. Sci. 1 (1) (2011) 9–16.
[102] H.L. Rehm, J.S. Berg, L.D. Brooks, C.D. Bustamante, J.P. Evans, M.J. Landrum, … R.L. Nussbaum, ClinGen—the clinical genome resource, N. Engl. J. Med. 372 (23) (2015) 2235–2242.
[103] K. Wang, Q. Chen, Y. Feng, H. Yang, W. Wu, P. Zhang, … W. Du, Single nucleotide polymorphisms in CDKAL1 gene are associated with risk of gestational diabetes mellitus in the chinese population, J. Diabetes Res. 2019 (2019) 1–7.
[104] T. Mikeska, J.M. Craig, DNA methylation biomarkers: cancer and beyond, Genes 5 (3) (2014) 821–864.
[105] P. Hydbring, G. Badalian-Very, Clinical applications of microRNAs, F1000Research 2 (2013) 1–16.
[106] International HapMap Consortium, A second generation human haplotype map of over 3.1 million SNPs, Nature 449 (7164) (2007) 851.

[107] International SNP Map Working Group, A map of human genome sequence variation containing 1.42 million single nucleotide polymorphisms, Nature 409 (6822) (2001) 928.
[108] M.I. McCarthy, Genomics, type II diabetes, and obesity, N. Engl. J. Med. 363 (24) (2010) 2339–2350.
[109] T. Rankinen, A. Zuberi, Y.C. Chagnon, S.J. Weisnagel, G. Argyropoulos, B. Walts, ... C. Bouchard, The human obesity gene map: the 2005 update, Obesity 14 (4) (2006) 529–644.
[110] A.P. Morris, B.F. Voight, T.M. Teslovich, T. Ferreira, A.V. Segre, V. Steinthorsdottir, ... A. Mahajan, Large-scale association analysis provides insights into the genetic architecture and pathophysiology of type II diabetes, Nat. Genet. 44 (9) (2012) 981.
[111] S. Dias, C. Pheiffer, Y. Abrahams, P. Rheeder, S. Adam, Molecular biomarkers for gestational diabetes mellitus, Int. J. Mol. Sci. 19 (10) (2018) 2926.
[112] D. Welter, J. MacArthur, J. Morales, T. Burdett, P. Hall, H. Junkins, ... L. Hindorff, The NHGRI GWAS catalog, a curated resource of SNP-trait associations, Nucleic Acids Res. 42 (D1) (2013) D1001–D1006.
[113] P.V. Popova, A.A. Klyushina, L.B. Vasilyeva, A.S. Tkachuk, Y.A. Bolotko, A.S. Gerasimov, ... A.A. Kostareva, Effect of gene-lifestyle interaction on gestational diabetes risk, OncoTarget 8 (67) (2017) 112024.
[114] P.H. Andraweera, K.L. Gatford, G.A. Dekker, S. Leemaqz, R.W. Jayasekara, V.H. Dissanayake, ... C.T. Roberts, The INSR rs2059806 single nucleotide polymorphism, a genetic risk factor for vascular and metabolic disease, associates with pre-eclampsia, Reprod. Biomed. Online 34 (4) (2017) 392–398.
[115] G. Cho, J. Yim, Y. Choi, J. Ko, S.-H. Lee, Review of machine learning algorithms for diagnosing mental illness, Psychiatry Investig. 16 (4) (2019) 262.
[116] B. Meskó, Z. Drobni, É. Bényei, B. Gergely, Z. Győrffy, Digital health is a cultural transformation of traditional healthcare, mHealth 3 (2017) 1–8.
[117] F.E. Ayo, J.B. Awotunde, S.O. Folorunso, R.O. Ogundokun, P. Idoko, J.I. Adekunle, O.I. Dauda, A fuzzy based method for diagnosis of acne skin disease severity, i-manager's J. Pattern Recogn. 5 (2) (2018) 10.
[118] M. Eremia, C.-C. Liu, A.-A. Edris, Advanced Solutions in Power Systems: HVDC, FACTS, and Artificial Intelligences, John Wiley & Son, 2016.
[119] M. Bakator, D. Radosav, Deep learning and medical diagnosis: a review of literature, Multimodal Technol. Interaction 2 (3) (2018) 47.
[120] B. Mesko, The Role of Artificial Intelligence in Precision Medicine, Taylor & Francis, 2017.
[121] D.M. Mukhopadhyay, M.O. Balitanas, A. Farkhod, S.-H. Jeon, D. Bhattacharyya, Genetic algorithm: a tutorial review, Int. J. Grid Distributed Comput. 2 (3) (2009) 25–32.
[122] R.S. Parpinelli, H.S. Lopes, New inspirations in swarm intelligence: a survey, Int. J. Bio-Inspired Comput. 3 (1) (2011) 1–16.
[123] R. Poli, W.B. Langdon, N.F. McPhee, J.R. Koza, A Field Guide to Genetic Programming, Lulu.com, 2008.
[124] W. Zhang, M.-P. Jia, L. Zhu, X.-A. Yan, Comprehensive overview of computational intelligence techniques for machinery condition monitoring and fault diagnosis, Chinese J. Mech. Eng. 30 (4) (2017) 782–795.
[125] P.N.M. Dorantes, G.M. Mendez, Non-iterative radial basis function neural networks to quality control via image processing, IEEE Lat. Am. Trans. 13 (10) (2015) 3447–3451.
[126] J. Kung, D. Kim, S. Mukhopadhyay, On the impact of an energy-accuracy tradeoff in a digital cellular neural network for image processing, IEEE Trans. Comput. Aided Des. Integr. Circuits Syst. 34 (7) (2015) 1070–1081.

[127] A.W. Setiawan, A.B. Suksmono, T.R. Mengko, O.S. Santoso, Performance evaluation of color retinal image quality assessment in asymmetric channel VQ coding, Int. J. eHealth Med. Commun. 4 (3) (2013) 1–19.
[128] Z. Li, C. Yang, C.-Y. Su, J. Deng, W. Zhang, Vision-based model predictive control for steering of a nonholonomic mobile robot, IEEE Trans. Control Syst. Technol. 24 (2) (2015) 553–564.
[129] L. Porzi, S.R. Bulo, A. Penate-Sanchez, E. Ricci, F. Moreno-Noguer, Learning depth-aware deep representations for robotic perception, IEEE Robot. Autom. Lett. 2 (2) (2016) 468–475.
[130] V. Mitra, G. Sivaraman, H. Nam, C. Espy-Wilson, E. Saltzman, M. Tiede, Hybrid convolutional neural networks for articulatory and acoustic information based speech recognition, Speech Commun. 89 (2017) 103–112.
[131] S.M. Siniscalchi, D. Yu, L. Deng, C.-H. Lee, Exploiting deep neural networks for detection-based speech recognition, Neurocomputing 106 (2013) 148–157.
[132] G. Zhang, Y. Shen, Exponential synchronization of delayed memristor-based chaotic neural networks via periodically intermittent control, Neural Netw. 55 (2014) 1–10.
[133] A. Abbe, C. Grouin, P. Zweigenbaum, B. Falissard, Text mining applications in psychiatry: a systematic literature review, Int. J. Methods Psychiatr. Res. 25 (2) (2016) 86–100.
[134] L. Pereira, R. Rijo, C. Silva, R. Martinho, Text mining applied to electronic medical records: a literature review, Int. J. E-Health Med. Commun. 6 (3) (2015) 1–18.
[135] W. Sun, Z. Cai, Y. Li, F. Liu, S. Fang, G. Wang, Data processing and text mining technologies on electronic medical records: a review, J. Healthcare Eng. 2018 (2018) 1–9.
[136] F. Charfi, A. Kraiem, Comparative study of ECG classification performance using decision tree algorithms, Int. J. E-Health Med. Commun. 3 (4) (2012) 102–120.
[137] F. Jia, Y. Lei, J. Lin, X. Zhou, N. Lu, Deep neural networks: a promising tool for fault characteristic mining and intelligent diagnosis of rotating machinery with massive data, Mech. Syst. Signal Process. 72 (2016) 303–315.
[138] A.T. Azar, S. Vaidyanathan, Computational Intelligence Applications in Modeling and Control, Springer, 2015.
[139] V. Piuri, F. Scotti, M. Roveri, Computational intelligence in industrial quality control, in: Paper Presented at the IEEE International Workshop on Intelligent Signal Processing, 2005, 2005.
[140] X. Chen, Q. Zhang, Y. Sun, Model-based compensation and pareto-optimal trajectory modification method for robotic applications, Int. J. Precis. Eng. Manuf. 20 (7) (2019) 1–11.
[141] J. Ogbemhe, K. Mpofu, Towards achieving a fully intelligent robotic arc welding: a review, Ind. Robot. Int. J. 42 (5) (2015) 475–484.
[142] X. Wang, Y. Ding, Adaptive real-time predictive compensation control for 6-DOF serial arc welding manipulator, Chinese J. Mech. Eng. 23 (3) (2010) 361–366.
[143] Q. Zhang, J. Zhang, A. Chemori, X. Xiang, Virtual submerged floating operational system for robotic manipulation, Complexity (2018) 2018. 9528313 18 pphttps://doi.org/10.1155/2018/9528313.
[144] Q. Al-Shayea, G. El-Refae, S. Yaseen, Artificial neural networks for medical diagnosis using biomedical datasets, Int. J. Behav. Healthcare Res. 21 4 (1) (2013) 45–63.
[145] Q.K. Al-Shayea, Artificial neural networks in medical diagnosis, Int. J. Comput. Sci. Issues 8 (2) (2011) 150–154.
[146] S. Kamruzzaman, M. Islam, An algorithm to extract rules from artificial neural networks for medical diagnosis problems, arXiv (2010) 1009.4566.
[147] J.L. Patel, R.K. Goyal, Applications of artificial neural networks in medical science, Curr. Clin. Pharmacol. 2 (3) (2007) 217–226.

[148] R.D. Labati, A. Genovese, E. Munoz, V. Piuri, F. Scotti, G. Sforza, Computational intelligence for industrial and environmental applications, in: Paper Presented at the 2016 IEEE 8th International Conference on Intelligent Systems (IS), 2016.
[149] V. Piuri, Computational intelligence for industrial and environmental applications, in: Paper Presented at the 2017 IEEE International Conference on Computational Intelligence and Virtual Environments for Measurement Systems and Applications (CIVEMSA), 2017.
[150] S. Kar, S. Das, P.K. Ghosh, Applications of neuro-fuzzy systems: a brief review and future outline, Appl. Soft Comput. 15 (2014) 243–259.
[151] Z. Pezeshki, S.A.S. Ivari, Applications of BIM: a brief review and future outline, Arch. Comput. Methods Eng. 25 (2) (2018) 273–312.
[152] G.B. Huang, Q.Y. Zhu, C.K. Siew, Extreme learning machine: theory and applications, Neurocomputing 70 (1–3) (2006) 489–501.
[153] J. Schmidhuber, Deep learning in neural networks: an overview, Neural Netw. 61 (2015) 85–117.
[154] R. Bekkerman, M. Bilenko, J. Langford, Scaling Up Machine Learning: Parallel and Distributed Approaches, Cambridge University Press, 2011.
[155] Y. Bengio, Learning deep architectures for AI, Foundations Trends® Mach. Learn. 2 (1) (2009) 1–127.
[156] I. del Campo, R. Finker, M.V. Martinez, J. Echanobe, F. Doctor, A real-time driver identification system based on artificial neural networks and cepstral analysis, in: Paper Presented at the 2014 International Joint Conference on Neural Networks (IJCNN), 2014.
[157] G. Hinton, L. Deng, D. Yu, G. Dahl, A.-R. Mohamed, N. Jaitly, … B. Kingsbury, Deep neural networks for acoustic modeling in speech recognition, IEEE Sign Process. Mag. 29 (2012).
[158] Q.V. Le, Building high-level features using large scale unsupervised learning, in: Paper Presented at the 2013 IEEE International Conference on Acoustics, Speech and Signal Processing, 2013.
[159] G.E. Hinton, R.R. Salakhutdinov, Reducing the dimensionality of data with neural networks, Science 313 (5786) (2006) 504–507.
[160] F. Arthur, K.R. Hossein, Deep learning in medical image analysis: a third eye for doctors, J. Stomatol. Oral Maxillofac. Surg. 120 (4) (2019) 279–288.
[161] X. Liu, L. Faes, A.U. Kale, S.K. Wagner, D.J. Fu, A. Bruynseels, … C. Kern, A comparison of deep learning performance against health-care professionals in detecting diseases from medical imaging: a systematic review and meta-analysis, Lancet Digit. Health 1 (6) (2019) e271–e297.
[162] C. Gholipour, F. Rahim, A. Fakhree, B. Ziapour, Using an artificial neural networks (ANNs) model for prediction of intensive care unit (ICU) outcome and length of stay at hospital in traumatic patients, J. Clin. Diagn. Res. 9 (4) (2015) OC19.
[163] L. Huang, X. Liang, J. Zha, Analysis of factors influencing hospitalization costs for patients with lung cancer surgery based on the BP neural network, Chinese Med. Record English Ed. 2 (6) (2014) 237–241.
[164] G. Zheeng, C. Zhang, L. Li, Bringing business intelligence to healthcare informatics curriculum: a preliminary investigation, in: Paper Presented at the Proceedings of the 45th ACM Technical Symposium on Computer Science Education, 2014.
[165] K.W. Johnson, J.T. Soto, B.S. Glicksberg, K. Shameer, R. Miotto, M. Ali, … J.T. Dudley, Artificial intelligence in cardiology, J. Am. Coll. Cardiol. 71 (23) (2018) 2668–2679.
[166] C. Krittanawong, H. Zhang, Z. Wang, M. Aydar, T. Kitai, Artificial intelligence in precision cardiovascular medicine, J. Am. Coll. Cardiol. 69 (21) (2017) 2657–2664.

[167] J. Nahar, T. Imam, K.S. Tickle, Y.-P.P. Chen, Computational intelligence for heart disease diagnosis: a medical knowledge-driven approach, Expert Syst. Appl. 40 (1) (2013) 96–104.
[168] R. Gargeya, T. Leng, Automated identification of diabetic retinopathy using deep learning, Ophthalmology 124 (7) (2017) 962–969.
[169] S. Gubbi, P. Hamet, J. Tremblay, C. Koch, F. Hannah-Shmouni, Artificial intelligence and machine learning in endocrinology and metabolism: the dawn of a new Era, Front. Endocrinol. 10 (2019) 185.
[170] P. Hamet, J. Tremblay, Artificial intelligence in medicine, Metabolism 69 (2017) S36–S40.
[171] R. Al-Massri, Y. Al-Astel, H. Ziadia, D.K. Mousa, S.S. Abu-Naser, Classification prediction of SBRCTs cancers using artificial neural network, Int. J. Acad. Eng. Res. 2 (11) (2018) 1–7.
[172] A.R.M. Al-shamasneh, U.H.B. Obaidellah, Artificial intelligence techniques for cancer detection and classification: a review study, Eur. Sci. J. 13 (3) (2017) 342–370.
[173] X. Bargalló, G. Santamaría, M. del Amo, P. Arguis, J. Ríos, J. Grau, … M. Velasco, Single reading with computer-aided detection performed by selected radiologists in a breast cancer screening program, Eur. J. Radiol. 83 (11) (2014) 2019–2023.
[174] A.T. Watanabe, V. Lim, H.X. Vu, R. Chim, E. Weise, J. Liu, … C.E. Comstock, Improved cancer detection using artificial intelligence: a retrospective evaluation of missed cancers on mammography, J. Digit. Imaging 32 (4) (2019) 625–637.
[175] D.S. Kermany, M. Goldbaum, W. Cai, C.C. Valentim, H. Liang, S.L. Baxter, … F. Yan, Identifying medical diagnoses and treatable diseases by image-based deep learning, Cell 172 (5) (2018) 1122–1131.e1129.
[176] F. Wang, A. Preininger, AI in health: state of the art, challenges, and future directions, Yearb. Med. Inform. 28 (01) (2019) 016–026.
[177] J. De Fauw, J.R. Ledsam, B. Romera-Paredes, S. Nikolov, N. Tomasev, S. Blackwell, … D. Visentin, Clinically applicable deep learning for diagnosis and referral in retinal disease, Nat. Med. 24 (9) (2018) 1342.
[178] Ö. Çiçek, A. Abdulkadir, S.S. Lienkamp, T. Brox, O. Ronneberger, 3D U-Net: learning dense volumetric segmentation from the sparse annotation, in: Paper Presented at the International Conference on Medical Image Computing and Computer-Assisted Intervention, 2016.
[179] C. Combi, Editorial from the new editor-in-chief: artificial intelligence in medicine and the forthcoming challenges, Artif. Intell. Med. 76 (2017) 37.
[180] C. Combi, G. Pozzi, Clinical information systems and artificial intelligence: recent research trends, Yearb. Med. Inform. 28 (01) (2019) 083–094.
[181] N. Shahid, T. Rappon, W. Berta, Applications of artificial neural networks in health care organizational decision-making: a scoping review, PLoS One 14 (2) (2019) e0212356.
[182] F. Jiang, Y. Jiang, H. Zhi, Y. Dong, H. Li, S. Ma, … Y. Wang, Artificial intelligence in healthcare: past, present, and future, Stroke Vasc. Neurol. 2 (4) (2017) 230–243.
[183] N.S. Weingart, R.M. Wilson, R.W. Gibberd, B. Harrison, Epidemiology of medical error, BMJ 320 (7237) (2000) 774–777.
[184] M.L. Graber, N. Franklin, R. Gordon, Diagnostic error in internal medicine, Arch. Intern. Med. 165 (13) (2005) 1493–1499.
[185] B. Winters, J. Custer, S.M. Galvagno, E. Colantuoni, S.G. Kapoor, H. Lee, … P. Pronovost, Diagnostic errors in the intensive care unit: a systematic review of autopsy studies, BMJ Qual. Saf. 21 (11) (2012) 894–902.
[186] C.S. Lee, P.G. Nagy, S.J. Weaver, D.E. Newman-Toker, Cognitive and system factors contributing to diagnostic errors in radiology, Am. J. Roentgenol. 201 (3) (2013) 611–617.

[187] T.M. Maddox, J.S. Rumsfeld, P.R. Payne, Questions for artificial intelligence in health care, JAMA 321 (1) (2019) 31–32.
[188] T. Panch, H. Mattie, L.A. Celi, The "inconvenient truth" about AI in healthcare, NPJ Digit. Med. 2 (1) (2019) 1–3.
[189] F.S. Collins, H. Varmus, A new initiative on precision medicine, N. Engl. J. Med. 372 (9) (2015) 793–795.
[190] T. Panch, P. Szolovits, R. Atun, Artificial intelligence, machine learning, and health systems, J. Glob. Health 8 (2) (2018).
[191] C.J. Hoofnagle, B. van der Sloot, F.Z. Borgesius, The European Union general data protection regulation: what it is and what it means, Inf. Commun. Technol. Law 28 (1) (2019) 65–98.
[192] A. Shaban-Nejad, M. Michalowski, D.L. Buckeridge, Health Intelligence: How Artificial Intelligence Transforms Population and Personalized Health, Nature Publishing Group, 2018.
[193] M.W. Aslam, Z. Zhu, A.K. Nandi, Feature generation using genetic programming with comparative partner selection for diabetes classification, Expert Syst. Appl. 40 (13) (2013) 5402–5412.
[194] E. Bonifacio, Predicting type 1 diabetes using biomarkers, Diabetes Care 38 (6) (2015) 989–996.
[195] L. Cai, H. Wu, D. Li, K. Zhou, F. Zou, Type II diabetes biomarkers of human gut microbiota selected via iterative sure independent screening method, PLoS One 10 (10) (2015) e0140827.
[196] J. Ganesalingam, R. Bowser, The application of biomarkers in clinical trials for motor neuron disease, Biomark. Med. 4 (2) (2010) 281–297.
[197] J.B. Awotunde, O.E. Matiluko, O.W. Fatai, Medical diagnosis system using fuzzy logic, Afr. J. Comput. ICT 7 (2) (2014) 99–106.
[198] M.J. Selleck, M. Senthil, N.R. Wall, Making meaningful clinical use of biomarkers, Biomark. Insights 12 (2017) 1–7.
[199] P. Kandhasamy, R. Balamurugan, S. Kannimuthu, Stellar mass black hole for engineering optimization recent, in: Developments in Intelligent Nature-Inspired Computing, IGI Global, 2017, pp. 62–90.
[200] M. Schena, D. Shalon, R. Heller, A. Chai, P.O. Brown, R.W. Davis, Parallel human genome analysis: microarray-based expression monitoring of 1000 genes, Proc. Natl. Acad. Sci. 93 (20) (1996) 10614–10619.
[201] M. Pyingkodi, R. Thangarajan, Meta-analysis in autism gene expression dataset with biclustering methods using a random cuckoo search algorithm, Asian J. Res. Soc. Sci. Hum. 7 (2) (2017) 186–194.
[202] M. Schena, D. Shalon, R.W. Davis, P.O. Brown, Quantitative monitoring of gene expression patterns with a complementary DNA microarray, Science 270 (5235) (1995) 467–470.
[203] L.-Y. Chuang, C.-H. Yang, K.-C. Wu, C.-H. Yang, A hybrid feature selection method for DNA microarray data, Comput. Biol. Med. 41 (4) (2011) 228–237.
[204] D. Sisodia, D.S. Sisodia, Prediction of diabetes using classification algorithms, Procedia Comput. Sc. 132 (2018) 1578–1585.
[205] N. Sneha, T. Gangil, Analysis of diabetes mellitus for early prediction using optimal features selection, J. Big Data 6 (1) (2019) 13.
[206] Q. Shen, Z. Mei, B.X. Ye, Simultaneous genes and training samples selection by modified particle swarm optimization for gene expression data classification, Comput. Biol. Med. 39 (7) (2009) 646–649.
[207] D.L. Tong, A.C. Schierz, Hybrid genetic algorithm-neural network: feature extraction for unpreprocessed microarray data, Artif. Intell. Med. 53 (1) (2011) 47–56.

[208] L. Li, C.R. Weinberg, T.A. Darden, L.G. Pedersen, Gene selection for sample classification based on gene expression data: a study of sensitivity to the choice of parameters of the GA/KNN method, Bioinformatics 17 (12) (2001) 1131–1142.
[209] C.-S. Yang, L.-Y. Chuang, C.-H. Ke, C.-H. Yang, A hybrid feature selection method for microarray classification, IAENG Int. J. Comput. Sci. 35 (3) (2008) 1–3.
[210] C. Li, B. Singh, R. Graves-Deal, H. Ma, A. Starchenko, W.H. Fry, … M.P. Khan, A three-dimensional culture system identifies a new mode of cetuximab resistance and disease-relevant genes in colorectal cancer, Proc. Natl. Acad. Sci. 114 (14) (2017) E2852–E2861.
[211] M. Xiong, W. Li, J. Zhao, L. Jin, E. Boerwinkle, Feature (gene) selection in gene expression-based tumor classification, Mol. Genet. Metab. 73 (3) (2001) 239–247.
[212] C.C. Bojarczuk, H.S. Lopes, A.A. Freitas, Genetic programming for knowledge discovery in chest-pain diagnosis, IEEE Eng. Med. Biol. Mag. 19 (4) (2000) 38–44.
[213] S.I.-J. Chien, Y. Ding, C. Wei, Dynamic bus arrival time prediction with artificial neural networks, J. Transp. Eng. 128 (5) (2002) 429–438.
[214] I. De Falco, A. Della Cioppa, E. Tarantino, Discovering interesting classification rules with genetic programming, Appl. Soft Comput. 1 (4) (2002) 257–269.
[215] N. Altman, M. Krzywinski, The curse (s) of dimensionality, Nat. Methods 15 (2018) 399–400.
[216] M.W. Libbrecht, W.S. Noble, Machine learning applications in genetics and genomics, Nat. Rev. Genet. 16 (6) (2015) 321–332.
[217] C. Xu, S.A. Jackson, Machine Learning and Complex Biological Data, BioMed Central, 2019.
[218] J. Zou, M. Huss, A. Abid, P. Mohammadi, A. Torkamani, A. Telenti, A primer on deep learning in genomics, Nat. Genet. 51 (1) (2019) 12–18.
[219] D.R. Schrider, A.D. Kern, Supervised machine learning for population genetics: a new paradigm, Trends Genet. 34 (4) (2018) 301–312.
[220] J.D. Washburn, M.K. Mejia-Guerra, G. Ramstein, K.A. Kremling, R. Valluru, E.S. Buckler, H. Wang, Evolutionarily informed deep learning methods for predicting relative transcript abundance from DNA sequence, Proc. Natl. Acad. Sci. 116 (12) (2019) 5542–5549.
[221] C. Angermueller, H.J. Lee, W. Reik, O. Stegle, DeepCpG: accurate prediction of single-cell DNA methylation states using deep learning, Genome Biol. 18 (1) (2017) 67.
[222] C.-P. Lee, Y. Leu, A novel hybrid feature selection method for microarray data analysis, Appl. Soft Comput. 11 (1) (2011) 208–213.
[223] Y.V. Sun, S.L. Kardia, Imputing missing genotypic data of single-nucleotide polymorphisms using neural networks, Eur. J. Hum. Genet. 16 (4) (2008) 487.
[224] R. Poplin, P.-C. Chang, D. Alexander, S. Schwartz, T. Colthurst, A. Ku, … P.T. Afshar, A universal SNP and small-indel variant caller using deep neural networks, Nat. Biotechnol. 36 (10) (2018) 983.
[225] T. Ching, D.S. Himmelstein, B.K. Beaulieu-Jones, A.A. Kalinin, B.T. Do, G.P. Way, … M.M. Hoffman, Opportunities and obstacles for deep learning in biology and medicine, J. R. Soc. Interface 15 (141) (2018) 20170387.
[226] D.W. Hubbard, The Failure of Risk Management: Why It's Broken and How to Fix It, John Wiley & Sons, 2009.
[227] M. Milik, J. Skolnick, Insertion of peptide chains into lipid membranes: an off-lattice Monte Carlo dynamics model, Proteins Struct. Funct. Bioinform. 15 (1) (1993) 10–25.
[228] P. Ojeda, M.E. Garcia, A. Londoño, N.-Y. Chen, Monte Carlo simulations of proteins in cages: influence of confinement on the stability of intermediate states, Biophys. J. 96 (3) (2009) 1076–1082.

[229] D.L. McLeish, Monte Carlo Simulation and Finance, Working Paper 2004, pp. 1–329.
[230] D.L. McLeish, Monte Carlo Simulation and Finance, vol. 276, John Wiley & Sons, 2011.
[231] M.K. Khedr, Project Risk Management Using Monte Carlo Simulation, AACE International Transactions, 2006 RI21.
[232] K. Kurihara, N. Nishiuchi, Efficient Monte Carlo simulation method of GERT-type network for project management, Comput. Ind. Eng. 42 (2–4) (2002) 521–531.
[233] Y.H. Kwak, L. Ingall, Exploring Monte Carlo simulation applications for project management, Risk Manage. 9 (1) (2007) 44–57.
[234] Y.H. Kwak, L. Ingall, Exploring Monte Carlo simulation applications for project management, IEEE Eng. Manag. Rev. 37 (2) (2009) 83.
[235] B. McCabe, Construction engineering and project management III: Monte Carlo simulation for schedule risks, in: Paper Presented at the Proceedings of the 35th Conference on Winter Simulation: Driving Innovation, 2003.
[236] Z. Francis, S. Incerti, V. Ivanchenko, C. Champion, M. Karamitros, M. Bernal, Z. El Bitar, Monte Carlo simulation of energy-deposit clustering for ions of the same LET in liquid water, Phys. Med. Biol. 57 (1) (2011) 209.
[237] N. Rathore, J.J. de Pablo, Monte Carlo simulation of proteins through a random walk in energy space, J. Chem. Phys. 116 (16) (2002) 7225–7230.
[238] T. Schoonjans, V.A. Solé, L. Vincze, M.S. del Rio, K. Appel, C. Ferrero, A general Monte Carlo simulation of energy-dispersive X-ray fluorescence spectrometers—Part 6. Quantification through iterative simulations, Spectrochim. Acta B At. Spectrosc. 82 (2013) 36–41.
[239] T. Schoonjans, L. Vincze, V.A. Solé, M.S. del Rio, P. Brondeel, G. Silversmit, … C. Ferrero, A general Monte Carlo simulation of energy dispersive X-ray fluorescence spectrometers—part 5: polarized radiation, stratified samples, cascade effects, M-lines, Spectrochim. Acta B At. Spectrosc. 70 (2012) 10–23.
[240] M. Jahangirian, T. Eldabi, A. Naseer, L.K. Stergioulas, T. Young, Simulation in manufacturing and business: a review, Eur. J. Oper. Res. 203 (1) (2010) 1–13.
[241] M. Li, F. Yang, R. Uzsoy, J. Xu, A metamodel-based Monte Carlo simulation approach for responsive production planning of manufacturing systems, J. Manufact. Syst. 38 (2016) 114–133.
[242] F. Wu, J.-Y. Dantan, A. Etienne, A. Siadat, P. Martin, Improved algorithm for tolerance allocation based on Monte Carlo simulation and discrete optimization, Comput. Ind. Eng. 56 (4) (2009) 1402–1413.
[243] Z. Cao, Y. Wang, D. Li, Practical reliability analysis of slope stability by advanced Monte Carlo simulations in a spreadsheet, in: Probabilistic Approaches for Geotechnical Site Characterization and Slope Stability Analysis, Springer, Berlin, 2017, pp. 147–167.
[244] D. Griffiths, G. Fenton, Risk Assessment in Geotechnical Engineering, John Wiley & Sons, Inc, Hoboken, NJ, 2008.
[245] P. Marek, J. Brozzetti, M. Gustar, I. Elishakoff, Probabilistic assessment of structures using Monte Carlo simulations, Appl. Mech. Rev. 55 (2) (2002) B31–B32.
[246] S. Raychaudhuri, Introduction to Monte Carlo simulation, in: Paper Presented at the 2008 Winter Simulation Conference, 2008.
[247] J.L. Kuti, P.K. Dandekar, C.H. Nightingale, D.P. Nicolau, Use of Monte Carlo simulation to design an optimized pharmacodynamic dosing strategy for meropenem, J. Clin. Pharmacol. 43 (10) (2003) 1116–1123.
[248] N. Ren, J. Liang, X. Qu, J. Li, B. Lu, J. Tian, GPU-based Monte Carlo simulation for light propagation in complex heterogeneous tissues, Opt. Express 18 (7) (2010) 6811–6823.

[249] P.K. Romano, N.E. Horelik, B.R. Herman, A.G. Nelson, B. Forget, K. Smith, OpenMC: A state-of-the-art Monte Carlo code for research and development, in: Paper Presented at the SNA + MC 2013-Joint International Conference on Supercomputing in Nuclear Applications + Monte Carlo, 2014.

[250] M.U. Bartsch, How Much Is Enough? Monte Carlo Simulations of an Oil Stabilization Fund for Nigeria, International Monetary Fund, 2006.

[251] L.Y. Batan, G.D. Graff, T.H. Bradley, Techno-economic and Monte Carlo probabilistic analysis of microalgae biofuel production system, Bioresour. Technol. 219 (2016) 45–52.

[252] J.E. Bickel, R.B. Bratvold, From uncertainty quantification to decision making in the oil and gas industry, Energy Explor. Exploit. 26 (5) (2008) 311–325.

[253] F. Salvat, J.M. Fernández-Varea, E. Acosta, J.P. Sempau, A code system for Monte Carlo simulation of electron and photon transport, in: Proceedings of a Workshop/Training Course, OECD/NEA, 2001, pp. 5–7.

[254] I. Kawrakow, Accurate condensed history Monte Carlo simulation of electron transport. I. EGSnrc, the new EGS4 version, Med. Phys. 27 (3) (2000) 485–498.

[255] I. Kawrakow, Accurate condensed history Monte Carlo simulation of electron transport. II. Application to ion chamber response simulations, Med. Phys. 27 (3) (2000) 499–513.

[256] G. Rubino, B. Tuffin, Rare Event Simulation Using Monte Carlo Methods, John Wiley & Sons, 2009.

[257] S.K. Au, Z. Cao, Y. Wang, Implementing advanced Monte Carlo simulation under the spreadsheet environment, Struct. Saf. 32 (5) (2010) 281–292.

[258] C.L. Borges, D.M. Falcao, J.C.O. Mello, A.C. Melo, Composite reliability evaluation by sequential Monte Carlo simulation on parallel and distributed processing environments, IEEE Trans. Power Syst. 16 (2) (2001) 203–209.

[259] M. Cieslak, C. Lemieux, J. Hanan, P. Prusinkiewicz, Quasi-Monte Carlo simulation of the light environment of plants, Funct. Plant Biol. 35 (10) (2008) 837–849.

[260] M. Cieslak, C. Lemieux, P. Prusinkiewicz, Quasi-Monte Carlo simulation of the light environment of virtual plants, in: Paper Presented at the Proceedings of the 5th International Workshop on Functional-Structural Plant Models, 2007.

[261] M. Rodriguez, J. Sempau, L. Brualla, PRIMO: A graphical environment for the Monte Carlo simulation of Varian and Elekta linacs, Strahlenther. Onkol. 189 (10) (2013) 881–886.

[262] S.V. Cai, T.R. Famula, A.M. Oberbauer, R.S. Hess, Heritability and complex segregation analysis of diabetes mellitus in American Eskimo Dogs, J. Vet. Intern. Med. 33 (5) (2019) 1926–1934.

[263] E.S. Chiaka, M.B. Adam, Bayesian analysis via Markov chain Monte Carlo algorithm on logistic regression model, Global J. Pure Appl. Math. 15 (2) (2019) 191–205.

[264] C.K. Kramer, C. Ye, S. Campbell, R. Retnakaran, Comparison of new glucose-lowering drugs on risk of heart failure in type II diabetes: a network meta-analysis, JACC: Heart Failure 6 (10) (2018) 823–830.

[265] P. Rai, Y. Zhang, A. Stover, C. Iloabuchi, K. Kamal, M. Chiumente, Insulin delivery systems for type 1 diabetes mellitus—a comparison using a decision analysis modeling approach, Value Health 21 (2018) S168.

[266] D. Sokolović, J. Ranković, V. Stanković, R. Stefanović, S. Karaleić, B. Mekić, ... A. M. Veselinović, QSAR study of dipeptidyl peptidase-4 inhibitors based on the Monte Carlo method, Med. Chem. Res. 26 (4) (2017) 796–804.

[267] S.G. Sosa Rubi, L. Dainelli, I. Silva-Zolezzi, P. Detzel, S.E. y Sosa, E. Reyes-Muñoz, ... R. Lopez-Ridaura, Short-term health and economic burden of gestational diabetes mellitus in Mexico: A modeling study, Diabetes Res. Clin. Pract. 153 (2019) 114–124.

[268] A.L. Beam, I.S. Kohane, Big data and machine learning in health care, Jama 319 (13) (2018) 1317–1318.
[269] P. Mamoshina, A. Vieira, E. Putin, A. Zhavoronkov, Applications of deep learning in biomedicine, Mol. Pharm. 13 (5) (2016) 1445–1454.
[270] E.A. Ashley, A.J. Butte, M.T. Wheeler, R. Chen, T.E. Klein, F.E. Dewey, ... A.A. Morgan, Clinical assessment incorporating a personal genome, Lancet 375 (9725) (2010) 1525–1535.
[271] C.A.C. Montaez, P. Fergus, A.C. Montaez, A. Hussain, D. Al-Jumeily, C. Chalmers, Deep learning classification of polygenic obesity using genome-wide association study SNPs, in: 2018 International Joint Conference on Neural Networks (IJCNN), July, IEEE, 2018, pp. 1–8.
[272] T.A. Manolio, Bringing genome-wide association findings into clinical use, Nat. Rev. Genet. 14 (8) (2013) 549–558.
[273] Y. Wei, C.K. Tsang, X.S. Zheng, Mechanisms of regulation of RNA polymerase III-dependent transcription by TORC1, EMBO J. 28 (15) (2009) 2220–2230.
[274] G. Abraham, M. Inouye, Genomic risk prediction of complex human disease and its clinical application, Curr. Opin. Genet. Dev. 33 (2015) 10–16.
[275] J. Kim, J. Kim, M.J. Kwak, M. Bajaj, Genetic prediction of type 2 diabetes using deep neural network, Clin. Genet. 93 (4) (2018) 822–829.
[276] G.M. Clarke, C.A. Anderson, F.H. Pettersson, L.R. Cardon, A.P. Morris, K.T. Zondervan, Basic statistical analysis in genetic case-control studies, Nat. Protoc. 6 (2) (2011) 121–133.
[277] B.A. McKinney, D.M. Reif, M.D. Ritchie, J.H. Moore, Machine learning for detecting gene-gene interactions, Appl. Bioinform. 5 (2) (2006) 77–88.
[278] J. Chaki, N. Dey, Pattern analysis of genetics and genomics: a survey of the state-of-art, Multimed. Tools Appl. (2019) 1–32.
[279] A. Belorkar, L. Wong, GFS: fuzzy preprocessing for effective gene expression analysis, BMC Bioinformatics 17 (17) (2016) 540.
[280] A. Wagner, A. Regev, N. Yosef, Revealing the vectors of cellular identity with single-cell genomics, Nat. Biotechnol. 34 (11) (2016) 1145.
[281] C.J. Walsh, P. Hu, J. Batt, C.C.D. Santos, Microarray meta-analysis and cross-platform normalization: integrative genomics for robust biomarker discovery, Microarrays 4 (3) (2015) 389–406.
[282] M.H. Bailey, C. Tokheim, E. Porta-Pardo, S. Sengupta, D. Bertrand, A. Weerasinghe, ... P.K.S. Ng, Comprehensive characterization of cancer driver genes and mutations, Cell 173 (2) (2018) 371–385.
[283] T.A. Brown, Genomes 4, Garland Science, 2018.
[284] K.M. Weiss, J.D. Terwilliger, How many diseases does it take to map a gene with SNPs? Nat. Genet. 26 (2) (2000) 151–157.
[285] C.F. Wright, T.W. Fitzgerald, W.D. Jones, S. Clayton, J.F. McRae, M. Van Kogelenberg, ... A.P. Bevan, Genetic diagnosis of developmental disorders in the DDD study: a scalable analysis of genome-wide research data, Lancet 385 (9975) (2015) 1305–1314.
[286] L.M.Q. Abualigah, Feature Selection and Enhanced Krill Herd Algorithm for Text Document Clustering, Springer, Berlin, 2019, pp. 1–165.
[287] F.N. Esterhuysen, Development of a Simple Artificial Intelligence Method to Accurately Subtype Breast Cancers Based on Gene Expression Barcodes (Doctoral dissertation), University of the Western Cape, 2018.
[288] C. Laclau, V. Brault, Noise-free latent block model for high dimensional data, Data Min. Knowl. Disc. 33 (2) (2019) 446–473.
[289] G.E. Hinton, A practical guide to training restricted Boltzmann machines, in: Neural Networks: Tricks of the Trade, Springer, 2012, pp. 599–619.

[290] G.E. Hinton, S. Osindero, Y.-W. Teh, A fast learning algorithm for deep belief nets, Neural Comput. 18 (7) (2006) 1527–1554.
[291] X. Rong, Deepnet: deep learning toolkit R, (2014)R Package Version 0.2. https://cran.r-project.org/web/packages/deepnet/index.html.

CHAPTER TEN

An application of cypher query-based dynamic rule-based decision tree over suicide statistics dataset with Neo4j

S. Anjana and K. Lavanya
School of Computer Science & Engineering, VIT, Vellore, India

1. Introduction

Reports on the increased rate of suicides triggered the WHO to adopt Comprehensive Mental Health Action Plan 2013–2020 in the World Health Assembly. The principal intention of the action plan is to strengthen leadership and govern the mental health of people effectively [1]. According to the fact sheets published by WHO, the cause of death for nearly 800,000 people is by suicide and more people attempt suicide every year [2]. Government and private organizations all around the world are conducting suicide awareness programs. This chapter attempts to analyze the crude suicide rate data obtained from Global Health Observatory Data Repository. The data contains suicide rates from different countries and from different WHO regions along with the number of attributes that participate in the analysis process. In addition to that, it explains the suicide rate estimate for women and men separately in both WHO region and country, which gives further scope for the analysis. As the suicide rate is increasing drastically, it demands a stable analysis so that accurate prevention measurements can be taken. Most of the recent studies on suicide datasets have included age-based analysis. These analyses include classification of suicide data based on age and sex. This helped us to conclude that suicide occurs throughout the lifespan and is the second leading cause of death among 15–29 year olds globally. Likewise, more studies are required on the datasets to get a useful outcome. Hence, it requires multiple criteria for analysis on suicide data so that the awareness can be more precise. This chapter takes WHO region and country as standards for the investigation on suicide datasets. This gives an

Intelligent IoT Systems in Personalized Health Care
https://doi.org/10.1016/B978-0-12-821187-8.00010-1

insight on how suicide rates are distributed worldwide, which further helps in identifying the areas that require close attention regarding the matter.

In this chapter, analysis is performed by creating a rule-based decision tree on the dataset. Decision tree is a very important tool in machine learning, which is widely used for classification and regression problems. It utilizes a tree-like structure or a graph of decisions and visualizes the results of analytical decision analysis. The efficient visualization techniques make decision trees more suitable for the analysis. A rule-based decision tree is the result of learning from a set of decision rules that cover the data instances. The decision rules depend on the dataset and the analysis criteria. Rule-based decision trees calculate results based on the rules formed and the results are visually and explicitly represented in a graph model.

As the data is stored in different forms and structures, a graph database is a suitable option for creating the decision trees on the interlinked data. Neo4j is one among the popular graph databases that provide high scalability and reliability. The open-source graph database has full transactional property and written in Java programming language. It represents dataset in the form of nodes and relationships with attributes assigned to both of them in the key-value form [3]. Neo4j not only helps to represent data but also allows easy retrieval of data from the graph using labels, which is one of the necessary requirements in rule-based decision trees. Neo4j creates nodes for each entity in the dataset with appropriate attributes given in the dataset. Dependencies between the nodes are represented by using relationships. Relationships in the decision trees are important as they participate in the traversal to create dynamic rule-based decision trees. Engineering rule-based decision trees require search. To make the search easier, each data node in the dataset is provided with an index using the indexing method. Indexing is a way to optimize performance of a database by minimizing the number of searches. The inbuilt indexing mechanism provided by Neo4j makes the task easier. This chapter also gives an insight on the importance of indexing by evaluating the response time of the algorithm with and without indexing of nodes. The rule-based decision tree is generated from the data model that contains the indexed data nodes and relationships.

The features of Neo4j make the implementation of the decision tree convenient [4]. Traversal is one of the most attractive features of graph databases. Neo4j supports traversal and provides plenty of tree-based algorithms to make the traversal more efficient. In rule-based decision trees, evaluation of rules occurs dynamically in each node and the result of evaluation helps in traversing the different levels of data model. The cypher query language used in Neo4j is easy to understand and implement, which gives the user an optimized storage and access. The cypher query language provides optimized performance with the least number of codes. Since rule-based decision trees are far different from traditional decision trees, traversal through the nodes is difficult. However, the cypher query language provides a user-friendly interface to create and implement dynamic rule-based decision trees. Neo4j helps to model the dynamic rule-based decision tree and assist in evaluating the rule dynamically, which further aids in an efficient analysis of the suicide dataset. Most of the analysis is visualized in pie charts or bar charts. In this chapter, analysis results are displayed in a tree structure with the help of Neo4j.

2. Literature review

Majority of the suicide rates data analysis are performed on the basis of age and income. As per WHO, one person dies every 40 s due to suicide. This study also suggests that suicide rates are high in high-income countries. As the study shows a warning increase in suicide rate, it requires more in-depth analysis on the data. Different perceptions toward the analysis provide more solutions to the problem. In this chapter, the study criteria are WHO regions and country. Analyzing medical data is often cumbersome as it contains accumulated large quantities of information about patients and their medical conditions. But in the case of suicide rate data the availability and consistency of data is deficient. Precise input data always provide accurate results. Automated systems are best for medical diagnosis because it would enhance medical care, reduce human errors, and reduce costs. An automated system makes use of a variety of machine learning algorithms for the implementation of the system. Purushottam et al. describe a heart disease prediction system using the decision tree. Tools such as WEKA and KEEL are used for data analysis and to study classification decision rules, respectively [5]. Likewise, rule-based decision trees are used for fault diagnosis of water quality monitoring devices. The chapter explains that decision trees have great potential for fault diagnosis of online water quality devices than any other algorithms [6]. Permanasari and Nurlayli elaborate how to analyze the cardiotocogram data for fetal distress determination using decision trees. The results show that decision trees produce minimum misclassification errors [7]. Decision trees are used mainly in the medical diagnosis domain. Albu explains how logical inference to decision tree results in medical diagnosis. The chapter analyzes many diagnosis systems that use decision trees and verify that decision trees offer better results than other machine learning algorithms applied in medical decision-making [8]. A set of decision rules forms a rule-based decision tree, which is used for classification and regression. Rule-based decision trees completely depend on the rules. Decision trees formed from the rules are easier to restructure because restructuring requires changes in rules not in the decision trees [9]. The new learning method for decision trees with rules outperforms the ID3 method. Abdelhalim and Traore compare the performance of RBDT 1 with other existing learning methodologies for decision trees [10]. Hence, decision trees are suitable tools for medical analysis purposes. Other than medical analysis, social networking in the health-care domain is booming as it engages health-care professionals in knowledge transfer to masses or to themselves. In addition, the limitations of RDBMS in implementing the social media for HSN platform are overcome with the help of Neo4j graph database [11]. Other than classification and regression, Neo4j helps to represent data progression. This practice is best suitable for the medical industry to represent disease progression [12]. Neo4j is the graph platform that helps in depicting complex relationships among entities. This graph database can create a framework for tensor ontologies. Here, a multilayered graph is used to represent the multiple connections between vertex pairs showing the coexistence of entities in the system [13]. Lu et al. describe how Neo4j aids in storing and processing mass data with the complex and connected attributes. The dataset used to elaborate the study is a film dataset, which contains data related to movies. The selected film dataset which contains the details such as the name of the film, director etc. is a multiconnected dynamic data. These multiconnected dynamic data are represented efficiently in Neo4j according to the film website to meet the user's requirements and interest [14]. Neo4j has always been a very good tool to represent connected data. Neo4j outperforms other databases in implementing social networking and recommendation systems [15]. With the help of cypher query language, Neo4j is used in recommendation systems. Dharmawan and Sarno explain how Neo4j helps in book recommendation using cypher queries. The graph database gives an easy way to represent the metadata of books and the cypher language provides an easy way to communicate with the graph. According to the user's interest, recommendations are made with the aid of cypher query language [16]. In another paper, Stark et al. describe another recommendation system with Neo4j, which is a drug recommendation system for migraine. The scalable database helps to store the details of patients, drugs, and other aforementioned elements. The flexible data model and other features of Neo4j make it more suitable for the recommendation system [17]. Johnpaul and Mathew try to describe the approach of NoSql query design and analysis on different datasets over Neo4j. This chapter also explains the importance of NoSQL database, and describes how the features of Neo4j fit modern database systems. It also provides information about the performance of cypher query language [18]. A detailed study on the architecture and internal mechanism of the Neo4j graph database points out that the query performance of the database solemnly depends on the application [19]. Drakopoulos proposes harmonic centrality with the structural ranking presentation, which is executed easily in a short time. Directed graph ranking is necessary for the vertices to get result and if the graph contains large number of vertices and nodes then it is a big challenge to design an algorithm and implement for vertex ranking. In the twitter network, to overcome network oblivious tensor fusion methodology is implemented with Neo4j and tensor toolbox [20]. Applications that involve dependent data make use of the inherent features of the Neo4j database. For example, application to improve the semantic queries [21] or lighting and failure information from different enterprise application system (EAPs) uses Neo4j for better performance [22].

Previous studies have shown that analysis of suicide rate data is mostly done using classification algorithms. Machine learning algorithms are applied to medical data to discover hidden knowledge. Also, decision trees are one among the most used algorithms for medical data evaluation. This chapter suggests a rule-based decision method to evaluate the suicide rate data with WHO region and country as standards. And the results are displayed in tree structure than charts. Neo4j can efficiently generate and evaluate a rule-based decision tree. The easy retrieval policy of the Neo4j database helps the node level rule evaluation simpler for rule-based decision trees. In addition, Neo4j can provide results in a visually convincing manner.

3. Proposed system

The main objective of the chapter is to analyze the estimated suicide rates so that it helps in the implementation of prevention action plan for suicides by giving the awareness in a more precise way. The mental health statistics shows that it varies according to sex, age, and region. So a detailed analysis is required to ensure that the prevention method implemented hits the reality of the actual problem. The workflow of the project starts with preprocessing of data and loading the data into the Neo4j database. An index for each of the data node in the database has been provided by using the indexing method for easier search requirements. The next step is finding the relationship between the nodes and creating the decision tree model using the relationships. Neo4j helps to represent the connected dataset. According to the analysis criteria, rules are generated and rules are evaluated in each node using the rule-based decision tree using cypher. The rules are in the form of cypher queries. The formation of the rules completely depends on the analysis criteria and the data. Rules have been used to traverse through the dataset while dynamically creating the rule-based decision tree. Representation of the dynamically formed rule-based decision trees are done by Neo4j (Fig. 1).

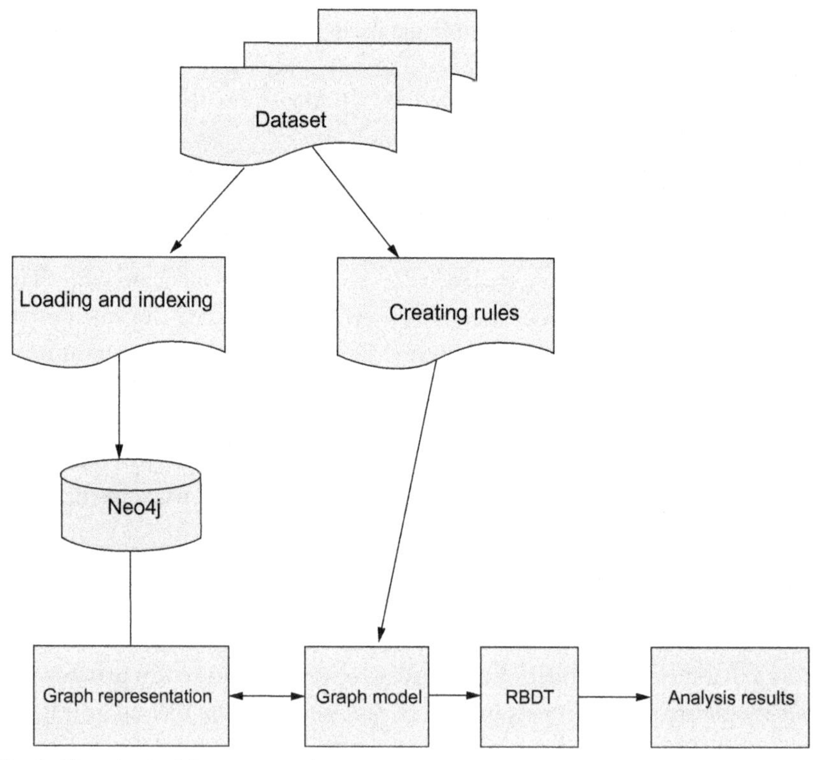

Fig. 1 Flowchart of the proposed system.

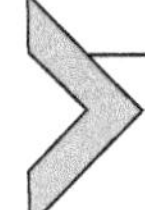

4. Methodology

4.1 Dataset description

The analysis was carried out using three datasets. Datasets are collected from the Global Health Observatory Repository provided by WHO. The data consists of the suicide rate estimates by country (dataCountry), suicide rate estimates by region based on sex (dataRegion), and suicide rate estimates by WHO region (dataWHO). A root node is created additionally in order to satisfy the tree structure of the dataset. The dataset of suicide rate estimates by country has seven attributes, which include the WHO region, country, sex, and age-standardized suicide rates (per 100,000 population) in the years 2016, 2015, 2010, and 2000. The dataset contains a total of 551 entities. The dataset of suicide rate estimates by WHO region based on sex has 7 attributes, which includes the WHO region, sex, and crude suicide rates (per 100,000 population) in the years 2016, 2015, 2010, 2005, and 2000. This dataset contributes 23 entities to the existing dataset. The dataset suicide rate estimates by WHO region has age-standardized suicide rates (per 100,000 population) in the year 2016 for both sexes as attributes. The third dataset has seven entities where each entity corresponds to a WHO region. Relationships are created between the different dataset nodes to model the data in a tree format starting with root node followed by dataWHO, dataRegion, and dataCountry datasets, respectively (Fig. 2).

4.2 Preprocessing the dataset

The raw data collected from WHO repository has been preprocessed before the actual analysis process. Since the data nodes in the dataset cannot be identified uniquely using a single attribute, a primary key called id has been mapped to every node in the dataset initially. The dataset of the suicide rate estimates by country does not depict which WHO region that country belongs to; hence, a new attribute is added to the dataset named the WHO region. A study is conducted to find the WHO region of each country in the dataset and added those values to the existing dataset. The new dataset is more suitable for generating the decision tree as we can create relationships between the dataset of the suicide rate estimates by country and suicide rate estimates by WHO region.

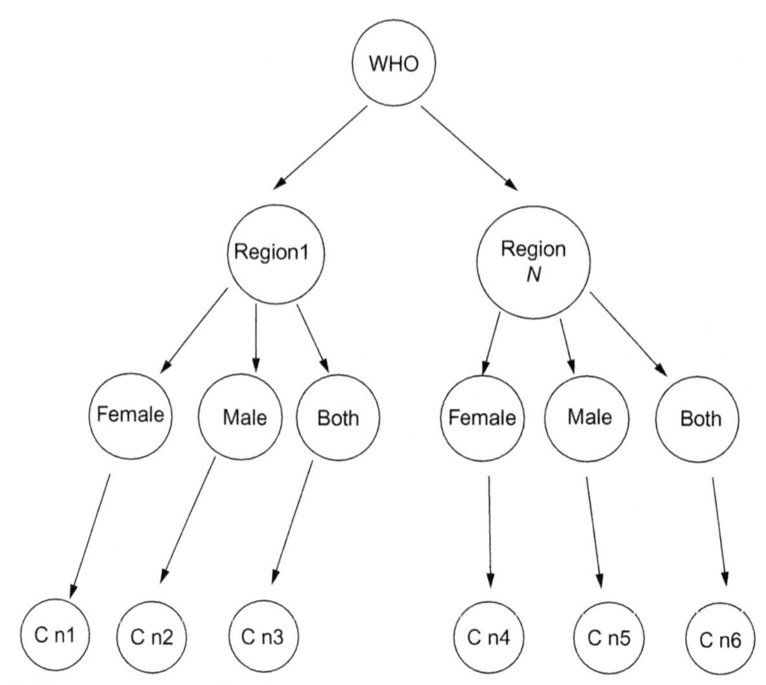

Fig. 2 Managing process flows in the dataset.

4.3 Algorithm

The analysis on suicide rate dataset is performed by generating the dynamic rule-based decision tree. A tree has many analogies in real life, and it has influenced a wide area of machine learning, covering both classification and regression. The dynamic rule-based decision tree is developed by evaluating a set of rules on suicide rate datasets. To create an optimized rule-based decision tree in terms of time, a two-level processing is performed on the datasets. Preprocessing the dataset by adding additional attributes to the existing one made the entities unique and powerful. The preprocessed datasets are indexed using the composite index method where multiple attributes are employed to spawn the indices for the nodes. Rules are generated based on the analysis criteria. Preprocessing and indexing helps in searching, matching, and traversing the dataset while making the rule-based decision tree. Rules are the important components required for initiating the rule-based decision tree. Rules are made based on analysis criteria. Since the tree generated from the dataset has three levels, a minimum of two rules are required to process the dataset. Traversing through the tree is by evaluating the rules at each level of the tree. The rule-based decision tree expands to the

1. Start

2. Preprocess the dataset.

3. Load the datasets dataWHO, dataRegion and dataCountry.

4. Create nodes for each dataset entities.

5. For each nodes in the dataset,

    5.1Create index.

6.Create relationships between the datasets.

    6.1 Create relationship between Root node and nodes in dataWHO.

    6.2 For each node1 in dataWHO, Create a relationship with node 2 in dataRegion WHERE node1.WHOregion = node2.WHOregion.

    6.3 For each node1 in dataRegion, Create a relationship with node 2 in dataCountry WHERE node1.WHOregion = node2.WHOregion.

7. Create Rules required for analysis in R.

8.For each level Li in the tree,

    8.1 Evaluate Ri

    8.2 If(Evaluation = TRUE)

        8.2.1 Traverse to next level.

    8.3 Else

        8.3.1 Stop

9.Return the rule based decision tree.

10.Stop

Fig. 3 Dynamic decision tree algorithm Neo4j.

next level only when the evaluation at the current level is successful. Finally, a rule-based decision tree has prompted with the required data (Fig. 3).

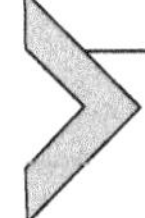

5. Working with Neo4j

5.1 Loading the database

The datasets undergo preprocessing prior to import. The preprocessed dataset needs to be imported to Neo4j repository before loading in to the Neo4j graph database. After loading the repository the datasets are loaded in to the Neo4j database by creating nodes for each entity in the datasets (Fig. 4).

After loading the dataset in to the Neo4j database, 578 nodes have formed and shown as Fig. 5.

```
CREATE (Root:WHO{id:0})

WITH count(*) as dummy

LOAD CSV WITH HEADERS FROM "file:///dataWHO.csv" AS row CREATE (n:dataWHO) SET n = row,n.id =
(row.id),n.WHOregion = (row.WHOregion),n.sa2016 = (row. sa2016)

WITH count(*) as dummy

LOAD CSV WITH HEADERS FROM "file:///dataRegion.csv" AS row CREATE (n:dataRegion) SET n = row,n.id
= (row.id),n.WHOregion = (row.WHOregion),n.Sex = (row. Sex),n.d2016 = (row. d2016),n.d2015 = (row.
d2015),n.d2010 = (row. d2010),n.d2005 = (row. d2005),n.d2000 = (row. d2000)

WITH count(*) as dummy

LOAD CSV WITH HEADERS FROM "file:///dataCountry.csv" AS row CREATE (n:dataCountry) SET n =
row,n.WHOregion = (row.WHOregion),n.Country= (row.Country),n.Sex = (row.Sex),n.sr2016 =
(row.sr2016),n.sr2015 = (row.sr2015),n.sr2010 = (row.sr2010),n.sr2000 = (row.sr2000)
```

Fig. 4 Query to load dataset.

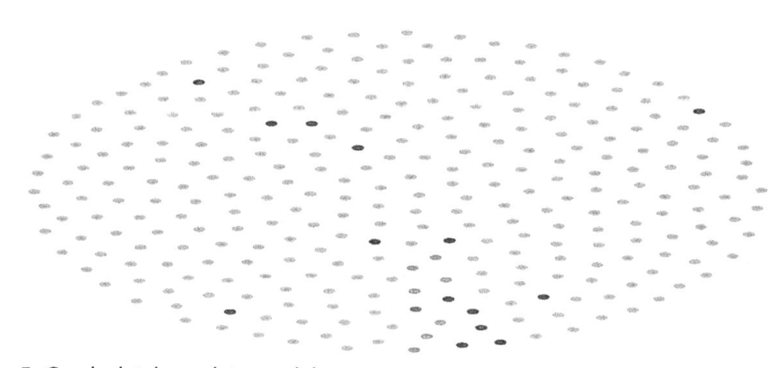

Fig. 5 Graph database data model.

5.2 Indexing

Indexing is a technique to optimize performance of a database by minimizing the number of accesses required when a query is processed. An index or database index is a data structure, which is used to quickly locate and access the data in a database. Neo4j supports indexing and provides index to the data nodes present in the Neo4j database. An index on a node based on multiple properties are created for the data nodes in the system. A composite index is produced using all the properties assigned for particular nodes. As most of the nodes in the dataset consist of similar characteristics and values, a composite indexing is necessary to make the indices of the nodes unique and easily accessible. CREATE INDEX command along with the attributes creates index for the nodes (Fig. 6).

```
CREATE INDEX ON:WHO(id)
CREATE INDEX ON:dataWHO(id,sa2016)
CREATE INDEX ON:dataRegion(id,WHOregion,Sex,d2000,d2005,d2010,d2015,d2016)
CREATE INDEX ON:dataCountry(id,Sex,WHOregion,sr2016,sr2015,sr2010,sr2000)
CALL db.indexes
```

Fig. 6 Neo4j index creation.

5.3 Creating relationships

After loading and indexing the datasets, the next step is creating relationship between the dependent data nodes. As the database consists of three datasets, in order to model the datasets in a tree format, relationships have been created between the nodes of different datasets. A root node was created to act as the start node of the tree. From the tree node, a relation HAS_REGION is started which ends to the nodes in dataWHO dataset (Fig. 7).

Seven relationships that are formed in the database between the root node and the nodes in dataWHO dataset as the dataWHO contains seven WHO region details (Fig. 8).

Similarly, another relationship suicide _rate_based_on_sex is initiated from the nodes in dataWHO to the nodes in the dataRegion dataset. Here a constraint is added to create the relation. A relation is created only if the nodes belong to the same WHO region (Fig. 9).

The cypher query gives rise to 21 relationships in the database (Fig. 10).

Another relation has implemented between the nodes in the dataRegion dataset and nodes in dataCountry dataset. The relation suicide _rate_based_on_Country provides connection between the datasets by checking the WHO region in both datasets (Fig. 11).

About 1647 relations have formed at the end of the relationship creation. The connected data structure is now ready for traversal and to imply rule-based decision trees. This completes the data modeling of the dataset (Fig. 12).

```
//relation1
MATCH (Root:WHO),(WHO:dataWHO) CREATE UNIQUE (Root)-[R1:HAS_REGION]->(WHO)
```

Fig. 7 Creation of Relation1.

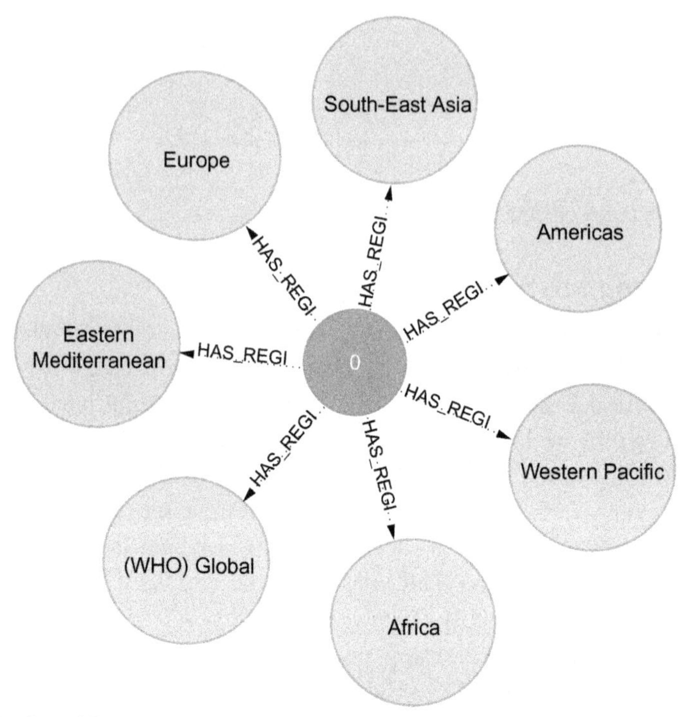

Fig. 8 Real-world WHO region representation.

```
//relation 2

MATCH (WHO:dataWHO),(Region:dataRegion) WHERE WHO.WHOregion = Region.WHOregion CREATE
UNIQUE (WHO)-[R2:suicide _rate_based_on_sex]->( Region)
```

Fig. 9 Creation of Relation 2.

5.4 Rule-based decision tree

Rule-based decision tree is a machine learning methodology for learning decision tree from a set of declarative decision rules. The decision rules are evaluated on the data instances present to generate the rule-based decision tree. One of the advantages of the rule-based decision tree is decision trees, which can be generated from a stable or dynamic set of rules, and it can be done on demand to produce accurate results. For rule-based decision trees, rules are the key factor that gives the desired output. The rules used by the rule-based decision tree could be generated either by an expert or induced directly from a rule induction method or indirectly by extracting them from a decision tree. Here we followed the second method, which

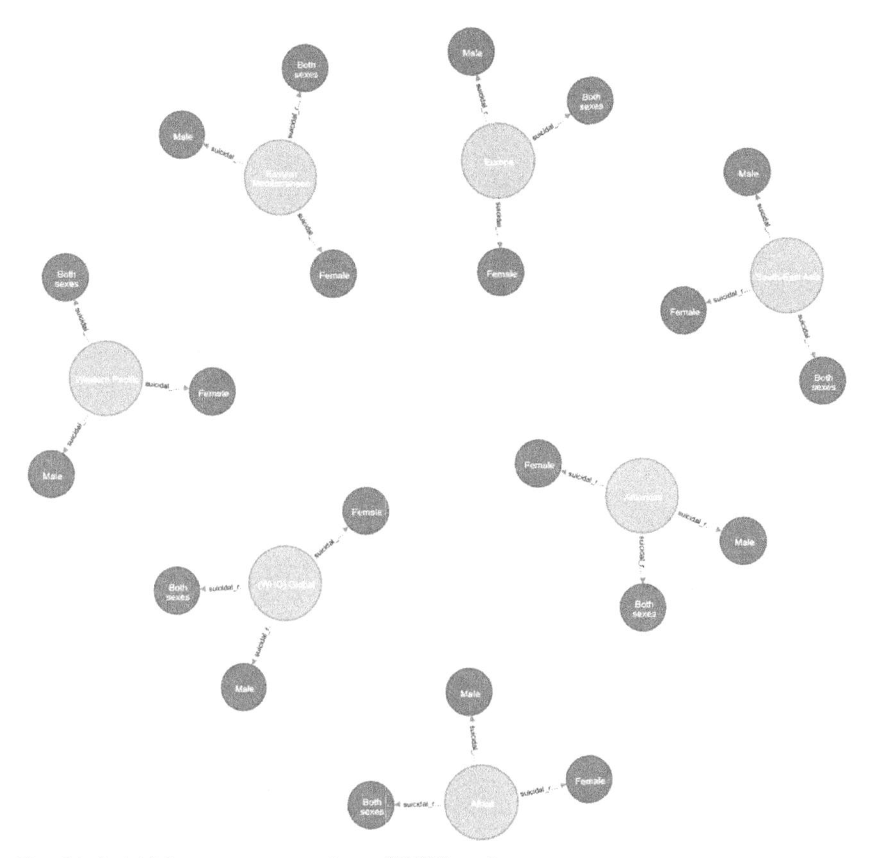

Fig. 10 Suicidal rate representation of WHO region.

```
//relation 3 MATCH (Region:dataRegion),(Country: dataCountry) WHERE Region.WHOregion =
Country.WHOregion CREATE (Region)-[R3:suicide _rate_based_on_Country]->(Country)
```

Fig. 11 Creation of Relation3.

is indirectly extracting them from a decision tree. Rules have relation to the analysis requirement. Based on the demands, rules are generated from the existing decision tree and are evaluated dynamically to produce the necessary outcome.

This chapter analyzes the mental health dataset, which contains the suicide rates of WHO regions and countries. To create a rule-based decision tree from the dataset we need to create a set of rules that are applicable for the dataset. Consider a scenario where we need to analyze the suicide_rate in 2016 which is >5.0 of females which are in the South-East

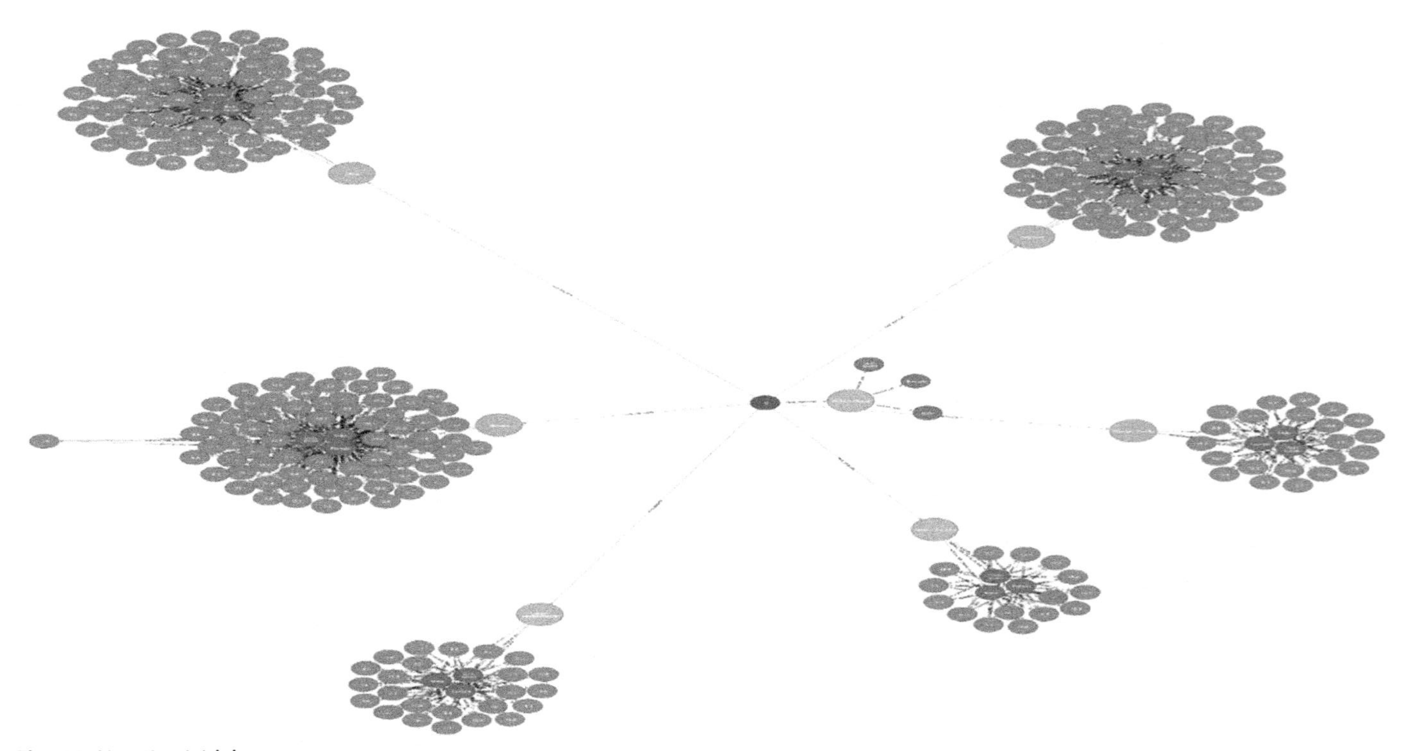

Fig. 12 Neo4j suicidal trees.

Rule 1: WHOregion is South east Asia region

Rule 2: Sex is female

Rule 3: suicide_rate in the year 2016 greater than 5.0

Fig. 13 Dynamic decision tree rules.

Asia region. To classify the data instances based on the requirement, we need to create rules. Rules are created for each level in the tree, based on the scenario. The rules generated from the above scenario are WHOregion, which is the South-East Asia region, sex is female and suicide_rate in 2016 is >5.0 (Fig. 13).

The following code snippet includes the above rules resulting in the desired rule-based decision tree (Fig. 14).

The traversal begins starting from the root node. In the next level, rule 1 is evaluated for the South-East Asia region. The rule is evaluated in all nodes of the current level. If the rule is evaluated successful at any one of the nodes, those nodes are selected for further traversal. In the next level, the second rule is assessed where the sex is female. As done for the previous level, the nodes where sex is female are selected for more travel through the tree. In the next level, the next rule is suicide rate in 2016 which is >5.0. This rule is checked for all nodes in the next level. The nodes that are found to be true for the rules are selected to form the rule-based decision tree (Fig. 15).

The data instances have been classified based on the rules at each level of the tree to result in the rule-based decision tree, which provides scope for analysis.

Consider another scenario to find the countries in all the WHO regions whose suicide rate in 2016 and 2015 are >5.0. To categorize the dataset based on the requirements, rules have been created by analyzing the requirements and the decision tree. Here rules such as all region, suicide rate in 2015 >5.0, and suicide rate in 2016 >5.0 has been formed explicitly from the decision tree (Fig. 16).

```
//RBDT#1

MATCH(root:WHO) MATCH(root)-[rel:HAS_REGION]->(WHO:dataWHO) WHERE WHO.WHOregion='South-
East Asia' MATCH(WHO)-[rel1:suicide _rate_based_on_sex]->(region:dataRegion) WHERE region.Sex='Female'
MATCH(region)-[rel2:suicide _rate_based_on_Country]->(Country:dataCountry) WHERE Country.Sex ='Female'
OR Country.sr2016 >='5.0'

RETURN root,rel,WHO,rel1,region,rel2,Country
```

Fig. 14 Query for rule-based decision trees.

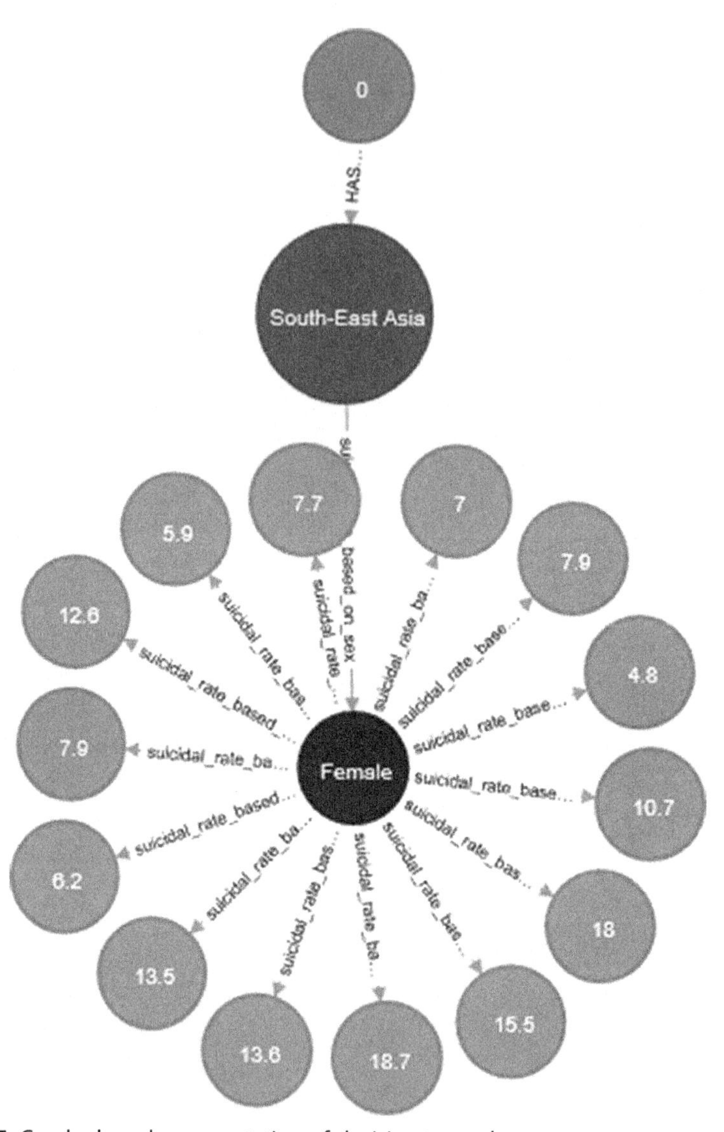

Fig. 15 Gender-based representation of decision tree rules.

Rule 1: suicide rate in the year 2015 greater than 5.0 and suicide rate in the year 2016 greater than 5.0

Fig. 16 Creation of rule for suicidal rate year wise.

```
//RBDT#2 MATCH(root:WHO)

MATCH(root)-[rel:HAS_REGION]->(WHO:dataWHO) MATCH(WHO)-[rel1:suicide _rate_based_on_sex]->(region:dataRegion) WHERE region.Sex='Both sexes' MATCH(region)-[rel2:suicide _rate_based_on_Country]->(Country:dataCountry) WHERE Country.sr2015 >='5.0' AND Country.sr2016 >='5.0'

RETURN root,rel,WHO,rel1,region,rel2,Country
```

Fig. 17 Query for year-wise suicidal rate using the decision tree.

The following code snippet includes the above rules and results in the desired rule-based decision tree (Fig. 17).

The rules evaluated at each level of the decision tree starting from the root node. Since the requirement is applicable to all the regions, no checks are given to the region level. Here the rules are applied to the nodes at the bottom level to find the countries based on the rules (Fig. 18).

6. Performance evaluation

The preprocessing and indexing procedures performed on the dataset contributed positively, while arranging the dataset in the tree pattern. Preprocess of the dataset helped to create the connected data. Meanwhile, indexing helped to optimize the time to connect data. Fig. 19 shows the difference in response time while creating the relations between the nodes. It clearly shows that the response time in creating relationship between the nodes with indexing is remarkably lower than the response time in creating relationship between the nodes without indexing. Indexing helps to minimize the response time.

Likewise, evaluation of response time for creating the rule-based decision tree with and without indexing shows the huge time gap and clearly states the capability of indexing in reducing the search and processing time. Fig. 20 shows the response time for creating two rule-based decision trees with and without index to the nodes.

7. Conclusion

WHO considers prevention of suicide as a public health priority. This chapter analyzes the suicide rate data with the tree type visualization taking WHO region and country as standards for evaluation. The graph database Neo4j is best suitable for representing the connected and semi-structured dataset. The creation and processing of the suicide rate dataset is made easy by using the decision tree algorithm. The analysis on suicide rate dataset

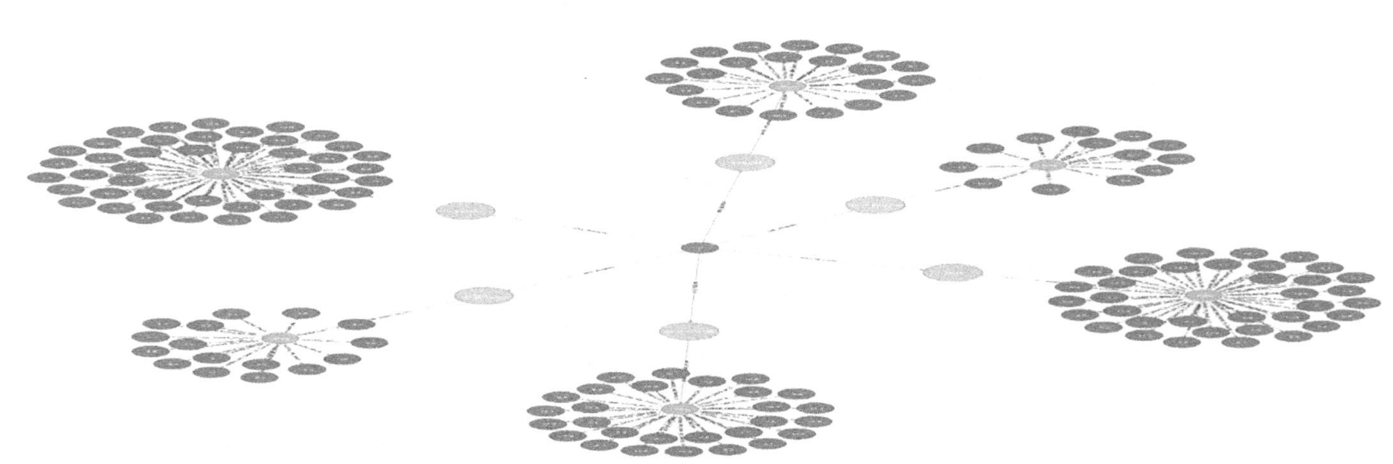

Fig. 18 Suicide data year wise based on the decision tree.

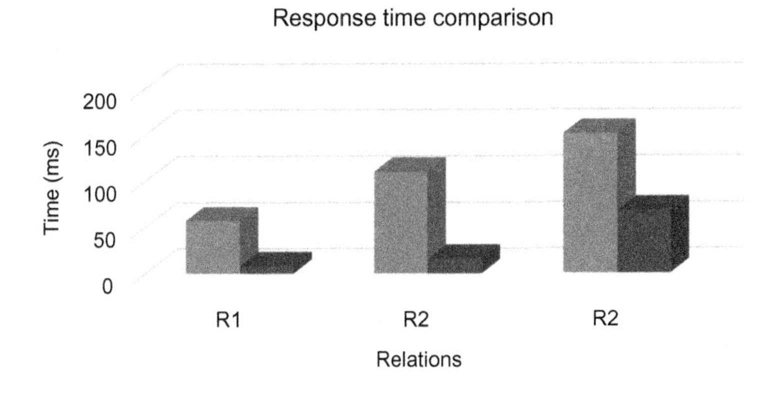

Fig. 19 Response time comparison for creating relationships.

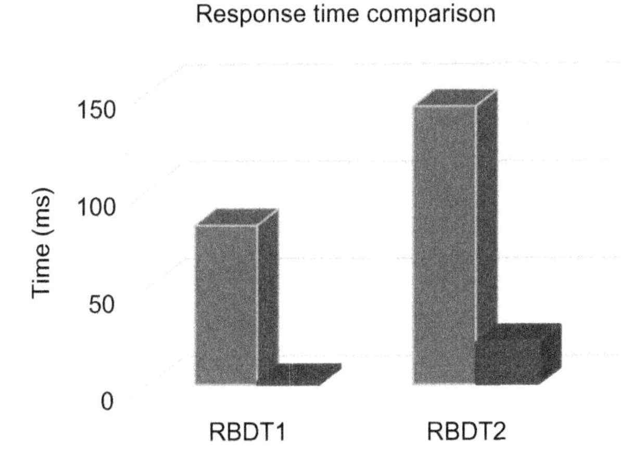

Fig. 20 Response time comparison for creating RBDT.

needs to be done accurately and dynamically in order to understand the trend in the dataset. The study shows the distribution of suicide rate country wise which comes under WHO regions. As from the tree, it is clearly visible that more countries under Africa region reported suicide. Table 1 shows the WHO region-wise data distribution.

Likewise, more data can be extracted over WHO region and country based on different rules. Neo4j is the best suitable database to represent such complex and multiconnected dataset. The rule-based decision tree

Table 1 Neo4j-based rule formation and results.

		No. of countries reported suicide				
S. No.	Rule	Africa	America	Eastern Mediterranean	South-East Asia	Western Pacific
1	Suicide rate in 2015 >5.0 and suicide rate in 2016 >5.0	37	22	16	13	29

algorithm plays a vital role in analyzing and managing the dataset. According to the analysis requirements, rules have generated from the external tree and are evaluated at each level of the tree to provide the precise rule-based decision tree. Evaluating the rules and traversal through the tree based on the evaluation became easier as Neo4j database has graph characteristics. With the help of Neo4j, rule-based decision trees can be generated on demand in a short period. The comparison of response time for creating the rule-based decision tree with and without indexing the nodes accurately shows the importance of indexing. Indexing optimizes the whole process by reducing the search and evaluation time and decreases the time by up to 85%.

References

[1] Comprehensive mental health action plan 2013–2020. Available from: https://www.who.int/mental_health/action_plan_2013/en/.

[2] Suicide. Available from: https://www.who.int/news-room/fact-sheets/detail/suicide.

[3] What is Graph Database. Available from: https://neo4j.com/developer/graph-database/.

[4] Why Neo4j? Top Ten Reasons. Available from: https://neo4j.com/top-ten-reasons/.

[5] Purushottam, K. Saxena, R. Sharma, Efficient heart disease prediction system using decision tree, in: International Conference on Computing Communication & Automation (ICCCA 2015), 2015, pp. 72–77.

[6] S. Liu, L. Xu, Q. Li, X. Zhao, D. Li, Fault diagnosis of water quality monitoring devices based on multiclass support vector machines and rule-based decision trees, IEEE Access 6 (c) (2018) 22184–22195.

[7] A.E. Permanasari, A. Nurlayli, Decision tree to analyze the cardiotocogram data for fetal distress determination, in: Proceeding—2017 International Conference on Sustainable Information Engineering and Technology (SIET 2017), vol. 2018-January, 2018, pp. 459–463.

[8] A. Albu, From logical inference to decision trees in medical diagnosis, in: 2017 E-Health and Bioengineering Conference (EHB 2017), 2017, pp. 65–68.

[9] A. Abdelhalim, I. Traore, Y. Nakkabi, Creating decision trees from rules using RBDT-1, Comput. Intell. 32 (2) (2016) 216–239.

[10] A. Abdelhalim, I. Traore, A new method for learning decision trees from rules, in: 8th International Conference on Machine Learning and Applications (ICMLA 2009), no. Mvd, 2009, pp. 693–698.

[11] A. Celesti, A. Buzachis, A. Galletta, G. Fiumara, M. Fazio, M. Villari, Analysis of a NoSQL graph DBMS for a hospital social network, in: Proceeding—IEEE Symposium on Computers and Communications, vol. 2018-June, 2018, pp. 1298–1303.
[12] H. Memarzadeh, N. Ghadiri, S.P. Zarmehr, A graph database approach for temporal modeling of disease progression, in: 2018 8th International Conference on Computer and Knowledge Engineering (ICCKE 2018), no. ICCKE, 2018, pp. 293–297.
[13] G. Drakopoulos, Tensor fusion of social structural and functional analytics over Neo4j, in: IISA 2016—7th International Conference on Information, Intelligence, Systems and Applications, 2016, pp. 1–6.
[14] H. Lu, Z. Hong, M. Shi, Analysis of film data based on Neo4j, in: Proceeding—16th IEEE/ACIS International Conference on Computer and Information Science (ICIS 2017), 2017, pp. 675–677.
[15] A. Virk, R. Rani, Recommendations using graphs on Neo4j, in: 2018 International Conference on Inventive Research in Computing Applications, no. ICIRCA, 2018, pp. 133–138.
[16] I.N.P.W. Dharmawan, R. Sarno, Book recommendation using Neo4j graph database in BibTeX book metadata, in: Proceeding—2017 3rd International Conference on Science in Information Technology: Theory and Application of IT for Education, Industry, and Society in Big Data Era (ICSITech 2017), vol. 2018-January, 2018, pp. 47–52.
[17] B. Stark, C. Knahl, M. Aydin, M. Samarah, K.O. Elish, BetterChoice: a migraine drug recommendation system based on Neo4J, in: 2017 2nd IEEE International Conference on Computational Intelligence and Applications (ICCIA 2017), vol. 2017-January, 2017, pp. 382–386.
[18] C.I. Johnpaul, T. Mathew, A Cypher query based NoSQL data mining on protein datasets using Neo4j graph database, in: 2017 4th International Conference on Advanced Computing and Communication Systems (ICACCS 2017), 2017, pp. 4–9.
[19] H. Huang, Z. Dong, Research on architecture and query performance based on distributed graph database Neo4j, in: 2013 3rd International Conference on Consumer Electronics, Communications and Networks (CECNet 2013)—Proceeding, 2013, pp. 533–536.
[20] G. Drakopoulos, A. Kanavos, Tensor-based document retrieval over Neo4j with an application to PubMed mining, in: IISA 2016—7th International Conference on Information, Intelligence, Systems & Applications, 2016, pp. 1–6.
[21] P. Kivikangas, M. Ishizuka, Improving semantic queries by utilizing UNL ontology and a graph database, in: Proceeding—IEEE 6th International Conference on Semantic Computing (ICSC 2012), 2012, pp. 83–86.
[22] Y. Ma, Z. Wu, L. Guan, B. Zhou, R. Li, Study on the relationship between transmission line failure rate and lightning information based on Neo4j, in: POWERCON 2014—2014 International Conference on Power System Technology Towards Green, Efficient and Smart Power System Proceeding, no. POWERCON, 2014, pp. 474–479.

CHAPTER ELEVEN

Exploring the possibilities of security and privacy issues in health-care IoT

R. Kishore[a] **and K. Kumar**[b]
[a]Sri Sivasubramaniya Nadar College of Engineering, Chennai, Tamilnadu, India
[b]Vellore Institute of Technology, Vellore, Tamilnadu, India

1. Introduction

Internet of Things (IoT) is a network of smart, self-configuring devices embedded with sensors and actuators that use diverse networking technologies to connect objects in the physical world to the web. IoT-enabled devices are designed and developed for a wide range of applications such as building and home automation, smart city, smart grid, smart agriculture, transportation, military, and health care, etc. Health-care industry particularly is expected to reach greater heights due to increasing health awareness among the people and increased access to insurance facilities. On the other hand, health-care services are becoming costlier and out of reach to a section of the population thereby a large section of the society will be at high risk as the number of chronic diseases is increasing with simultaneous aging of major population. Development of IoT-enabled devices, systems, and applications made health-care pocket friendly and easily accessible. Currently, the entire world is experiencing a pandemic situation due to coronavirus disease, COVID 19 outbreaks [1], wherein nearly 2.5 million positive cases were identified as on date [2] and 0.18 million deaths. In such pandemic situations, to avoid community spread, doctors nowadays are available for video consultation in most parts of the globe with the help of applications developed for such purpose. In the near future, it is expected that such growing technology would move the medical checks from hospital to patient's residence, so-called remote patient monitoring systems that would avoid unnecessary hospitalization if correctly diagnosed and will help a large section of society who cannot afford for such facilities. With such developments in IoT-based health-care systems and services, it is expected

Intelligent IoT Systems in Personalized Health Care
https://doi.org/10.1016/B978-0-12-821187-8.00011-3

that the quality of treatment will be improved thereby improving the health of patients.

In spite of the massive potentiality of IoT in health care, the complete communication infrastructure is weak from the security point of view and may result in loss of privacy for the patients. Security and privacy with respect to health-care data are the most important requirements that would bring remarkable challenges and openings to manufacturers, developers, service providers, and consumers. The most important security issues influencing the developing health-care IoT systems arise out of the security flaws present in the technologies used in building health-care IoT for relaying health-care data. Most of the IoT devices are constrained devices. A constrained device possesses limited power, limited memory, limited computation and communication capability, and is low cost. The communication in health-care IoT is wireless and hence vulnerable to various attacks. Typically in health-care IoTs, the patient's physiological parameters are sensed by the devices called body sensors. Body sensors relay the measured physiological parameters and the related events to the cloud that carries data analyses. Privacy in health-care IoT systems should be protected at every stage starting from the device till data analyses at the cloud which helps to disclose patient's health-related sensitive information. The privacy of users and security of patient's health-related information poses new challenges that need to be addressed in the health-care IoTs. The objective would be to identify the security and privacy challenges in health-care IoT and discuss possible solutions for constrained environments. In such environments there is a possibility of an attack in two different ways:

1. Attack in the network between the gateway and the devices or between the gateway and the cloud server
2. Attack between the cloud server and the end user (uses smartphone or any other external consoles)

Some of the attacks which are likely to occur in health-care IoT environment are spoofing attack, eavesdropping attack, data manipulation, malware Infection, and power analysis attack. To address the abovementioned attacks, the types of required security services such as confidentiality, data integrity, authentication, and secured firmware updates are to be defined properly. The most important aspect is to protect the cryptographic keys. The objective of the chapter is to explore the possibilities of various security and privacy issues possible in such scenarios. The security challenges in IoT health-care systems will be discussed in details. As the attacks occurring in

IoT devices are very difficult for the end users to detect, prevent, and apply corrective measures, appropriate security mechanisms to address the security challenges will be explained in detail.

On the other hand, various wireless communication protocols such as Bluetooth, Zigbee, LoRaWAN, etc. are emerging in IoT devices apart from WiFi. As the IoT devices are connected via such protocols and due to increase in the usage of such devices in a variety of applications, security breaches are found to be critical and the number of attacks is also increasing. Hence the type of attacks related to protocol usage in health-care scenarios will be discussed. Finally, as IoT devices run on various microcontrollers with proprietary real-time operating system (RTOS), field firmware upgradation of these devices is very essential by the manufacturers to understand the threats. Since the threats related to microcontrollers vary with the different types of controllers used in various devices, the different types of threats related to different controllers will be addressed. Over and above, in this proposed chapter, we would like to discuss the existing cryptographic algorithms or mechanisms available to address the security issues in IoT health-care devices and discuss the possible solutions. The challenge is to identify the limitations and explore the possibilities for further progress in IoT enabled health-care systems.

The rest of the chapter is organized as follows. In Section 2, a conceptual framework of IoT-based health-care system, communication, and protocol architectures are presented. Section 3 gives an introduction to applications of IoT in health-care. Section 4 elaborates upon the challenges involved in IoT health-care. Section 5 gives an insight into security and privacy issues in IoT health-care and proposes a possible solution in terms of a security model. Section 6 summarizes the discussions and possibilities for further developments.

2. IoT health-care framework

IoT-based health-care system enables real-time monitoring of patients with expected medical emergencies like heart attack, hypotension, hypertension, respiratory disorder, etc., via connected smart devices. The IoT devices capture the required physiological parameters of the patient under monitoring and forward the collected health data to the cloud via the Smartphone application to which the IoT devices are connected. The physiological parameters such as heart rate, blood pressure level, oxygen level, etc. stored in the cloud are shared with the concerned physician for remote monitoring. Center of Connected Health Policy (CCHP), a nonprofit organization, in its case study found that there was nearly 50% reduction in 30-day readmission rate because of such remote monitoring on patients with heart failure severity [3].

Fig. 1 depicts the conceptual framework for effective remote patient monitoring with heart disease. The IoT device shown in Fig. 1 uses noninvasive technique to capture heart rate measurements exactly at the ulnar side of the forearm in the wrist. The sensor will be embedded within the strap of a watch and will be strapped in the wrist. The IoT device is connected to the smartphone application. The device can capture the heart rate and use the data connection of the smartphone to transfer the heart rate data to the cloud. The data stored in the cloud can be shared with the concerned doctor or even to the insurance company. At the cloud server, machine learning and training can be used to look for patterns in the data that is collected from an individual and make better decisions in the future and alert the individual or the concerned physician regarding the possible medical emergency. This allows the device to learn automatically without human

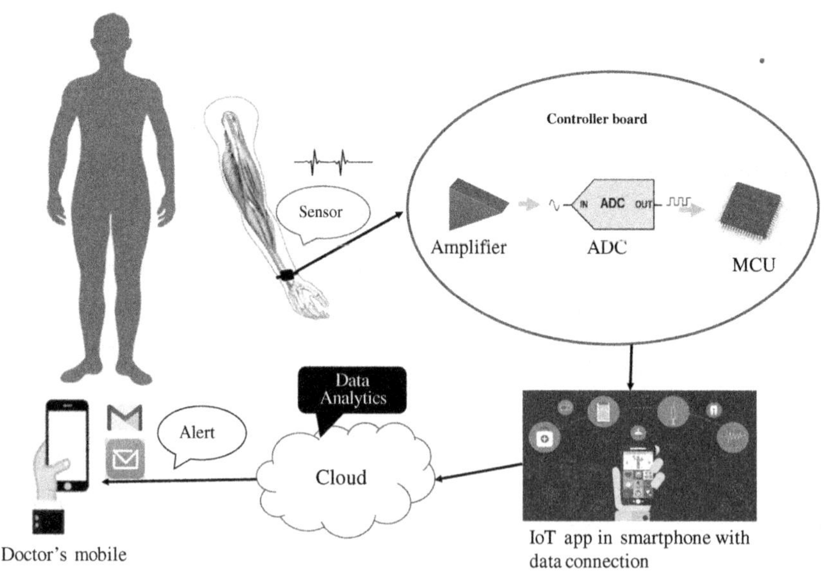

Fig. 1 Conceptual framework for remote patient monitoring and health care via wearable.

intervention or assistance and adjust actions accordingly. Whenever the individual is stressed, it gets reflected on the data that is stored in the cloud and the same thing can be intimated to the individual or the physician by generating a chart of the heart rate data over a period.

Fig. 2 depicts the communication architecture framework for IoT health-care system. IoT-based health-care system can completely automate patient care with developing new technologies and health-care facilities. Protocols such as Bluetooth, 802.11, 802.16, 802.15.4, Zigbee, etc., can be used to transfer the physiological parameters captured by the IoT device to the cloud as depicted in Fig. 2. The health data stored in the cloud can be shared with the concerned physician thereby reducing unnecessary visits and improving resource management efficiently. Such IoT-enabled systems are capable of performing data analyses over the cloud and thereby storing only the final reports in the form of graphs or bar charts instead of raw data. Machine learning, deep learning, and Artificial Intelligence can be used to look for patterns in the data that is collected from an individual and make better predictions and alert the concerned physician regarding the possible medical emergency. Hence IoT-based health-care systems provide time alerts with better accuracy thereby improving the efficiency of remote patient monitoring. On the other hand, IoT-based health-care systems

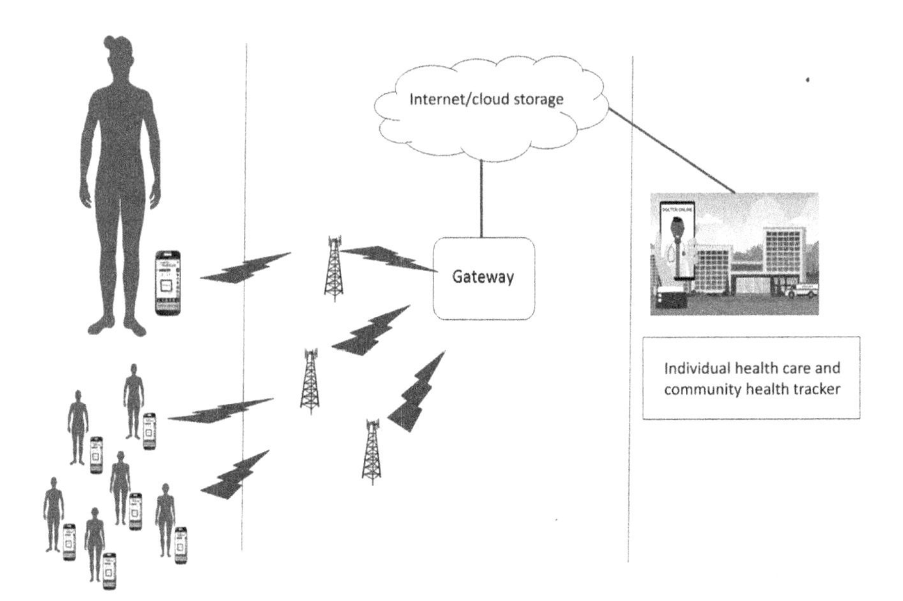

Fig. 2 Communication architecture framework for IoT health-care system.

collect a huge amount of data about the patient's health history that can be used to support medical research in terms of statistical study. Most of the IoT-based application follows five-layer protocol architecture model, in which Link and PHY layer supports 802.11/802.15.1/802.15.4/802.16/, Network layer supports IPv4/IPv6/6LoWPAN, transport layer supports TCP/UDP, and application layer supports http/COAP/MQTT/WebSockets [4]. As discussed in Ref. [5] IoT health-care system devices use IPv6 and 6LoWPAN systems for data transmission over 802.15.4 protocol. Data are replied back by the 802.15.4 enabled devices with user datagram protocol (UDP) at the transport layer.

3. IoT health-care applications

The innovation in the field of IoT is expected to enable a wide variety of applications across the globe. The various IoT health-care applications currently used by the doctors, patients, insurance companies, medical research institutes, and various other organizations that are directly or indirectly participating in patient's health care are real-time body temperature measurements, heart rate measurements, noninvasive glucose level measurements, blood pressure monitoring, continuous monitoring of blood oxygen levels, IoT-enabled fully automated systems for physically challenged people etc. [5]. Nowadays health-care industries talk about ingestible sensors that can monitor the medication level in the body round the clock, thereby send an alert in case of any irregularities and hearing aids that are connected to a smartphone with Bluetooth technology. Halo Neuroscience, a Silicon Valley firm and Thync, a venture-backed start-up had come up with gadgets in the form of headsets that send low-level electric current to the targeted region of the brain thereby improving the mood [6, 7]. The rise in IoT has come up with devices with computer vision and artificial intelligence (AI) technology that reads the environment and helps the visually impaired patient to detect obstacles and navigate accordingly [3]. The most important facility that should be noted in all the applications introduced so far is the availability of health-care data. The health-care data collected from the patients are stored in the cloud and analyzed, thereby complete statistical details about the patients in the form of charts or bar graphs can be made available for the physicians irrespective of the date, time and venue, that avoid manual maintenance and makes doctor's work easier than before.

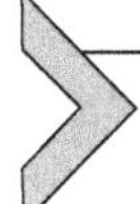

4. Challenges

4.1 Security and privacy

As the collected health-care data is stored in the cloud for further processing and sharing, the most important challenge faced by IoT health-care system is security and privacy issues. There is a possibility that the hackers steal the personal information of patients and as well as the information pertaining to top physicians with which they can create fake identities to buy certain important drugs and even medical emergency equipment. Another possible cyberattack is that the hackers can steal the patient's personal data and can initiate a fraudulent claim with the concerned insurance company. Data protection becomes an important challenge with respect to IoT health-care systems as the amount of data involved is huge and sensitive too.

4.2 Lack of standardization

As per the current practical scenario, no unified standard exists; solutions that are available are basically domain specific for particular application vertical. The different types of devices introduced into the market by different manufacturers, when interconnected, pose real challenges as different devices follow different protocols and standards which in turn complicate the process of data collection, processing, and aggregation. The lack of nonuniform standards and data formats may adversely affect the scalability of IoT health-care systems to a greater extent. Developing unified standards may seem impossible due to domain knowledge differences, but an effort to develop interoperability is a challenging task [8].

4.3 Accuracy in decision-making

Most of the IoT health-care applications require continuous monitoring of patients round the clock. Apart from that the number of devices involved in measuring the health-care data is several numbers based on the requirement and hence the amount of data involved is huge. As the devices follow different communication protocols, data processing and data aggregation are becoming difficult. This, in turn, affects the accuracy in decision-making by doctors [3]. The issue aggravates with the number of devices.

4.4 Device type

Another challenging task is to decide the technology to develop miniaturized IoT devices that need to be embedded in some form with the patients to collect the physiological parameters. Most such devices are constrained in terms of power, memory, computation, and communication capabilities. Deciding appropriate embedded parts and integrating them into the required IoT device is a challenging task. The choice should be such that it should not harm the patients under observation.

4.5 Software development and maintenance

Software application development involves four stages namely: setup, development, debugging and testing, and publishing. In the development of IoT health-care applications, the participation of concerned medical experts in terms of sharing their experience and suggestions will ensure better quality. At the same time, for the application to be sustainable, there should be frequent updates based on the current advances in the medical science, new types of diseases and disorders, methods of early detection of such diseases, and posing new diagnostic challenges. Additionally, there is a requirement of customized computing platforms for IoT health care with suitable application program interfaces (APIs), disease-oriented libraries and appropriate frameworks as the requirements are highly demanding and sophisticated [5].

4.6 Cost-effective

Actual requirement is, the IoT-based health care must be low-cost facilitating affordability by a common man. Achieving low-cost IoT health-care technologies pose a new challenge to most of the developed countries.

5. Security and privacy issues in IoT health-care

Smart IoT devices in health-care systems provide many advantages but they pose a lot of challenges with respect to security and privacy issues related to health-related data of the patients.

5.1 Privacy issues

In health-care systems, the complete details of the patients under observation are maintained in the form of health records. The most important requirement in such scenarios is, all these health records must be kept highly confidential, and otherwise, this can become a target for cyberattacks leading to the disclosure of private data resulting in privacy issues. Since the communication of such patient's personal data in IoT health-care systems is done wirelessly, the attackers can inject eavesdropping attack by stealing information that the smart IoT devices, mobile phones, and computers exchange over the network. Therefore protecting the data exchanged over the network becomes an important requirement. Different countries have different laws to protect patient's health data. But the users of such systems are not very much aware of what is happening to the health data collected from them. For example, people from different age group and different work culture, use fitness wearable devices and think that the physiological parameters collected are protected under certain rules and regulations, but the reality is quite different [9]. In addition to all these privacy issues, there is a possibility that the attackers may steal location details of such patients as all these devices are connected to the smartphone application via the data connection.

5.2 Security requirements and challenges

Following are the general security requirements with respect to the IoT health-care systems:

- **Confidentiality:** Providing privacy and secrecy of health-care information to prevent unauthorized access.
- **Integrity:** Ensuring that message is not altered during the transition.
- **Authentication:** Authenticating the participating connected devices before granting a limited resource, or revealing information.
- **Flexibility:** Security technique chosen must work in any kind of environment and it must support the addition of devices, in general, it must be flexible.
- **Scalability:** Techniques chosen must support scalability, which forms an important issue in IoT-enabled systems, i.e., the technique must be applicable for both smaller networks and larger ones.

One of the important challenges in IoT-based systems is to provide very good security with limited resources. The challenges in providing security for IoT health-care systems are [10]:

- First challenge is that there must be a good compromise between minimization of resource consumption and maximization of security level. At the same time, resource constraints should be taken care.

- Security mechanism hosted on a device is dependent on the capabilities and constraints of sensor node hardware.
- Third challenge is that the lack of predefined topology facilitates attackers for passive eavesdropping, active interfering, leaking of secret information, interfering message, impersonating nodes, etc.
- Fourth challenge is that communication is through radio, Wire-based security scheme impractical for such applications [11].
- Fifth challenge is that the topology is not fixed. Variations in the number of connected devices are expected to be arbitrary fashion. Failures may be permanent or intermittent.
- Sixth challenge is that the overall cost should be as low as possible.

Constraints in IoT health-care systems can be defined in three different ways as node constraints, network constraints, and physical limitations. The constraints and the corresponding challenges are:

- **Limited energy**: IoT devices have limited battery life and hence a simple key establishment technique consuming less computation power must be used.
- **Limited communication bandwidth**: Due to bandwidth constraints the key establishment techniques must allow only small-sized data to be transferred at a time.
- **Limited memory**: The selected security technique must use less memory at the same time it should provide a higher level of security.
- **Limited communication range**: Limited energy supply also limits the communication range. To overcome this, a sort of network processing should be performed where a selected set of nodes takes care of aggregating and forwarding the data to cloud.

5.3 Security threats in IoT health-care systems

Considering the sensitivity of the health data exchanged among the connected devices in IoT health-care systems, attacks are possible on different levels from devices, to applications and cloud. Interoperability of IoT devices leads to vulnerabilities and the related challenges [9]. The following are the important security threats possible in the case of IoT health-care systems:

(i) Denial of Service (DoS) attacks

DoS attacks cause communications links to be lost or unavailable. Such attacks threaten health-care service availability, network functionality, and device responsibility. The system loses its basic functionality to implement the requirements. In may happen in different forms from devices to protocols and storage.

Jamming is a popular Denial of Service attack where the adversaries interfere with the communication frequencies of the sensor nodes. Here the adversaries select few jamming nodes from the entire networks. The possible solution, in this case, is the proper choice of spread spectrum technique. But the complexity involved with respect to computation and the cost involved will be huge [10]. To be more specific, the adversary targets a single system with several compromised devices and crashes the entire framework making data unavailable, successfully injecting distributed denial of service (DDoS) attacks [9].

(ii) Tampering

In this case, the adversaries can get access, and become successful in compromising some IoT devices that can be used as a portal to steal patient's sensitive health data that is termed "Medjacking" by TrapX, a solution provider for security issues during June 2015. Additionally, the adversary can get access to the security credentials of the users. One of the best defense mechanism for such an attack is manufacturing devices with self-destruction capability whereby devices vaporize memory contents in case of intrusion and this may prevent leakage of information. Another solution is proposing an efficient fault-tolerant protocol that maintains proper functioning of the network even if some nodes are compromised. But exploring the possibilities of implementing such mechanisms in IoT-based systems pose a new challenge.

(iii) Modification and fabrication

As mentioned in the previous attack, once the adversary succeeds in compromising few IoT devices, he/she can access the sensitive health data and modify them misleading the other entities involved in the IoT network. On the other hand, the hacker can inject false messages into the network and confuse the other connected entities, introducing fabrication attack. The best defenses against this type of attacks are making use of proper encryption and authentication techniques.

(iv) Replay attack

The adversary after successfully compromising the IoT device can replay the existing messages at some later point of time thereby threatening data freshness. The best defenses against this type of attacks are making use of proper time stamps and cryptographic nonce.

(v) Unauthorized data access

Each IoT-based health-care system application will have a huge number of users and the data involved in such a scenario is huge. In such scenarios preventing unauthorized access to the health data and related resources becomes an important issue. The best defenses against this type of attacks are making use of appropriate access control and authentication mechanisms.

(vi) Hardware and Software Compromise

In such cases, the attackers succeed in tampering IoT devices and configure the compromised devices with malicious codes. In addition to that an attacker explores the software vulnerabilities and malfunction IoT devices [5]. As discussed earlier, one of the best defense mechanisms for such an attack is manufacturing devices with self-destruction capability whereby devices vaporize memory contents in case of intrusion and this may prevent leakage of information.

(vii) Exploiting the vulnerabilities of protocol standards

The attackers may exploit the possible vulnerabilities with the defined protocol standards for injecting malicious activities into the networks. For example, If MAC sublayer of IEEE 802.15.4 standard use CSMA/CA with RTS/CTS, in which RTS/CTS packets are used to reserve channel access to transmit data. There is a possibility that an adversary intentionally injects RTS packets continuously to a targeted node by ignoring CTS reply packets thereby flooding the link of the targeted node [10]. Such nodes are often termed as malicious nodes or self-sacrificing nodes. The best solution is that a node can limit itself in accepting connections from the same identity. A particular node will not accept more than fixed number connections from the same identity and there should be a proper selection of this threshold.

(viii) Cyber Security Attacks

As the IoT-based health-care systems rely on the cloud for storage and analyses, the cybercriminals can inject attacks such as hidden https tunnels, hidden DNS tunnels, and ransomware and botnet attacks. In hidden https tunnels, the attackers use the https tunnels meant for actual data communication for its own command and control that looks like normal traffic over long periods of time [12]. On the other hand, the cybercriminals can insert malicious software or stole health-care information into DNS queries and responses that can even bypass the firewalls. Ransomware is a form of malware that encrypts the victim's information and demands a ransom to restore access to data, whereas botnets are created by infecting systems with malicious software and make those systems slave to the botnet creator. Currently, it is found that the rate of occurrence of both ransomware and botnets in IoT health-care systems is less when compared to other applications [12].

5.4 Possible solutions

Following security mechanisms can be considered at the stage of design, development, production, and deployment of IoT health-care devices:

(i) Encryption

Encryption ensures confidentiality. The sensitive data can be protected by creating secure links. Appropriate encryption algorithms can be used to create required certificates that can be used to exchange health-care data securely.

(ii) Authentication

Manufacturers need to issue certificates for IoT devices for verifying the identity of the authorized users in the network.

(iii) Integrity

IoT health-care data exchanged wirelessly must be signed with the created certificates so that data is not altered during the transition that ensures data integrity.

Suppose the IoT health-care framework uses the devices Xbee module mounted on an appropriate microcontroller, the security features in Xbee include 128-bit AES encryption, two security keys namely network key and link key that can be preconfigured or obtained during joining, and support for a trust center. The security features available with Xbee satisfies the security requirements such as message integrity, confidentiality, and authentication. In such scenarios, all the participating devices can be predistributed with the link key and the coordinator node can be made responsible for selecting a network key for encryption and distributing the network key encrypted by the link key to the joining devices. Data transmissions are always encrypted with the network key that is hop to hop, and can optionally be end-to-end encrypted with the Application Support Sublayer (APS) link key [13]. On the other hand, three cryptographic layers are defined in Waspmote encryption libraries namely link layer, application layer, and secure web server connection, if Waspmote is chosen as the microcontroller. As per the link layer, all the participating nodes in the network share a common predistributed key that can be used to encrypt the data using AES 128. This process can be carried out using specific hardware integrated in the same Zigbee radio, thereby the efficiency of energy consumption of devices can be taken care. In this kind of setup, if a malicious node sends a message, the message will be discarded in the first hop itself and hence link layer itself will be able to ensure that no third party devices can get connected to the network thus providing efficient access control. Apart from access control, Waspmote encryption libraries can also provide solutions to common security issues such as authentication, confidentiality, integrity, and repudiation with Waspmote hash files, Waspmote AES files, and Waspmote RSA files, respectively. Hence acceptable security level can be achieved with the device used in the IoT health-care framework, which is in turn dependent on the secured storage of keys.

Suppose, if a device without predistributed keys wishes to join the network, at least a single unprotected key must be sent to enable encrypted communication. This one-time exchange of unprotected key may end up with sniffing attack [14] that would lead to compromise security of the whole network. On the other hand, most of the device type mentioned in such application scenarios is not tamper-resistant and hence there is a possibility that an attacker succeeds in accessing the secret keys and other privileged information. The solution for the first case is to use only preconfigured keys as discussed earlier and in the second case, logic needs to be developed to erase the security credentials if tampering is detected. The resilience of the framework can be further increased by proposing logic to periodically change the network key in regular intervals of time so that known plaintext attack can be avoided.

Exploring the possibility of the second level of security by developing lightweight digital signature algorithm using Waspmote hash files, Waspmote AES files, and Waspmote RSA files available with Waspmote encryption libraries will be a great challenge [15]. Here the signature can be generated using the private key of the signer and the signature is attached along with the hash code of the message before transmission. The receiver has to verify the signature using the public key of the signer. Once the signature is verified, the link will be established between the two devices and the message will be transmitted. The challenge here will be making the entire process lightweight suitable for constrained environments by choosing a complex number computed from a small prime number within a specific range for computing signature parameters instead of large prime numbers to avoid computational overhead.

6. Summary

In this chapter, we have discussed the possible communication and protocol architecture of IoT health-care systems. Overview of various applications of IoT in health care is discussed along with the challenges involved. It is understood that due to sensitive data handling, security and privacy become an important challenge in the design, development, and deployment of IoT health-care systems. The chapter also explored the privacy issues, security requirements and challenges, security attacks, and a short discussion about the possible solutions. It is understood that security factor must be taken care at various levels starting from devices to networks and finally cloud. Confidentiality, integrity, and authentication are the required services for maintaining security in IoT devices. Extensive care should be taken to design a proper security technique that suits IoT systems. A trade-off between the requirements and resources of IoT determines which security mechanism should be employed for a particular application.

References

[1] World Health Organization, Health Topics—Coronavirus, https://www.who.int/health-topics/coronavirus#tab=tab_1, 2020 (Accessed 22 April 2020).
[2] Worldometer, COVID-19 Coronavirus Pandemic, https://www.worldometers.info/coronavirus/, 2020 (Accessed 22 April 2020).
[3] Peerbits, Internet of Things in Healthcare: Applications, Benefits and Challenges, https://www.peerbits.com/blog/internet-of-things-healthcare-applications-benefits-and-challenges.html, 2020 (Accessed 23 April 2020).
[4] A. Bahga, V. Madisetti, Internet of Things—A Hands-on Approach, Universities Press, 2015.
[5] S.M.R. Islam, D. Kwak, M.H. Kabir, M. Hossain, K. Kwak, The Internet of things for health care: a comprehensive survey, IEEE Access 3 (2015) 678–708.
[6] Thync, Breakthrough Bioelectronic Products, https://www.thync.com, 2020 (Accessed 24 April 2020).
[7] Halo Neuroscience, How Halo Sport Works?, https://www.haloneuro.com, 2020 (Accessed 24 April 2020).
[8] H. Zhou, The Internet of Things in the Cloud—A Middleware Perspective, CRC Press—Taylor & Francis Group, 2013.
[9] Sciforce, Ensuring Privacy and Security in Healthcare IoT, https://medium.com/sciforce/ensuring-privacy-and-security-in-the-healthcare-iot-7b97549d629c, 2020 (Accessed 24 April 2020).
[10] H. Kumar, D. Sarma, A. Kar, Security threats in wireless sensor networks, IEEE Aerosp. Electron. Syst. Mag. 23 (6) (2008) 39–45.
[11] H. Kumar, D. Sarma, A. Kar, Security threats in wireless sensor networks, in: IEEE International Carnahan Conferences Security Technology, Oct. 2006, pp. 243–251.
[12] Tripwire, Cyber Security Challenges in Healthcare IoT Device, https://www.tripwire.com/state-of-security/security-data-protection/iot/cyber-security-healthcare-iot/, 2020 (Accessed 21 April 2020).
[13] Digi International, Zigbee RF Modules User Guide, XBee/XBee-PRO Zigbee RF Modules User Guide Datasheet 90000976, https://www.digi.com/resources/documentation/digidocs/pdfs/90000976.pdf, 2018 (Revised June 2017) (Accessed 02 April 2020).
[14] T. Zillner, ZigBee exploited: the good, the bad and the ugly, Magdebg. J. Sicherheitsforschung 12 (2016) 699–704. Retrieved September 2, 2016, from http://www.sicherheitsforschung-magdeburg.de/publikationen/journal.html.
[15] Libelium, Waspmote Encryption Libraries Programming Guide, Document version: v7.0-02/2017 http://www.libelium.com/downloads/documentation/encryption_libraries_guide_eng.pdf, 2017 (Accessed 03 April 2020).

Subject Index

Note: Page numbers followed by *f* indicate figures, *t* indicate tables, and *b* indicate boxes.

B

C

D

J

K

L

M

N

O

P

Q

R

S

Printed in the United States
By Bookmasters